Mathematics, Substance and Surmise

Ernest Davis • Philip J. Davis

Editors

Mathematics, Substance and Surmise

Views on the Meaning and Ontology of Mathematics

 Springer

Editors
Ernest Davis
Department of Computer Science
New York University
New York, NY, USA

Philip J. Davis
Department of Applied Mathematics
Brown University
Providence, RI, USA

ISBN 978-3-319-21472-6   ISBN 978-3-319-21473-3   (eBook)
DOI 10.1007/978-3-319-21473-3

Library of Congress Control Number: 2015954083

Mathematics Subject Classification (2010): 01Axx, 00A30, 03-02, 11-04, 11Y-XX, 40-04, 65-04, 68W30, 97Mxx, 03XX, 97C30.

Springer Cham Heidelberg New York Dordrecht London
© Springer International Publishing Switzerland 2015

Printed on acid-free paper

Springer International Publishing AG Switzerland is part of Springer Science+Business Media (www.springer.com)

Contents

Introduction

Ernest Davis

Mathematics discusses an enormous menagerie of mathematical objects: the number 28, the regular icosahedron, the finite field of size 169, the Gaussian distribution, and so on. It makes statements about them: 28 is a perfect number, the Gaussian distribution is symmetric about its mean. Yet it is not at all clear what kind of entity these objects *are*. Mathematical objects do not seem to be exactly like physical entities, like the Eiffel Tower; nor like fictional entities, like Hamlet; nor like socially constructed entities, like the English language or the US Senate; nor like structures arbitrarily imposed on the world, like the constellation Orion. Do mathematicians *invent* mathematical objects; or *posit* them; or *discover* them? Perhaps objects *emerge* of themselves, from the sea of mathematical thinking, or perhaps they "come into being as we probe" as suggested by Michael Dummett [1].

Most of us who have done mathematics have at least the strong impression that the truth of mathematical statements is independent both of human choices, unlike truths about Hamlet, and of the state of the external world, unlike truths about the planet Venus. Though it has sometimes been argued that mathematical facts are just statements that hold by definition, that certainly doesn't *seem* to be the case; the fact that the number of primes less than N is approximately $N/\log(N)$ is certainly not in any way an *obvious* restatement of the definition of a prime number. Is mathematical knowledge fundamentally different from other kinds of knowledge or is it simply on one end of a spectrum of certainty?

Similarly, the truth of mathematics—like science in general, but even more strongly—is traditionally viewed as independent of the quirks and flaws of human society and politics. We know, however, that math has often been used for political purposes, often beneficent ones, but all too often to justify and enable oppression

E. Davis (✉)
Department of Computer Science, Courant Institute of Mathematical Sciences,
New York University, New York, NY, USA
e-mail: davise@cs.nyu.edu

© Springer International Publishing Switzerland 2015
E. Davis, P.J. Davis (eds.), *Mathematics, Substance and Surmise*,
DOI 10.1007/978-3-319-21473-3_1

and cruelty.[1] Most scientists would view such applications of mathematics as scientifically unwarranted; avoidable, at least in principle; and in any case irrelevant to the validity of the mathematics in its own terms. Others would argue that "the validity of the mathematics in its own terms" is an illusion and the phrase is propaganda; and that the study of mathematics, and the placing of mathematics on a pedestal, carry inherent political baggage. "Freedom is the freedom to say that two plus two makes four" wrote George Orwell, in a fiercely political book whose title is one of the most famous numbers in literature; was he right, or is the statement that two plus two makes four a subtle endorsement of power and subjection?

Concomitant with these general questions are many more specific ones. Are the integer 28, the real number 28.0, the complex number $28.0+0i$, the 1×1 matrix [28], and the constant function $f(x) = 28$ the same entity or different entities? Different programming languages have different answers. Is "the integer 28" a single entity or a collection of similar entities; the signed integer, the whole number, the ordinal, the cardinal, and so on? Did Euler mean the *same thing* that we do when he wrote an integral sign? For that matter, do a contemporary measure theorist, a PDE expert, a numerical analyst, and a statistician all mean the same thing when they use an integral sign?

Such questions have been debated among philosophers and mathematicians for at least two and a half millennia. But, though the questions are eternal, the answers may not be. The standpoint from which we view these issues is significantly different from Hilbert and Poincaré, to say nothing of Newton and Leibnitz, Plato and Pythagoras, reflecting the many changes the last century has brought. Mathematics itself has changed tremendously: vast new areas, new techniques, new modes of thought have opened up, while other areas have been largely abandoned. The applications and misapplications of mathematics to the sciences, technology, the arts, the humanities, and society have exploded. The electronic computer has arrived and has transformed the landscape. Computer technology offers a whole collection of new opportunities, new techniques, and new challenges for mathematical research; it also brings along its own distinctive viewpoint on mathematics.

The past century has also seen enormous growth in our understanding of mathematics and mathematical concepts as a cognitive and cultural phenomenon. A great deal is now known about the psychology and even the neuroscience of basic mathematical ability; about mathematical concepts in other cultures; about mathematical reasoning in children, in pre-verbal infants, and in animals.

Moreover the larger intellectual environment has altered, and with it, our views of truth and knowledge generally. Works such as Kuhn's analysis of scientific progress and Foucault's analysis of the social aspects of knowledge have become part of the general intellectual currency. One can decide to reject them, but one cannot ignore them.

---

[1]The forthcoming book *Weapons of Math Destruction* by Cathy O'Neil studies how modern methods of data collection and analysis can feed this kind of abuse.

This book

The seventeen essays in this book address many different aspects of the ontology and meaning of mathematics from many different viewpoints. The authors include mathematicians, philosophers, computer scientists, cognitive psychologists, sociologists, historians, and educators. Some attack the ontological problem head on and propose a specific answer to the question, "What is a mathematical object?"; some attack it obliquely, by discussing related aspects of mathematical thought and practice; and some argue that the question is either useless or meaningless.

It would be both unnecessary and foolhardy to attempt to summarize the individual chapters. But it may be of interest to note some common themes:

- The history of math, mathematical philosophy, and mathematical practice. (Avigad, Bailey & Borwein, Gillies, Gray, Lützen, O'Halloran, Martin & Pease, Ross, Stillwell, Verran). Among scientists, mathematicians are particularly aware of the history of their field; and philosophers sometimes seem to be aware of little else. Many different aspects and stages in the evolution of thinking about mathematics and its objects are traced in these different essays.

- The real line and Euclidean geometry (Bailey & Borwein, Berlinski, E. Davis, Gillies, Gray, Lützen, Stillwell). These essays in the collection touch on many different mathematical theories and objects, but the problems associated with $\mathbb{R}$ and $\mathbb{R}^n$ seem particularly prominent.

- The role of language (Avigad, Azzouni, Gray, O'Halloran, Piantadosi, Ross, Sinclair). On the one hand, mathematics itself seems to be something like a language; Ross discusses the view that mathematics is a universal or ideal language. On the other hand, a question like "Do mathematical objects exist?" is itself a linguistic expression; and it can be argued that difficulties of answering it derive from illusions about the precision or scope of language. Sinclair argues that we may be using the wrong language for mathematics; rather than thinking of mathematical entities as nouns, we should be thinking of them as verbs. Martin and Pease's essay focuses on the related issue of communication in mathematical collaboration.

- The mathematics of the 21st century (Avigad, Bailey & Borwein, Martin & Pease, Sinclair). Several of our authors look forward to a broadening of the conceptualization and the practice of mathematics in the coming century. The answers to the questions "What is mathematics?" and "What are mathematical objects?" may change, in directions that have recently emerged in the mathematical community.

- Applications. E. Davis considers the applications of geometry to robotics. Bailey and Borwein discuss numerous applications of mathematical simulation including space travel, planetary dynamics, protein analysis, and snow crystals. Verran considers the (mis)applications of statistics to policy. In the opposite direction, Berlinski discusses the difficulty of making precise the sense in which mathematics can be applied at all to physics or to any other non-mathematical domain.

- Psychology: How people think about mathematics. This is front and center in Rips, but it is just below the surface in *all* the essays. Arguably, that is the real subject of this book.

Why a multidisciplinary^2 collection?

In the last few decades, universities, research institutions, and funding agencies have made a large, deliberate effort to encourage interdisciplinary research. There is a good reason for this. On the one hand, there is much important research that requires the involvement of multiple disciplines. On the other hand, overwhelmingly, the institutions of science and scholarship—departments, academic programs, journals, conferences, and so on—are set up along disciplinary lines. As a result, it can often be hard for good interdisciplinary work to get published; for researchers to get promoted, tenured, and recognized; and for students to get trained. Therefore, it is both necessary and highly important for the powers that be to counteract this tendency by energetically welcoming and promoting interdisciplinary research.

However this laudable effort has often been both taken too far and trivialized. The word "multidisciplinary" and its many near synonyms^3 have often become mindless mantras, particularly among university administrators. At times they have become terms of purely generic praise, indiscriminately applied to any research, however, narrow in scope. There has been some healthy reaction against this (e.g., [2]), but in general, the fad is still in full swing.

Since this academic trend is both so important and also so faddish and so often overhyped, it is wise to be initially both welcoming and skeptical of each new manifestation. A collection like this one raises two natural question in that regard. First: The existence of mathematical objects and the truth of mathematical statements are clearly within the purview of the philosophy of mathematics. In fact, they are central questions in the philosophy of mathematics, and there is a large philosophical literature on the subject. So why should one suppose that other disciplines have anything to contribute to the question?

Second: Each of the authors in this collection is an expert in their own discipline, and primarily publishes their work in journal articles and books addressed to other experts in their discipline. Jody Azzouni publishes articles addressed to philosophers in *Synthese, Philosophia Mathematica,* etc.; Lance Rips and Steve Piantadosi publish articles addressed to psychologists in *Psychological Science,*

---

^{2}This book is, strictly speaking, *multidisplinary* rather than *interdisciplinary.* That is, it brings together multiple disciplines in a single volume, but does not reflect any very strong integration of these.

^{3}These include "cross-disciplinary," "extradisciplinary," "hyperdisciplinary," "interdisciplinary," "metadisciplinary," "neo-disciplinary," "omnidisciplinary," "pandisciplinary," "pluridisciplinary," "polydisciplinary," "postdisciplinary," "supradisciplinary," "superdisciplinary," and "transdisciplinary." The reader may Google for specific citations.

Cognition, etc.; and so on down the line. Contrary to the cult of interdisciplinarity, this kind of specialized communication to a limited audience is not regrettable; it is fruitful and essentially inevitable, given the degree of expertise needed to read an original technical research article in any given field. What do we actually expect to accomplish by putting all these disparate viewpoints together between the covers of a book? Will the result be anything more than nineteen specialists all talking past one another?

The essays in this book are themselves the best answer to the first question. Manifestly, each of the disciplines represented here does have its own distinctive viewpoints and approaches to the nature of mathematical objects, mathematical truths, and mathematical knowledge, and brings to bear considerations that the other disciplines tend to discount or overlook. The relations between large cardinal theory and real analysis that John Stillwell explains certainly bear on questions of mathematical ontology; so, in a different way, does the psychological evidence that Lance Rips discusses. I will not say that the question of the nature of mathematics is too important to be left to the philosophers of mathematics; but it is, perhaps, too protean and too elusive.

As regards the second question, one has to keep ones expectations modest, at least in the short term. We do not expect any dramatic direct interdisciplinary cross-fertilizations to emerge here. Lance Rips will not find that David Bailey and Jon Borwein's computational verification of intricate identities in real analysis give him the key he needs for his studies of the cognitive development of mathematical understanding; nor vice versa. The most that one hopes for is a slight broadening of view, a partial wakening from dogmatic slumbers. As a scientist spends years struggling to apply her own favorite methods to her own chosen problems, it can be easy to lose sight of the fact that there are not only other answers, but even other questions. Seeing the very different way in which other people look at your subject matter is valuable, even though their answers or their questions are only occasionally directly useful, because they shine an indirect light that leads you to your own alternative paths.

Further in the future, though, perhaps we can look forward to a deeper integration of these points of view. In particular, as mentioned above, all of these essays engage with the question of how people think about mathematics; and therefore it is reasonable to suppose that a complete theory of that would explain how people think about mathematics, from basic counting to inaccessible cardinals and beyond, might draw on and combine all these kinds of considerations. This collection is perhaps a small step toward that ultimate overarching vision.

Gaps

There are some regrettable gaps in our collection. On the mathematical side: There is no discussion of probability, which, it seems to me, raises important and difficult questions at least of interpretation, if not of ontology. Verran's chapter deals incisively with the uses of statistics, but we have no discussion of the theory or the history of statistics. We are thin in algebra; Gray's and Lützen's essays discuss 18th and 19th century developments, but the great accomplishments of the 20th century go almost unmentioned; we have no one to tell us in what sense the monster group exists. On the disciplinary side: we have no one from the natural sciences and no one from the arts. It is not so easy to collect all the contributors for a book that one might wish for.

Web site

There is a web site for the book at http://www.cs.nyu.edu/faculty/davise/ MathOntology/ with supplementary materials.

Order of chapters

The chapters have been ordered so as to maximize the mean similarity between consecutive chapters, subject to the constraint that the chapter by Martin and Pease came first, since that seemed like a good starting point. Details can be found at the web site.

Excitements

In the final analysis, perhaps the best claim that mathematical objects have on existence is the excitement that they provoke in their devotees. Of all the wonderful material in this book, perhaps my favorite is a short anecdote that Bailey and Borwein tell of Paul Erdős (a variant on steroids of the well-known story of Ramanujan and cab #1729.)

Shortly before his death, Paul Erdős was shown the form

$$\sum_{k=1}^{n} \frac{2n^2}{k^2} \prod_{i=1}^{n-1} \frac{(4k^4 + i^4)}{\prod_{\substack{i=1 \\ i \neq k}}^{n}(k^4 - i^4)} = \binom{2n}{n}$$

at a meeting in Kalamazoo. He was ill and lethargic, but he immediately perked up, rushed off, and returned 20 minutes later saying excitedly that he had no idea how to prove [the formula], but that if proven it would have implications for the irrationality of $\zeta(3)$ and $\zeta(7)$. (Somehow in 20 minutes, an unwell Erdős had intuited backwards our whole discovery process.)

Similarly, the best justification for raising the question, "Do mathematical objects exist?" is this collection of fascinating and insightful responses that the question has elicited; even among those authors who have rejected the question as meaningless. Speaking personally, few things in my professional life have given me more pleasure than editing this book with my father.[4] It was really thrilling to open each email with a new chapter from another author, and see the wonderful stone soups that they had concocted starting with our simple-minded question. If our readers share that pleasure and excitement, then the book is a success.

Acknowledgements We thank all the authors for their wonderful contributions and their encouragement. Particular thanks to Jeremy Avigad, Jon Borwein, and Gary Marcus for much helpful feedback and many valuable suggestions.

References

1. M. Dummett, 'Truth', *Proceedings of the Aristotelian Society*, n.s. **69**, 1959, 141–162.
2. J.A. Jacobs, *In defense of disciplines: Interdisciplinarity and specialization in the research university*, U. Chicago, 2013.

---

[4]Philip Davis and Ernest Davis are father and son, in case you were wondering.

Hardy, Littlewood and *polymath*

Ursula Martin and Alison Pease

Abstract In the early twenty-first century the *polymath* experiments saw some of the most distinguished mathematicians in the world work together on significant research problems, writing down what they were doing on a blog for all to see as they went along. They drew widespread attention as they offered an unusual opportunity to see mathematics in progress. In this paper we contrast *polymath* with a famous collaboration from the early twentieth century, that of the Cambridge mathematicians G H Hardy and J E Littlewood. We look at the collaborations, and the institutions and structures that enabled them, as a contribution to understanding how collaboration enables mathematical advance.

In the early twenty-first century the *polymath* experiments saw some of the most distinguished mathematicians in the world work together on significant research problems, writing down what they were doing on a blog for all to see as they went along. They drew widespread attention as they offered an unusual opportunity to see mathematics in progress. In this paper we contrast *polymath* with a famous collaboration from the early twentieth century, that of the Cambridge mathematicians G H Hardy and J E Littlewood. We look at the collaborations, and the institutions and structures that enabled them, as a contribution to understanding how collaboration enables mathematical advance.

U. Martin (✉)
Department of Computer Science, University of Oxford, Wolfson Building,
Parks Road, Oxford OX1 3QD, UK
e-mail: ursula.martin@cs.ox.ac.uk

A. Pease
School of Computing, University of Dundee, Queen Mother Building, Balfour Street,
Dundee DD1 4HN, Scotland, UK
e-mail: a.pease@dundee.ac.uk

© Springer International Publishing Switzerland 2015
E. Davis, P.J. Davis (eds.), *Mathematics, Substance and Surmise*,
DOI 10.1007/978-3-319-21473-3_2

1   Hardy and Littlewood

Godfrey Harold Hardy (1877–1947) and John Edensor Littlewood (1885–1977) established the UK as a leading centre of mathematical analysis. Between 1912 and 1948 they published nearly 100 joint papers in analysis and number theory: Hardy published around 300 altogether and Littlewood about half that number.

They had a formidable reputation for joint work, which was less usual at the time than today. Harald Bohr remarked "Nowadays, there are only three really great English mathematicians: Hardy, Littlewood, and Hardy-Littlewood", and the eminent number theorist Landau wrote "The mathematician Hardy-Littlewood was the best in the world, with Littlewood the more original genius and Hardy the better journalist."

Their contribution to UK mathematics went well beyond joint papers, taking the UK from something of a backwater of mathematical research to a leading position in the rapidly developing fields of analysis, number theory and the theory of functions.

Both were undergraduates at Cambridge, where mathematical education at the time consisted of coaching for the Tripos, highly competitive and narrow university examinations, with the goal of finishing top of the class and being celebrated as "Senior Wrangler".

Hardy's approach to mathematics was transformed by reading Jordan's "Cours d'analyse", which he said opened his eyes to "what mathematics really meant", and through his writing and teaching he brought to the UK the rigour of proof-oriented continental mathematics in the style of Borel, Lebesgue and Landau. His 1908 textbook 'Pure Mathematics' was the first exposition in English of analysis in this style: it is renowned for its clarity, and remains in print to this day.

Both helped many students develop successful mathematical careers. Titchmarsh describes "conversation classes" where "Mathematicians of all nationalities and ages were encouraged to hold forth on their own work, and the whole thing was conducted with a delightful informality that gave ample scope for free discussion after each paper". They navigated the politics to find their proteges positions in Cambridge (the most prized of which, only open to men, were in their own college, Trinity) and elsewhere, and, for the most successful, Fellowships of the Royal Society. Hardy in particular put considerable time into the London Mathematical Society. Since its founding in 1865, it had become the UK's learned society for mathematics, with its membership reflecting a shift in mathematics research, as a largely amateur community was replaced by mathematicians employed in universities. It had developed an international role in exchange of mathematical ideas, with its journals, famous for their rigorous standards of refereeing, exchanged with mathematical societies around the world, and its monthly meetings in London providing an opportunity for face-to-face discussions. Hardy was president twice; Littlewood claimed as president to have never said anything other than 'I declare this meeting closed'. Hardy was also active in the International Congress of Mathematicians, and to opening up communication and organising visits to and from overseas mathematicians, in contrast to the isolation of British mathematics in the previous century.

To look back at this collaboration is to look back at a bygone age of gentlemanly politeness, board and lodging and after-dinner conversation in the splendid surroundings of Trinity College, handwritten letters and notes delivered by a college servant, or sent by ship across the world, and apparently abundant free time for watching cricket in Cambridge, or mountaineering during the lengthy vacations.

Hardy saw proof as a collaborative act of understanding, a teacher transmitting a route map to a student. Writing in the philosophy journal Mind in 1929 he responds to contemporary debates on logic and foundation:

> I have myself always thought of a mathematician as in the first instance an observer, a man who gazes at a distant range of mountains and notes down his observations. His object is simply to distinguish clearly and notify to others as many different peaks as he can. There are some peaks which he can distinguish easily, while others are less clear. He sees A sharply, while of B he can obtain only transitory glimpses. At last he makes out a ridge which leads from A, and following it to its end he discovers that it culminates in B. B is now fixed in his vision, and from this point he can proceed to further discoveries. In other cases perhaps he can distinguish a ridge which vanishes in the distance, and conjectures that it leads to a peak in the clouds or below the horizon. But when he sees a peak he believes that it is there simply because he sees it. If he wishes someone else to see it, he points to it, either directly or through the chain of summits which led him to recognise it himself. When his pupil also sees it, the research, the argument, the proof is finished.[1]
>
> The analogy is a rough one, but I am sure that it is not altogether misleading. If we were to push it to its extreme we should be led to a rather paradoxical conclusion; that there is, strictly, no such thing as mathematical proof; that we can, in the last analysis, do nothing but point; that proofs are what Littlewood and I call gas, rhetorical flourishes designed to affect psychology, pictures on the board in the lecture, devices to stimulate the imagination of pupils. This is plainly not the whole truth, but there is a good deal in it. The image gives us a genuine approximation to the processes of mathematical pedagogy on the one hand and of mathematical discovery on the other; it is only the very unsophisticated outsider who imagines that mathematicians make discoveries by turning the handle of some miraculous machine.

Hardy's Mathematician's Apology, published in 1940, is a sombre essay on the beauty and purpose of mathematics—emphasising its nature as "A map or picture, the joint product of many hands, a partial and imperfect copy (yet exact so far as it extends) of a section of mathematical reality". He describes his collaborations with Littlewood and Ramanujan.

> All my best work since then has been bound up with theirs, and it is obvious that my association with them was the decisive event of my life. I still say to myself when I am depressed, and find myself forced to listen to pompous and tiresome people, 'Well, I have done one the thing you could never have done, and that is to have collaborated with both Littlewood and Ramanujan on something like equal terms.'

---

[1] The wording calls to mind Odell's graphic description of the last sighting of his fellow climbers Mallory and Irvine, both known to Hardy, in their attempt on Everest in 1924: "In the sudden clearing of the atmosphere above me I saw the whole summit ridge and finally the peak of Everest unveiled".

Littlewood's Mathematician's Miscellany is a somewhat more light-hearted collection of reflections—he asserts "A good mathematical joke is better, and better mathematics, than a dozen mediocre papers." He describes their collaborative process in some detail

> [Hardy] took a sensual pleasure in 'calligraphy', and it would have been a deprivation if he didn't make the final copy of a joint paper. (My standard role in a joint paper was to make the logical skeleton, in shorthand - no distinction between r and r^2, 2π and 1, etc., etc. But when I said 'Lemma 17' it stayed Lemma 17.)

and elsewhere describing how Hardy unabbreviated the shorthand and added "what we used to call 'gas'." Hardy's deprivation was perhaps not all Littlewood claimed— as Hardy wrote to Bertrand Russell in 1919

> I wish you could find some tactful way of stirring up Littlewood to do a little writing. Heavens knows I am conscious of my huge debt to him. But the situation which is gradually stereotyping itself is very trying for me. It is that, in our collaboration, he will contribute ideas and ideas only: and that all the tedious part of the work has to be done by me. If I don't, it simply isn't done, and nothing would ever get published.

Dame Mary Cartwright (1900–1998) was a former student of Hardy and collaborator of both, the first female FRS in mathematics, and subsequently Mistress of Girton College Cambridge[2]. In a series of papers written when she was over 80 she drew upon Cambridge's archives of Hardy's and Littlewood's manuscripts, and her own memories and notes, to analyse their collaboration. Much of this was by letter: although their rooms in Trinity College were only a few minutes apart, they would often communicate through notes delivered by a servant. They were frequently in different places—from 1915–1919 Littlewood was away on war service; from 1920 to 1931 Hardy worked in Oxford; and in any case outside the short Cambridge terms (24 weeks a year in all) both travelled extensively, whether to work with overseas mathematicians or on lengthy holidays. According to Cartwright they rarely, if ever sat down together to write mathematics, and never discussed at a blackboard, though they did talk to each frequently to discuss critical points and to establish notation, which Littlewood disliked changing as he did a great deal of work in his head. Littlewood, who seems rarely to have bought paper and used all sorts of scraps, including the backs of exam papers and committee minutes, would write terse notes. Hardy, who prided himself on his hand-writing, and bought high-quality new paper (though he was careful to use both sides) would write out a draft and return it to Littlewood alongside the original for Littlewood to check. This way of working sometime led to confusion as when Hardy writes "Also return the scrap of your writing, since otherwise I shall have a letter of yours, not yet exhausted, with a mysterious hole in the middle and shall be confused when I have forgotten what it contained. I have various MSS of yours with irritating holes of this kind in them, and am always wondering if the hole didn't contain just the point."

---

[2]In the mid-twentieth century Cambridge had two women-only colleges and the rest, around two dozen, were all male.

A typical collaboration was on the paper, "A maximal theorem with function theoretic applications", which appeared in Acta Mathematica in July 1930. The main results are motivated by a maximisation theorem explained in a footnote: "The arguments used ... are indeed mostly of the type which are intuitive to a student of cricket averages. A batsman's average is increased by his playing an innings greater than his present average; if his average is increased by playing an innings x, it is further increased by playing next an innings $y > x$; and so forth". The 36 pages of the paper contain 27 theorems, and thanks to the clear exposition (or copious amounts of "gas") is still very readable: all the proofs are clearly set out, with plenty of surrounding text to motivate them, explain why results are best possible and other possible generalisations and proof techniques will not work, and so on.

The published papers reflect Hardy and Littlewood's correspondence, as they discuss hypotheses, difficulties, and alternative ways of arranging and synthesising material, or postponing some of it to future papers or collaborations with others. Ideas pass back and forth with requests from one or the other to give their opinion on a proposed piece of work or to "Please check" drafts. Both seem at ease asking for help or acknowledging mistakes: for example, Hardy writes "Over this I have, at various times, spent many fruitless hours. I seem to remember your saying that you once thought you had it, but succeeded in evading it. If so, you must apply have been at one time (what I never have) in apparent possession of adequate ideas & I wish you'd tell me roughly how you were arguing" or elsewhere "Here's the hell of a mess: it seems that the proofs ... are just fallacious" and "But I have the feeling that I must have missed something more obvious" or even "Are we not both being asses...". In an echo of the philosophical issues in the 1929 Mind paper Hardy distinguishes between 'checking' and 'understanding', writing of one of Littlewood's drafts that "It is a masterpiece of ingenuity, but very difficult (to "understand", not to check.)"

Bohr's 1947 article summarised what he called Hardy and Littlewood's Axioms of Collaboration, presented by Cartwright as:

1. When one wrote to the other, it was completely indifferent whether what they wrote was right or wrong.
2. When one received a letter from the other, he was under no obligation to read it, let alone answer it.
3. Although it did not really matter if they both simultaneously thought about the same detail, still it was preferable that they should not do so.
4. It was quite indifferent if one of them had not contributed the least bit to the contents of a paper under their common name.

While she is in agreement with their overall tone, Cartwright's analysis suggests that the Axioms themselves, and parts of the 'Mathematicians Apology', were somewhat rhetorical and at odds with the evidence of the letters, and the kindness of both men to students and collaborators: she suggests the word "indifferent" perhaps should be read as something akin to "tolerant of". She reported that in her own collaborations, Littlewood "would not let me put his name to any paper not actually written by him. I had to say it was based on joint work with him". Unusually for senior scientists of the time, Hardy and Littlewood seem to be have been generous with crediting the contributions of their students in their papers, either through footnotes or through naming them as co-authors.

In 1913 Hardy responded to a letter from the unknown Indian mathematician, Ramanujan, who had previously written to several of Hardy's more senior colleagues without getting a reply. This led to Ramanujan coming to Cambridge, and another remarkable collaboration, and five joint papers, earning Ramanujan an FRS and subsequently a fellowship of Trinity College (though the college's inability to provide his vegetarian diet meant he ate alone in his rooms). Ramanujan had an extraordinary mathematical intuition and was able to perceive remarkable combinatorial identities and asymptotic results. Working with Hardy he learned contemporary standards of exposition and proof, which he had not previously encountered. Hardy's mastery of European analytic function theory coupled with Ramaunjan's insight was, to quote Littlewood, "a singularly happy collaboration of two men, of quite unlike gifts, in which each contributed the best, most characteristic, and most fortunate work that was in him".

A profound result providing an exact formula for $p(n)$, the number of partitions of an integer, was also remarkable for the contribution of Percy McMahon. A former soldier and an instructor at a military college, McMahon wrote on combinatory analysis in the older tradition of English mathematics, and was famous for his calculating ability. Hardy and Ramanujan sought his help in calculating the first 200 values of $p(n)$, and as they acknowledge in the paper:

> To Major MacMahon in particular we owe many thanks for the amount of trouble he has taken over very tedious calculations. It is certain that, without the encouragement given by the results of these calculations, we should never have attempted to prove theoretical results at all comparable in precision with those which we have enunciated.

2   polymath

Timothy Gowers, a successor to Littlewood in the Rouse Ball chair at Cambridge, was awarded a Fields Medal in 1998 for work combining functional analysis and combinatorics, in particular his proof of Szemerdi's theorem. Gowers has characterised himself as a problem-solver rather than a theory-builder, drawing attention to the importance of problem solvers and problem solving in understanding and developing broad connections and analogies between topics not yet amenable to precise unifying theories. He writes articulately on his blog about many topics connected with mathematics, education and open science, and used this forum to launch his experiments in online collaborative proof which he called "*polymath*".

In a blog post on 27th January 2009 he asked "Is massively collaborative mathematics possible", suggesting that "If a large group of mathematicians could connect their brains efficiently, they could perhaps solve problems very efficiently as well", and proposing some ground rules, which we reproduce in an appendix. The post attracted 203 comments from around the globe, exploring philosophical and practical aspects of working together on a blog to solve problems, and a few days later he launched the first experiment. The problem chosen was to find a new proof of the density version of the "Hales Jewett Theorem", replacing the previously known very technical proof with a more accessible combinatorial argument which, it was hoped, would also open the door to generalisations of the result. Over the next seven weeks, 27 people contributed around 800 comments—around 170,000 words in all—with the contributors ranging from high-school teacher Jason Dyer to Gowers's fellow Fields Medallist Terry Tao. On March 10, 2009 Gowers was able to announce a new combinatorial proof of the result, writing "If this were a conventional way of producing mathematics, then it would be premature to make such an announcement—one would wait until the proof was completely written up with every single i dotted and every t crossed—but this is blog maths and we're free to make up conventions as we go along".

The result was written up as a conventional journal paper, with the author given as "D H J Polymath"—identifying the actual contributors requires some detective work on the blog—and published on the arxiv in 2009, and in the Annals of Mathematics in 2012. The journal version explains the process

> Before we start working towards the proof of the theorem, we would like briefly to mention that it was proved in a rather unusual "open source" way, which is why it is being published under a pseudonym. The work was carried out by several researchers, who wrote their thoughts, as they had them, in the form of blog comments at http://gowers.wordpress.com. Anybody who wanted to could participate, and at all stages of the process the comments were fully open to anybody who was interested."

A typical extract from the blog (Figure 1) shows the style of interaction. Participants, in line with the ground rules, were encouraged to present their ideas in an accessible way, to put forward partial ideas that might be wrong—"better to have had five stupid ideas than no ideas at all", to test out ideas on other participants before doing substantial work on them, and to treat other participants with respect. As the volume of comments and ideas grew, it became apparent that the blog structure made it hard for readers to extract the thread of the argument and keep up with what was going on, without having to digest everything that had been previously posted, and in future experiments a leader took on the task of drawing together the threads from time to time, identifying the most appropriate next direction, and restarting the discussion with a substantial new blog post.

By 2015 there had been nine endeavours in the *polymath* sequence, and a number of others in similar style. Not all had achieved publishable results, with some petering out through lack of participation, but all have left the record of their partial achievements online for others to see and learn from—a marked contrast to partial proof attempts that would normally end up in a waste-basket. The *minipolymath* series applied the *polymath* model to problems drawn from the International Mathematics Olympiad, in an attempt to learn more about the approach—problems typically took hours rather than months to be solved—and the accessible nature of the problem allowed a much greater range of participants.

polymath 8, launched in the summer of 2013, was motivated by Yitang Zhang's proof of a result about bounded gaps between primes. The twin primes conjecture

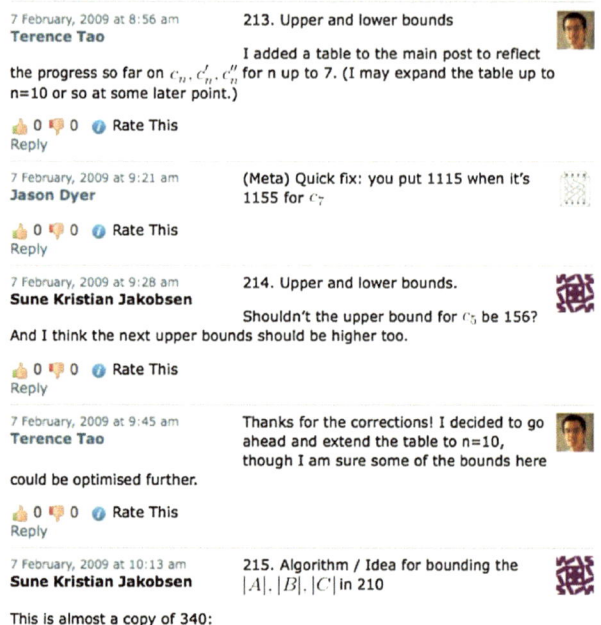

Fig. 1 An extract from the *polymath* blog for the proof of the Density Hales Jewett theorem

states that there are infinitely many pairs of primes that differ by 2: for example 3, 5; 11; 13 and so on. Zhang proved that there is a number K such that infinitely many pairs of primes differ by at most K, and showed that K is less than 70,000,000. After various discussions on other blogs, Tao formally launched the project, to improve the bound on K, on 13th June 2013. The first part of the project, *polymath* 8a, concluded with a bound of 4,680, and a research paper, also put together collaboratively, appeared on the arxiv in February 2014. The second part, *polymath* 8b, combined this with techniques independently developed in parallel by James Maynard, to reach a bound of 246, with a research paper appearing on the arxiv in July 2014. The participants also used Tao's blog to seek input for a retrospective paper reflecting on the experience, which appeared in the arxiv in September 2014.

One immediate concern was the scoping of the enquiry so as not to intimidate or hamper individuals working on their own on this hot topic: it was felt that this was more than countered by providing a resource for the mathematical community that would capture progress, and provide a way to pull together what would otherwise be many independent tweaks. The work was well suited to the *polymath* approach: the combination of Tao's leadership and the timeliness of the problem made it easy to recruit participants; the bound provided an obvious metric of progress and maintained momentum; and it naturally fell into five components forming what Tao called a "factory production line". The collaborative approach allowed people to learn new material, and brought rapid sharing of expertise, in particular knowledge of the literature, and access to computational and software skills.

Tao himself explained how he was drawn into the project, even though it interrupted another big project a piece of work that he expects to take some years, by the ease of making simple improvements to Zhang's bound, and summed up by saying "All in all, it was an exhausting and unpredictable experience, but also a highly thrilling and rewarding one." The time commitment was indeed intense—for example, a typical thread " *polymath*8b, II: Optimising the variational problem and the sieve" started on 22 November 2013, and ran for just over two weeks until 8th December. The initial post by Tao runs to about 4000 words—it is followed by 129 posts, of which 36, or just under a third, are also by Tao.

Tao and other participants were motivated above all by the kudos of solving a high-profile problem, in a way that was unlikely had they worked individually, but also by the excitement of the project, and the enthusiasm of the participants and the wider community. They reported enjoying working in this new way, especially the opportunity to work alongside research stars, the friendliness of the other participants, and their tolerance of errors, and the way in which the problem itself and the *polymath* format provided the incentive of frequent incremental progress, in a way not typical of solo working.

Participants needed to balance the incentives for participation against other concerns. Chief among these was the time commitment: participants reported the need for intense concentration and focus, with some working on it at a "furious pace" for several months; some feeling that the time required to grasp everything that was happening on the blog make *polymath* collaborations more, rather than less, time consuming than traditional individual work or small-group collaboration; and some feeling that the fast pace was deterring participants whose working style was slower and more reflective.

Pure mathematicians typically produce one or two journal papers a year, so, particularly for those who do not yet have established positions, there will be concerns that a substantial investment of time in *polymath* might damage their publication record. While such a time commitment would normally be worth the risk for a high-profile problem that has the likely reward of a good publication, the benefits are less clear-cut when the paper is authored under a group pseudonym (D H J Polymath), with the list of participants given in a linked wiki. As a participant remarked " *polymath* was a risk for those who did not have tenure". On the other hand, in a fast moving area, participants may feel that incorporating their ideas into the collective allows them to make a contribution that they would not have achieved with solo work, or that engaging in this way is better than being beaten to a result and getting no credit, especially if participation in a widely read blog is already adding to their external reputation.

An additional risk for those worried about their reputation can be that mistakes are exposed for ever in a public forum: pre-tenure mathematician Pace Nielsen was surprised that people were "impressed with my bravery" and would advise considering this issue before taking part. Rising star James Maynard observed: "It was very unusual for me to work in such a large group and so publicly—one really needed to lose inhibitions and be willing to post ideas that were not fully formed (and potentially wrong!) online for everyone to see."

Those reading the *polymath* 8 sites went well beyond the experts—with an audience appreciating the chance to see how mathematics was done behind the scenes, or as Tao put it "How the sausage is made". At its height it was getting three thousand hits a day, and even readers who knew little mathematics reported the excitement of checking regularly and watching the bounds go down. All the members of a class of number theory students at a summer school on Bounded Gaps admitted to following *polymath*. Perhaps typical was Andrew Roberts, an undergraduate who thanked the organisers for such an educational resource and reported "reading the posts and following the 'leader-board' felt a lot like an academic spectator sport. It was surreal, a bit like watching a piece of history as it occurred. It made the mathematics feel much more alive and social, rather than just coming from a textbook. I don't think us undergrads often get the chance to peak behind closed doors and watch professional mathematicians 'in the wild' like this, so from a career standpoint, it was illuminating". David Roberts, an Australian educator who used *polymath* in his classes to show students how the things they were learning were being used in cutting-edge research, reported "For me personally it felt like being able to sneak into the garage and watch a high-performance engine being built up from scratch; something I could never do, but could appreciate the end result, and admire the process". The good manners remarked upon by the expert participants extended to less-well informed users, with questions and comments from non-experts generally getting a polite response, often from Tao himself, and a few more outlandish comments, such as claims of a simple proof, being ignored except for a plethora of down-votes.

The experiments have attracted widespread attention, in academia and beyond. Gowers had worked closely with physicist Michael Nielson in designing the wiki and blog structure to support *polymath*, and in an article in Nature in 2009 the pair reflected on its wider implications, a theme developed further in Nielsen's 2012 book, Reinventing Discovery, and picked up by researchers in social and computer science analysing the broader phenomenon of open science enabled by the internet.

Like the participants, the analysts remarked on the value of the *polymath* blogs for capturing the records of how mathematics is done, the kinds of thinking that goes into the production of a proof, such as experimenting with examples, computations and concepts, and showing the dead ends and blind alleys. As Gowers and Nielson put it, "Who would have guessed that the working record of a mathematical project would read like a thriller?" The working records provide data for online ethnography, allowing analysis of what is involved in the creation of a proof—for example in the *minipolymath* projects as little as 30% of the comments were actual proof steps, with the rest being analysis of examples, concepts and so on.

Although *polymath* is often described as "crowdsourced science", the crowd is a remarkably small and expert one. The analogy has often been drawn with open source software projects—however these are typically organised in a much more modular and top-down fashion than is possible in developing a mathematical proof, where many ideas and strands will be interwoven in a manner, as Nielsen comments, much more akin to a novel.

These studies of crowdsourcing help explain the success of *polymath* and other attempts at open collaboration in mathematics. Mathematicians have well-established shared standards for exposition and argument, making it easy to resolve disputes. As the proof develops, the blog provides a shared cognitive space and short-term working memory. The ground rules allow for a dynamic division of labour, and encourage a breakdown into smaller subtasks thus reducing barriers to entry and increasing diversity. This diversity of skills is encouraged, and a later stage might be to design some form of rating or ranking—an "attention architecture"— so that participants can more easily find the subtask where they might be able to contribute, and more readily explore past contributions.

A striking aspect of *polymath* is that senior figures in the field are prepared to try such a bold experiment, to think though clearly for themselves what the requirements are, and to take a "user centred" view of the design. For example, it was suggested that participants might use a platform such as github, designed to support distributed version control, to make the final stage, collaborating on a paper, more straightforward: Tao's response was to prioritise user accessibility: "One thing I worry about is that if we use any form of technology more complicated than a blog comment box, we might lose some of the participants who might be turned off by the learning curve required". On the whole *polymath* has proved successful in solving problems amenable to the approach, with the added benefit of presenting to the public a new way of mathematics. The failures have been through failures to find the necessary mathematical breakthrough, rather than to any obvious failures of the *polymath* format.

3 Conclusions

To study the practice of mathematics is to study not only the theorems produced, but also what the individuals and institutions involved in their creation do to bring them about. Insofar as mathematical results require readers as well as writers, and mathematical knowledge accretes by building on the work of others, mathematics has always been a collaborative endeavour. To quote David Hume, writing in 1739

> There is no Algebraist nor Mathematician so expert in his science, as to place entire confidence in any truth immediately upon his discovery of it, or regard it as any thing, but a mere probability. Every time he runs over his proofs, his confidence encreases; but still more by the approbation of his friends; and is rais'd to its utmost perfection by the universal assent and applauses of the learned world.

The institutions surrounding mathematics provide trusted and well-established mechanisms for accrediting and disseminating knowledge through journals, meetings and opportunities for informal exchange, and rely on the collaboration of mathematicians for their success and authority, complementing the professionalisation of mathematics research and education, through employment of mathematicians by universities. They enable development and transmission of shared concepts and methods: not just the mathematical objects under consideration, but also expectations of rigour, proof and exposition; notions of what constitutes significant research questions; and expectations of how a professional mathematician should conduct themselves—for example in matters of citation or co-authorship.

Cartwright, Hardy, Littlewood, McMahon and Ramanujan collaborated at a time when these mechanisms were becoming fixed, provided through leading mathematics departments in Europe and the USA, and emerging learned societies patterned on the London Mathematical Society. One hundred years later collaborating pure mathematicians, the *polymath* collaborators included, still have as their ultimate goal research papers that meet traditional refereeing standards and are published in the conventional way. The goals of scholars are still a mixture of intellectual satisfaction and professional recognition, with then, as now, success measured by proving significant results and publishing them in significant journals, or the additional recognition of well-known prizes. A modern-day Cartwright or Ramanujan would take it for granted that men and women could apply for the same jobs, and expect to find something to eat when at work, and have their applications judged on objective criteria—though aware that the added kudos of a letter of reference from a major player or a highly visible online presence could do no harm.

Then, as now, successful collaboration requires a shared understanding of the mathematical objects under consideration, shared notions of rigour, proof and exposition, shared expectations of how professional mathematicians should conduct themselves, and above all that the collaborators have collectively the skills needed to solve the problem.

For modern collaborators the most striking difference is the speed of communication and the quantity and availability of information. Thanks to email, github, latex and skype mathematicians can collaborate at a distance, share hand-written drafts and work together on the final version of documents; blogs and facebook have replaced attendance at monthly meetings of the London Mathematical Society as a means of catching up with the latest scientific gossip; and the calculating skills of McMahon are as redundant as those of the blacksmith. Papers are available immediately they are written through the arxiv, rather than having to wait for them to cross the ocean in printed form and appear on the shelves of a library, and scholars can readily search, find and access current and historical literature. It has been argued that this flattening is providing a homogenising effect that has negative aspects—in that uniform expectations and standards are crowding out the distinctive approaches which might come from individuals who have not been through a standard education system, and further that the decline of distinct schools of mathematics, such as those that flourished in the former Soviet union, is leading to a narrowing not just of techniques and skills, but also of intellectual approach, restricting the kinds of mathematical questions that are asked.

All of this is true of any twenty-first century collaboration. Perhaps what is most remarkable and distinctive about the *polymath* experiments is the public display and sharing which was not possible in the past. An early aspiration for *polymath*—that the format would allow contributors to take part in small pieces of the discussion— was not really fulfilled as participants found that the fast pace required intense concentration and focus, and to leave and return was not really possible as it required too much effort to catch up with progress in the interim. However *polymath* succeeded in opening up the process of mathematical discovery to a wide audience from a variety of backgrounds, and the ground rules, and the involvement of senior figures on the discussions, reinforced in a very open and public way expectations of how mathematical collaboration should be conducted.

A number of participants and possible participants remarked on their distaste for the fast pace of *polymath* discussions, which requires people to work quickly and not mind making mistakes in public, and wondered whether this is really a necessary condition for collaboration, or merely a way for people who are comfortable with this to gain status in the community. One feels that G H Hardy, who hated the highly competitive nineteenth century Cambridge mathematical tripos, and was well known for his reticence with strangers, would have agreed.

Appendix—the *polymath* Ground Rules

These ground rules are reproduced from Gowers's Weblog post of 27 January 2009

1. The aim will be to produce a proof in a top-down manner. Thus, at least to start with, comments should be short and not too technical: they would be more like feasibility studies of various ideas.
2. Comments should be as easy to understand as is humanly possible. For a truly collaborative project it is not enough to have a good idea: you have to express it in such a way that others can build on it.
3. When you do research, you are more likely to succeed if you try out lots of stupid ideas. Similarly, stupid comments are welcome here. (In the sense in which I am using 'stupid', it means something completely different from 'unintelligent'. It just means not fully thought through.)
4. If you can see why somebody else's comment is stupid, point it out in a polite way. And if someone points out that your comment is stupid, do not take offence: better to have had five stupid ideas than no ideas at all. And if somebody wrongly points out that your idea is stupid, it is even more important not to take offence: just explain gently why their dismissal of your idea is itself stupid.
5. Don't actually use the word 'stupid', except perhaps of yourself.
6. The ideal outcome would be a solution of the problem with no single individual having to think all that hard. The hard thought would be done by a sort of super-mathematician whose brain is distributed amongst bits of the brains of lots of interlinked people. So try to resist the temptation to go away and think about something and come back with carefully polished thoughts: just give quick reactions to what you read and hope that the conversation will develop in good directions.

7. If you are convinced that you could answer a question, but it would just need a couple of weeks to go away and try a few things out, then still resist the temptation to do that. Instead, explain briefly, but as precisely as you can, why you think it is feasible to answer the question and see if the collective approach gets to the answer more quickly. (The hope is that every big idea can be broken down into a sequence of small ideas. The job of any individual collaborator is to have these small ideas until the big idea becomes obvious and therefore just a small addition to what has gone before.) Only go off on your own if there is a general consensus that is what you should do.

8. Similarly, suppose that somebody has an imprecise idea and you think that you can write out a fully precise version. This could be extremely valuable to the project, but don't rush ahead and do it. First, announce in a comment what you think you can do. If the responses to your comment suggest that others would welcome a fully detailed proof of some substatement, then write a further comment with a fully motivated explanation of what it is you can prove, and give a link to a pdf file that contains the proof.

9. Actual technical work, as described in 8, will mainly be of use if it can be treated as a module. That is, one would ideally like the result to be a short statement that others can use without understanding its proof.

10. Keep the discussion focused. For instance, if the project concerns a particular approach to a particular problem (as it will do at first), and it causes you to think of a completely different approach to that problem, or of a possible way of solving a different problem, then by all means mention this, but don't disappear down a different track.

11. However, if the different track seems to be particularly fruitful, then it would perhaps be OK to suggest it, and if there is widespread agreement that it would in fact be a good idea to abandon the original project (possibly temporarily) and pursue a new one—a kind of decision that individual mathematicians make all the time—then that is permissible.

12. Suppose the experiment actually results in something publishable. Even if only a very small number of people contribute the lion's share of the ideas, the paper will still be submitted under a collective pseudonym with a link to the entire online discussion.

Further reading

Hardy and Littlewood

Mathematical Proof, G H Hardy, Mind 38 (1929), 1–25
A Mathematician's Apology, G H Hardy, Cambridge University Press 1940
Littlewood's Miscellany, edited by Bela Bollobas, Cambridge University Press 1986

Some Hardy-Littlewood Manuscripts, M L Cartwright, Bulletin of the London Mathematical Society 13 (1981), 273–300

Later Hardy and Littlewood Manuscripts, M L Cartwright, Bulletin of the London Mathematical Society 17 (1985), 318–390

Partnership and Partition: A Case Study of Mathematical Exchange within the London Mathematical Society, Adrian Rice, Philosophia Scientiae, 19 (2015) 5–24

The polymath initiative

Massively collaborative mathematics, Timothy Gowers and Michael Nielsen, Nature, 461 (2009) 879–881

The *polymath* Blog http://polymathprojects.org

The *polymath* wiki http://michaelnielsen.org/polymath1

"Is massively collaborative mathematics possible?", Gowers's Weblog, Retrieved 2015-3-16. https://gowers.wordpress.com/2009/01/27/is-massively-collaborative-mathematics-possible/

Reinventing discovery, Michael Nielsen, Princeton 2012

D H J Polymath, "The 'bounded gaps between primes' *polymath* project: A retrospective analysis", Newsletter of the European Mathematical Society, 94 (2014) 13–23

Acknowledgements The first author acknowledges support from the UK Engineering and Physical Sciences Research Council under grant EP/K040251.

Experimental computation as an ontological game changer: The impact of modern mathematical computation tools on the ontology of mathematics

David H. Bailey and Jonathan M. Borwein

Abstract Robust, concrete and abstract, mathematical computation and inference on the scale now becoming possible should change the discourse about many matters mathematical. These include: what mathematics is, how we know something, how we persuade each other, what suffices as a proof, the infinite, mathematical discovery or invention, and other such issues.

1 Introduction

Like almost every other field of fundamental scientific research, mathematics (both pure and applied) has been significantly changed by the advent of advanced computer and related communication technology. Many pure mathematicians routinely employ symbolic computing tools, notably the commercial products *Maple* and *Mathematica*, in a manner that has become known as *experimental mathematics*: using the computer as a "laboratory" to perform exploratory experiments, to gain insight and intuition, to discover patterns that suggest provable mathematical facts, to test and/or falsify conjectures, and to numerically confirm analytically derived results.

Applied mathematicians have adopted computation with even more relish, in applications ranging from mathematical physics to engineering, biology, chemistry, and medicine. Indeed, at this point it is hard to imagine a study in applied mathematics that does not include some computational content. While we should distinguish "production" code from code used in the course of research, the methodology used by applied mathematics is essentially the same as the experimental approach in pure

D.H. Bailey (✉)
Lawrence Berkeley National Laboratory (retired), Berkeley, CA 94720, USA
e-mail: david@davidhbailey.com

J.M. Borwein
CARMA, The University of Newcastle, University Drive,
Callaghanm, NSW 2308, Australia
e-mail: Jonathan.Borwein@newcastle.edu.au

© Springer International Publishing Switzerland 2015
E. Davis, P.J. Davis (eds.), *Mathematics, Substance and Surmise*,
DOI 10.1007/978-3-319-21473-3_3

mathematics: using the computer as a "laboratory" in much the same way as a physicist, chemist, or biologist uses laboratory equipment (ranging from a simple test tube experiment to a large-scale analysis of the cosmic microwave background) to probe nature [51, 53].

This essay addresses how the emergence and proliferation of experimental methods has altered the doing of mathematical research, and how it raises new and often perplexing questions of what exactly is mathematical truth in the computer age. Following this introduction, the essay is divided into four sections[1]: (i) experimental methods in pure mathematics, (ii) experimental methods in applied mathematics, (iii) additional examples (involving a somewhat higher level of sophistication), and (iv) concluding remarks.

2 The experimental paradigm in pure mathematics

By *experimental mathematics* we mean the following *computationally-assisted* approach to mathematical research [24]:

1. Gain insight and *intuition*;
2. *Visualize* mathematical principles;
3. *Discover* new relationships;
4. *Test* and especially *falsify* conjectures;
5. *Explore* a possible result to see if it *merits* formal proof;
6. *Suggest* approaches for formal proof;
7. *Replace* lengthy hand derivations;
8. *Confirm* analytically derived results.

We often call this 'experimental mathodology.' As noted in [24, Ch. 1], some of these steps, such as to gain insight and intuition, have been part of traditional mathematics; indeed, experimentation need not involve computers, although nowadays almost all do. In [20–22], a more precise meaning is attached to each of these items. In [20, 21] the focus is on pedagogy, while [22] addresses many of the philosophical issues more directly. We will revisit these items as we continue our essay. With regards to item 5, we have often found the computer-based tools useful to tentatively confirm preliminary lemmas; then we can proceed fairly safely to see where they lead. If, at the end of the day, this line of reasoning has not led to anything of significance, at least we have not expended large amounts of time attempting to formally prove these lemmas (for example, see Section 2.6).

With regards to item 6, our intending meaning here is more along the lines of "computer-assisted" or "computer-directed proof," and thus is distinct from

---

[1]We borrow heavily from two of our recent prior articles [9] ('pure') and [8] ('applied'). They are reused with the permission of the *American Mathematical Society* and of *Princeton University Press*, respectively.

methods of computer-based *formal proof*. On the other hand, such methods have been pursued with significant success lately, such as in Thomas Hales' proof of the Kepler conjecture [36], a topic that we will revisit in Section 2.10. With regards to item 2, when the authors were undergraduates they both were taught to use calculus as a way of making accurate graphs. Today, good graphics tools (item 2) exchanges the cart and the horse. We graph to see structure that we then analyze [20, 21].

Before turning to concrete examples, we first mention two of our favorite tools. They both are essential to the attempts to find patterns, to develop insight, and to efficiently falsify errant conjectures (see items 1, 2, and 4 in the list above).

Integer relation detection. Given a vector of real or complex numbers x_i, an *integer relation algorithm* attempts to find a nontrivial set of integers a_i, such that $a_1 x_1 + a_2 x_2 + \cdots + a_n x_n = 0$. One common application of such an algorithm is to find new identities involving computed numeric constants.

For example, suppose one suspects that an integral (or any other numerical value) x_1 might be a linear sum of a list of terms $x_2, x_3, \ldots, x_n$. One can compute the integral and all the terms to high precision, typically several hundred digits, then provide the vector $(x_1, x_2, \ldots, x_n)$ to an integer relation algorithm. It will either determine that there is an integer-linear relation among these values, or provide a lower bound on the Euclidean norm of any integer relation vector (a_i) that the input vector might satisfy. If the algorithm does produce a relation, then solving it for x_1 produces an experimental identity for the original integral. The most commonly employed integer relation algorithm is the "PSLQ" algorithm of mathematician-sculptor Helaman Ferguson [24, pp. 230–234], although the *Lenstra-Lenstra-Lovasz* (LLL) algorithm can also be adapted for this purpose. In 2000, *integer relation methods* were named one of the top ten algorithms of the twentieth century by *Computing in Science and Engineering*. In our experience, the rapid falsification of hoped-for conjectures (item #4) is central to the use of integer relation methods.

High-precision arithmetic. One fundamental issue that arises in discussions of "truth" in mathematical computation, pure or applied, is the question of whether the results are numerically reliable, i.e., whether the precision of the underlying floating-point computation was sufficient. Most work in scientific or engineering computing relies on either 32-bit IEEE floating-point arithmetic (roughly seven decimal digit precision) or 64-bit IEEE floating-point arithmetic (roughly 16 decimal digit precision). But for an increasing body of studies, even 16-digit arithmetic is not sufficient. The most common form of high-precision arithmetic is "double-double" (equivalent to roughly 31-digit arithmetic) or "quad-precision" (equivalent to roughly 62-digit precision). Other studies require very high precision—hundreds or thousands of digits.

A premier example of the need for very high precision is the application of integer relation methods. It is easy to show that if one wishes to recover an n-long integer relation whose coefficients have maximum size d digits, then both the input data and the integer relation algorithm must be computed using at somewhat more than nd-digit precision, or else the underlying relation will be lost in a sea of numerical noise.

Algorithms for performing arithmetic and evaluating common transcendental functions with high-precision data structures have been known for some time, although challenges remain. Computer algebra software packages such as *Maple* and *Mathematica* typically include facilities for arbitrarily high precision, but for some applications researchers rely on internet-available software, such as the GNU multiprecision package. In many cases the implementation and high-level auxiliary tools provided in commercial packages are more than sufficient and easy to use, but 'caveat emptor' is always advisable.

2.1 Digital integrity, I

With regards to #8 above, we have found computer software to be particularly effective in ensuring the integrity of published mathematics. For example, we frequently check and correct identities in mathematical manuscripts by computing particular values on the left-hand side and right-hand side to high precision and comparing results—and then, if necessary, use software to repair defects—often in semi-automated fashion. As authors, we often know what sort of mistakes are most likely, so that is what we hunt for.

As a first example, in a study of "character sums" we wished to use the following result derived in [28]:

$$\sum_{m=1}^{\infty}\sum_{n=1}^{\infty}\frac{(-1)^{m+n-1}}{(2m-1)(m+n-1)^3} \tag{1}$$

$$\stackrel{?}{=} 4\operatorname{Li}_4\left(\frac{1}{2}\right) - \frac{51}{2880}\pi^4 - \frac{1}{6}\pi^2\log^2(2) + \frac{1}{6}\log^4(2) + \frac{7}{2}\log(2)\zeta(3).$$

Here $\operatorname{Li}_4(1/2)$ is a 4-th order polylogarithmic value. However, a subsequent computation to check results disclosed that whereas the left-hand side evaluates to $-0.872929289\ldots$, the right-hand side evaluates to $2.509330815\ldots$. Puzzled, we computed the sum, as well as each of the terms on the right-hand side (sans their coefficients), to 500-digit precision, then applied the PSLQ algorithm. PSLQ quickly found the following:

$$\sum_{m=1}^{\infty}\sum_{n=1}^{\infty}\frac{(-1)^{m+n-1}}{(2m-1)(m+n-1)^3} \tag{2}$$

$$= 4\operatorname{Li}_4\left(\frac{1}{2}\right) - \frac{151}{2880}\pi^4 - \frac{1}{6}\pi^2\log^2(2) + \frac{1}{6}\log^4(2) + \frac{7}{2}\log(2)\zeta(3).$$

In other words, in the process of transcribing (1) into the original manuscript, "151" had become "51." It is quite possible that this error would have gone undetected and uncorrected had we not been able to computationally check and correct such results. This may not always matter, but it can be crucial.

Along this line, Alexander Kaiser and the present authors [13] have developed some prototype software to refine and automate this process. Such semi-automated integrity checking becomes pressing when verifiable output from a symbolic manipulation might be the length of a Salinger novel. For instance, recently while studying expected radii of points in a hypercube [26], it was necessary to show existence of a "closed form" for

$$J(t) := \int_{[0,1]^2} \frac{\log(t + x^2 + y^2)}{(1 + x^2)(1 + y^2)} \, dx \, dy. \tag{3}$$

The computer verification of [26, Thm. 5.1] quickly returned a $100,000$-character "answer" that could be numerically validated very rapidly to hundreds of places (items #7 and #8). A highly interactive process reduced a basic instance of this expression to the concise formula:

$$J(2) - \frac{\pi^2}{8} \log 2 - \frac{7}{48} \zeta(3) + \frac{11}{24} \pi \, \mathrm{Cl}_2 \left(\frac{\pi}{6} \right) - \frac{29}{24} \pi \, \mathrm{Cl}_2 \left(\frac{5\pi}{6} \right), \tag{4}$$

where Cl_2 is the *Clausen function* $\mathrm{Cl}_2(\theta) := \sum_{n \geq 1} \sin(n\theta)/n^2$ (Cl_2 is the simplest non-elementary Fourier series). Automating such reductions will require a sophisticated simplification scheme with a very large and extensible knowledge base, but the tool described in [13] provides a reasonable first approximation. At this juncture, the choice of software is critical for such a tool. In retrospect, our *Mathematica* implementation would have been easier in *Maple*, due to its more flexible facility for manipulating expressions, while *Macsyma* would have been a fine choice were it still broadly accessible.

2.2 Discovering a truth

Giaquinto's [32, p. 50] attractive encapsulation

> In short, discovering a truth is coming to believe it in an independent, reliable, and rational way.

has the satisfactory consequence that a student can legitimately discover things already "known" to the teacher. Nor is it necessary to demand that each dissertation be absolutely original—only that it be independently discovered. For instance, a differential equation thesis is no less meritorious if the main results are subsequently found to have been accepted, unbeknown to the student, in a control theory journal a month earlier—provided they were independently discovered. Near-simultaneous independent discovery has occurred frequently in science, and such instances are likely to occur more and more frequently as the earth's "new nervous system" (Hillary Clinton's term in a policy address while Secretary of State) continues to pervade research.

Despite the conventional identification of mathematics with deductive reasoning, in his 1951 Gibbs lecture, Kurt Gödel (1906–1978) said:

> If mathematics describes an objective world just like physics, there is no reason why inductive methods should not be applied in mathematics just the same as in physics.

He held this view until the end of his life despite—or perhaps because of—the epochal deductive achievement of his incompleteness results.

Also, we emphasize that many great mathematicians from Archimedes and Galileo—who reputedly said *"All truths are easy to understand once they are discovered; the point is to discover them."*—to Gauss, Poincaré, and Carleson have emphasized how much it helps to "know" the answer beforehand. Two millennia ago, Archimedes wrote, in the Introduction to his long-lost and recently reconstituted *Method* manuscript,

> For it is easier to supply the proof when we have previously acquired, by the method, some knowledge of the questions than it is to find it without any previous knowledge.

Archimedes' *Method* can be thought of as a precursor to today's interactive geometry software, with the caveat that, for example, *Cinderella*, as opposed to *Geometer's SketchPad*, actually does provide proof certificates for much of Euclidean geometry.

As 2006 Abel Prize winner Lennart Carleson describes, in his 1966 speech to the International Congress on Mathematicians on the positive resolution of Luzin's 1913 conjecture (namely, that the Fourier series of square-summable functions converge pointwise a.e. to the function), after many years of seeking a counterexample, he finally decided none could exist. He expressed the importance of this confidence as follows:

> The most important aspect in solving a mathematical problem is the conviction of what is the true result. Then it took 2 or 3 years using the techniques that had been developed during the past 20 years or so.

In similar fashion, Ben Green and Terry Tao have commented that their proof of the *existence of arbitrarily long sequence of primes in arithmetic progression* was undertaken only after extensive computational evidence (experimental data) had been amassed by others [24]. Thus, we see in play the growing role that intelligent large-scale computing can play in providing the needed confidence to attempt a proof of a hard result (see item 4 at the start of Section 2).

2.3   Digital assistance

By *digital assistance*, we mean the use of:

1. *Integrated mathematical software* such as *Maple* and *Mathematica*, or indeed MATLAB and their open source variants such as SAGE and Octave.

2. *Specialized packages* such as CPLEX (optimization), PARI (computer algebra), SnapPea (topology), OpenFoam (fluid dynamics), Cinderella (geometry), and MAGMA (algebra and number theory).
3. *General-purpose programming languages* such as Python, C, C++, and Fortran-2000.
4. *Internet-based applications* such as: Sloane's Encyclopedia of Integer Sequences, the Inverse Symbolic Calculator,[2] Fractal Explorer, Jeff Weeks' Topological Games, or Euclid in Java.
5. *Internet databases and facilities* including Excel (and its competitors) Google, MathSciNet, arXiv, Wikipedia, Wolfram Alpha, MathWorld, MacTutor, Amazon, Kindle, GitHub, iPython, IBM's Watson, and many more that are not always so viewed.

A cross-section of Internet-based mathematical resources is available at http://www.experimentalmath.info. We are of the opinion that it is often more realistic to try to adapt resources being built for other purposes by much-deeper pockets, than to try to build better tools from better first principles without adequate money or personnel.

Many of the above tools entail data-mining in various forms, and, indeed, data-mining facilities can broadly enhance the research enterprise. The capacity to consult the Oxford dictionary and Wikipedia instantly within Kindle dramatically changes the nature of the reading process. Franklin [31] argues that Steinle's "exploratory experimentation" facilitated by "widening technology" and "wide instrumentation," as routinely done in fields such as pharmacology, astrophysics, medicine, and biotechnology, is leading to a reassessment of what legitimates experiment. In particular, a "local theory" (i.e., a specific model of the phenomenon being investigated) is not now a prerequisite. Thus, a pharmaceutical company can rapidly examine and discard tens of thousands of potentially active agents, and then focus resources on the ones that survive, rather than needing to determine in advance which are likely to work well. Similarly, aeronautical engineers can, by means of computer simulations, discard thousands of potential designs, and submit only the best prospects to full-fledged development and testing.

Hendrik Sørenson [49] concisely asserts that experimental mathematics—as defined above—is following a similar trajectory, with software such as *Mathematica*, *Maple*, and MATLAB playing the role of wide instrumentation:

> These aspects of exploratory experimentation and wide instrumentation originate from the philosophy of (natural) science and have not been much developed in the context of experimental mathematics. However, I claim that e.g. the importance of wide instrumentation for an exploratory approach to experiments that includes concept formation also pertain to mathematics.

---

[2]Most of the functionality of the ISC, which is now housed at http://carma-lx1.newcastle.edu.au:8087, is now built into the "identify" function of *Maple* starting with version 9.5. For example, the *Maple* command `identify(4.45033263602792)` returns $\sqrt{3}+e$, meaning that the decimal value given is simply approximated by $\sqrt{3} + e$.

In consequence, boundaries between mathematics and the natural sciences, and between inductive and deductive reasoning are blurred and becoming more so (see also [4]). This convergence also promises some relief from the frustration many mathematicians experience when attempting to describe their proposed methodology on grant applications to the satisfaction of traditional hard scientists. We leave unanswered the philosophically-vexing if mathematically-minor question as to whether genuine mathematical experiments (as discussed in [24]) truly exist, even if one embraces a fully idealist notion of mathematical existence. It surely *seems* to the two of us that they do.

2.4 Pi, partitions, and primes

The present authors cannot now imagine doing mathematics without a computer nearby. For example, characteristic and minimal polynomials, which were entirely abstract for us as students, now are members of a rapidly growing box of concrete symbolic tools. One's eyes may glaze over trying to determine structure in an infinite family of matrices including

$$M_4 = \begin{bmatrix} 2 & -21 & 63 & -105 \\ 1 & -12 & 36 & -55 \\ 1 & -8 & 20 & -25 \\ 1 & -5 & 9 & -8 \end{bmatrix} \quad M_6 = \begin{bmatrix} 2 & -33 & 165 & -495 & 990 & -1386 \\ 1 & -20 & 100 & -285 & 540 & -714 \\ 1 & -16 & 72 & -177 & 288 & -336 \\ 1 & -13 & 53 & -112 & 148 & -140 \\ 1 & -10 & 36 & -66 & 70 & -49 \\ 1 & -7 & 20 & -30 & 25 & -12 \end{bmatrix}$$

but a command-line instruction in a computer algebra system will reveal that both $M_4^3 - 3M_4 - 2I = 0$ and $M_6^3 - 3M_6 - 2I = 0$, thus illustrating items 1 and 3 at the start of Section 2. Likewise, more and more matrix manipulations are profitably, even necessarily, viewed graphically. As is now well known in numerical linear algebra, graphical tools are essential when trying to discern qualitative information such as the block structure of very large matrices, thus illustrating item 2 at the start of Section 2. See, for instance, Figure 1.

Equally accessible are many matrix decompositions, the use of Groebner bases, Risch's decision algorithm (to decide when an elementary function has an elementary indefinite integral), graph and group catalogues, and others. Many algorithmic components of a computer algebra system are today extraordinarily effective compared with two decades ago, when they were more like toys. This is equally true of extreme-precision calculation—a prerequisite for much of our own work [19, 23–25]. As we will illustrate, during the three decades that we have seriously tried to integrate computational experiments into research, we have experienced at least twelve Moore's law doublings of computer power and memory capacity, which when combined with the utilization of highly parallel clusters (with thousands of processing cores) and fiber-optic networking, has resulted in six to seven orders of magnitude speedup for many operations.

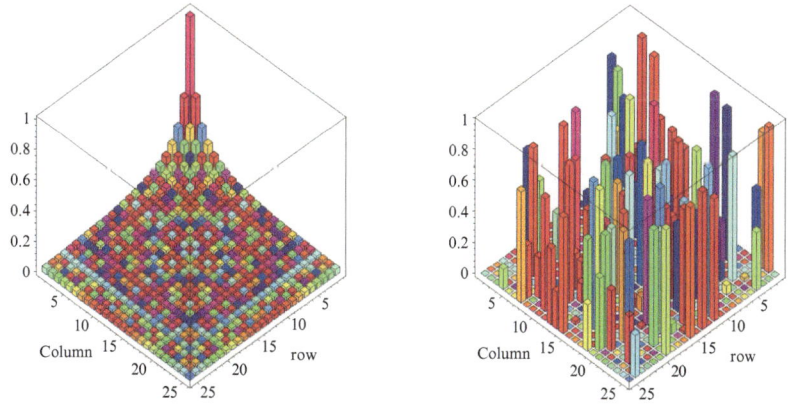

Fig. 1 Plots of a 25×25 Hilbert matrix (L) and a matrix with 50% sparsity and random $[0, 1]$ entries (R).

2.5 The partition function

Consider the number of *additive partitions*, $p(n)$, of a natural number, where we ignore order and zeroes. For instance, $5 = 4 + 1 = 3 + 2 = 3 + 1 + 1 = 2 + 2 + 1 = 2 + 1 + 1 + 1 = 1 + 1 + 1 + 1 + 1$, so $p(5) = 7$. The ordinary generating function (5) discovered by Euler is

$$\sum_{n=0}^{\infty} p(n)q^n = \prod_{k=1}^{\infty} \left(1 - q^k\right)^{-1}. \tag{5}$$

(This can be proven by using the geometric formula for $1/(1 - q^k)$ to expand each term and observing how powers of q^n occur.)

The famous computation by Percy MacMahon of $p(200) = 3972999029388$ at the beginning of the 20th century, done symbolically and entirely naively from (5) in *Maple* on a laptop, took 20 minutes in 1991 but only 0.17 seconds in 2010, while the many times more demanding computation

$$p(2000) = 4720819175619413888601432406799959512200344166$$

took just two minutes in 2009 and 40.7 seconds in 2014.[3] Moreover, in December 2008, the late Richard Crandall was able to calculate $p(10^9)$ in three seconds on his laptop, using the Hardy-Ramanujan-Rademacher 'finite' series for $p(n)$ along

---

[3]The difficulty of comparing timings and the growing inability to look under the hood (bonnet) in computer packages, either by design or through user ignorance, means all such comparisons should be taken with a grain of salt.

with FFT methods. Using these techniques, Crandall was also able to calculate the probable primes $p(1000046356)$ and $p(1000007396)$, each of which has roughly 35000 decimal digits.[4]

Such results make one wonder when easy access to computation discourages innovation: *Would Hardy and Ramanujan have still discovered their marvelous formula for p(n) if they had powerful computers at hand?*

2.6 Reciprocal series for π

Truly novel series for $1/\pi$, based on elliptic integrals, were discovered by Ramanujan around 1910 [11, 54]. One is:

$$\frac{1}{\pi} = \frac{2\sqrt{2}}{9801} \sum_{k=0}^{\infty} \frac{(4k)!\,(1103 + 26390k)}{(k!)^4 396^{4k}}. \tag{6}$$

Each term of (6) adds eight correct digits. Gosper used (6) for the computation of a then-record 17 million digits of π in 1985—thereby completing the first proof of (6) [24, Ch. 3]. Shortly thereafter, David and Gregory Chudnovsky found the following variant, which lies in the quadratic number field $Q(\sqrt{-163})$ rather than $Q(\sqrt{58})$:

$$\frac{1}{\pi} = 12 \sum_{k=0}^{\infty} \frac{(-1)^k\,(6k)!\,(13591409 + 545140134k)}{(3k)!\,(k!)^3\,640320^{3k+3/2}}. \tag{7}$$

Each term of (7) adds 14 correct digits. The brothers used this formula several times, culminating in a 1994 calculation of π to over four billion decimal digits. Their remarkable story was told in a prizewinning *New Yorker* article [48]. Remarkably, as we already noted earlier, (7) was used again in 2013 for the current record computation of π.

A few years ago Jésus Guillera found various Ramanujan-like identities for π, using integer relation methods. The three most basic—and entirely rational—identities are:

$$\frac{4}{\pi^2} = \sum_{n=0}^{\infty} (-1)^n r(n)^5 (13 + 180n + 820n^2) \left(\frac{1}{32} \right)^{2n+1} \tag{8}$$

$$\frac{2}{\pi^2} = \sum_{n=0}^{\infty} (-1)^n r(n)^5 (1 + 8n + 20n^2) \left(\frac{1}{2} \right)^{2n+1} \tag{9}$$

---

[4] See http://fredrikj.net/blog/2014/03/new-partition-function-record/ for a lovely description of the computation of $p(10^{20})$, which has over 11 billion digits and required knowing π to similar accuracy.

$$\frac{4}{\pi^3} \stackrel{?}{=} \sum_{n=0}^{\infty} r(n)^7 (1 + 14n + 76n^2 + 168n^3) \left(\frac{1}{8}\right)^{2n+1}, \qquad (10)$$

where $r(n) := (1/2 \cdot 3/2 \cdot \cdots \cdot (2n-1)/2)/n!$.

Guillera proved (8) and (9) in tandem, by very ingeniously using the Wilf-Zeilberger algorithm [46, 52] for formally proving hypergeometric-like identities [24, 25, 35, 54]. No other proof is known, and there seem to be no like formulae for $1/\pi^N$ with $N \geq 4$. The third, (10), is almost certainly true. Guillera ascribes (10) to Gourevich, who used integer relation methods to find it.

We were able to "discover" (10) using 30-digit arithmetic, and we checked it to 500 digits in 10 seconds, to 1200 digits in 6.25 minutes, and to 1500 digits in 25 minutes, all with naive command-line instructions in *Maple*. But it has no proof, nor does anyone have an inkling of how to prove it. It is not even clear that proof techniques used for (8) and (9) are relevant here, since, as experiment suggests, it has no "mate" in analogy to (8) and (9) [11]. Our intuition is that if a proof exists, it is more a verification than an explication, and so we stopped looking. We are happy just to "know" that the beautiful identity is true (although it would be more remarkable were it eventually to fail). It may be true for no good reason—it might just have no proof and be a very concrete Gödel-like statement.

There are other sporadic and unexplained examples based on other Pochhammer symbols, most impressively there is an unproven 2010 integer relation discovery by Cullen:

$$2^{11}/\pi^4 \stackrel{?}{=} \qquad (11)$$

$$\sum_{n=0}^{\infty} \frac{(\frac{1}{4})_n (\frac{1}{2})_n^7 (\frac{3}{4})_n}{(1)_n^9} (21 + 466n + 4340n^2 + 20632n^3 + 43680n^4) \left(\frac{1}{2}\right)^{12n}$$

2.7 π *without reciprocals*

In 2008 Guillera [35] produced another lovely pair of third-millennium identities—discovered with integer relation methods and proved with creative telescoping—this time for π^2 rather than its reciprocal. They are

$$\sum_{n=0}^{\infty} \frac{1}{2^{2n}} \frac{(x + \frac{1}{2})_n^3}{(x+1)_n^3} (6(n+x) + 1) = 8x \sum_{n=0}^{\infty} \frac{(\frac{1}{2})_n^2}{(x+1)_n^2}, \qquad (12)$$

and

$$\sum_{n=0}^{\infty} \frac{1}{2^{6n}} \frac{(x + \frac{1}{2})_n^3}{(x+1)_n^3} (42(n+x) + 5) = 32x \sum_{n=0}^{\infty} \frac{(x + \frac{1}{2})_n^2}{(2x+1)_n^2}. \qquad (13)$$

Here $(a)_n = a(a + 1) \cdots (a + n - 1)$ is the *rising factorial*. Substituting $x = 1/2$ in (12) and (13), he obtained respectively the formulae

$$\sum_{n=0}^{\infty} \frac{1}{2^{2n}} \frac{(1)_n^3}{\left(\frac{3}{2}\right)_n^3} (3n + 2) = \frac{\pi^2}{4}, \qquad \sum_{n=0}^{\infty} \frac{1}{2^{6n}} \frac{(1)_n^3}{\left(\frac{3}{2}\right)_n^3} (21n + 13) = 4 \frac{\pi^2}{3}.$$

2.8 Quartic algorithm for π

One measure of the dramatic increase in computer power available to experimental mathematicians is the fact that the record for computation of π has gone from 29.37 *million* decimal digits in 1986 to 12.1 *trillion* digits in 2013. These computations have typically involved Ramanujan's formula (6), the Chudnovsky formula (7), the Salamin-Brent algorithm [24, Ch. 3], or the Borwein quartic algorithm. The Borwein quartic algorithm, which was discovered by one of us and his brother Peter in 1983, with the help of a 16 Kbyte Radio Shack portable system, is the following: Set $a_0 := 6 - 4\sqrt{2}$ and $y_0 := \sqrt{2} - 1$, then iterate

$$y_{k+1} = \frac{1 - (1 - y_k^4)^{1/4}}{1 + (1 - y_k^4)^{1/4}},$$

$$a_{k+1} = a_k(1 + y_{k+1})^4 - 2^{2k+3} y_{k+1}(1 + y_{k+1} + y_{k+1}^2). \tag{14}$$

Then a_k converges *quartically* to $1/\pi$: each iteration approximately quadruples the number of correct digits. Twenty-two full-precision iterations of (14) produce an algebraic number that coincides with π to well more than 24 trillion places. Here is a highly abbreviated chronology of computations of π (based on http://en.wikipedia. org/wiki/Chronology_of_computation_of_pi).

- 1986: One of the present authors used (14) to compute 29.4 million digits of π. This required 28 hours on one CPU of the new Cray-2 at NASA Ames Research Center. Confirmation using the Salamin-Brent scheme took another 40 hours. This computation uncovered hardware and software errors on the Cray-2.
- Jan. 2009: Takahashi used (14) to compute 1.649 trillion digits (nearly 60,000 times the 1986 computation), requiring 73.5 hours on 1024 cores (and 6.348 Tbyte memory) of a Appro Xtreme-X3 system. Confirmation via the Salamin-Brent scheme took 64.2 hours and 6.732 Tbyte of main memory.
- Apr. 2009: Takahashi computed 2.576 trillion digits.
- Dec. 2009: Bellard computed nearly 2.7 trillion decimal digits (first in binary), using (7). This required 131 days on a single four-core workstation armed with large amounts of disk storage.
- Aug. 2010: Kondo and Yee computed 5 trillion decimal digits using (7). This was first done in binary, then converted to decimal. The binary digits were confirmed by computing 32 hexadecimal digits of π ending with position

4,152,410,118,610, using BBP-type formulas for π due to Bellard and Plouffe (see Section 2.9). Additional details are given at http://www.numberworld.org/misc_runs/pi-5t/announce_en.html. These digits appear to be "very normal."

- Dec. 2013: Kondo and Yee extended their computation to 12.1 trillion digits.[5]

Daniel Shanks, who in 1961 computed π to over 100,000 digits, once told Phil Davis that a billion-digit computation would be "forever impossible." But both Kanada and the Chudnovskys achieved that in 1989. Similarly, the intuitionists Brouwer and Heyting appealed to the "impossibility" of *ever* knowing whether and where the sequence 0123456789 appears in the decimal expansion of π, in defining a sequence $(x_k)_{k \in N}$ which is zero in each place except for a one in the n-th place where the sequence first starts to occur if at all.

This sequence converges to zero classically but was not then well-formed intuitionistically. Yet it was found in 1997 by Kanada, beginning at position 17387594880. Depending on ones ontological perspective, either nothing had changed, or the sequence had always converged but we were ignorant of such fact, or perhaps the sequence became convergent in 1997.

As late as 1989, Roger Penrose ventured, in the first edition of his book *The Emperor's New Mind*, that we likely will never know if a string of ten consecutive sevens occurs in the decimal expansion of π. Yet this string was found in 1997 by Kanada, beginning at position 22869046249.

Figure 2 shows the progress of π calculations since 1970, superimposed with a line that charts the long-term trend of Moore's Law. It is worth noting that whereas progress in computing π exceeded Moore's Law in the 1990s, it has lagged a bit in the past decade.

2.9 Digital integrity, II

There are many possible sources of errors in large computations of this type:

- The underlying formulas and algorithms used might conceivably be in error, or have been incorrectly transcribed.
- The computer programs implementing these algorithms, which necessarily employ sophisticated algorithms such as *fast Fourier transform*s to accelerate multiplication, may contain subtle bugs.
- Inadequate numeric precision may have been employed, invalidating some key steps of the algorithm.
- Erroneous programming constructs may have been employed to control parallel processing. Such errors are very hard to detect and rectify, since in many cases they cannot easily be replicated.

---

[5]See "12.1 Trillion Digits of Pi And we're out of disk space..." at http://www.numberworld.org/misc_runs/pi-12t/.

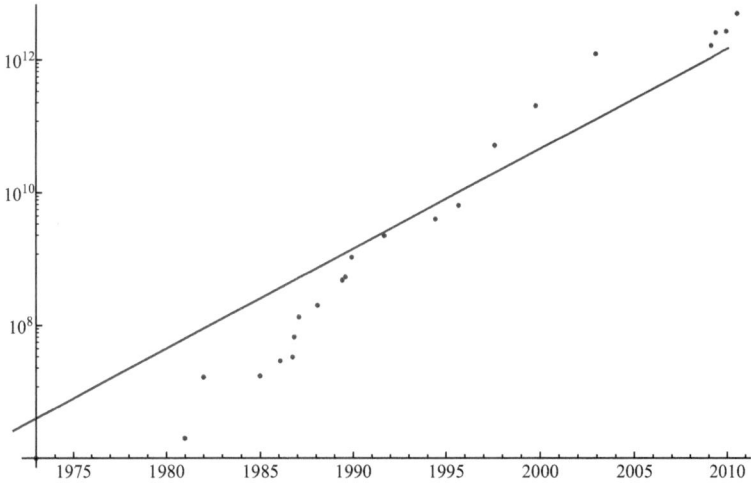

Fig. 2 Plot of π calculations, in digits (dots), compared with the long-term slope of Moore's Law (line).

- Hardware errors may have occurred in the middle of the run, rendering all subsequent computation invalid. This was a factor in the 1986 computation of π, as noted above.
- Quantum-mechanical errors may have corrupted the results, for example, when a stray subatomic particle interacts with a storage register [50].

So why should anyone believe the results of such calculations? The answer is that such calculations are always double-checked with an independent calculation done using some other algorithm, sometimes in more than one way. For instance, Kanada's 2002 computation of π to 1.3 trillion decimal digits involved first computing slightly over one trillion hexadecimal (base-16) digits, using (14). He found that the 20 hex digits of π beginning at position $10^{12} + 1$ are B4466E8D21 5388C4E014.

Kanada then calculated these same 20 hex digits directly, using the "BBP" algorithm [18]. The BBP algorithm for π is based on the formula

$$\pi = \sum_{i=0}^{\infty} \frac{1}{16^i} \left(\frac{4}{8i+1} - \frac{2}{8i+4} - \frac{1}{8i+5} - \frac{1}{8i+6} \right), \tag{15}$$

which was discovered via the PSLQ integer relation algorithm [24, pp. 232–234]. In particular, PSLQ discovered the formula

$$\pi = 4\,{}_2F_1 \left(\begin{matrix} 1, \frac{1}{4} \\ \frac{5}{4} \end{matrix} \middle| -\frac{1}{4} \right) + 2\tan^{-1}\left(\frac{1}{2}\right) - \log 5, \tag{16}$$

where $_2F_1 \left(\begin{array}{c} 1, \frac{1}{4} \\ \frac{5}{4} \end{array} \middle| -\frac{1}{4} \right) = 0.955933837\ldots$ is a Gaussian hypergeometric function. From (16), the series (15) almost immediately follows. The BBP algorithm, which is based on (15), permits one to calculate binary or hexadecimal digits of π beginning at an arbitrary starting point, without needing to calculate any of the preceding digits, by means of a simple scheme that requires only modest-precision arithmetic.

The result of the BBP calculation was B4466E8D21 5388C4E014. Needless to say, in spite of the many potential sources of error in both computations, the final results dramatically agree, thus confirming in a convincing albeit heuristic sense that both results are almost certainly correct. Although one cannot rigorously assign a "probability" to this event, note that the probability that two 20-long random hex digit strings perfectly agree is one in $16^{20} \approx 1.2089 \times 10^{24}$.

This raises the following question: *Which is more securely established*, the assertion that the hex digits of π in positions $10^{12} + 1$ through $10^{12} + 20$ are B4466E8D21 5388C4E014, or the final result of some very difficult work of mathematics that required hundreds or thousands of pages, that relied on many results quoted from other sources, and that (as is frequently the case) has been read in detail by only a relative handful of mathematicians besides the author? (See also [24, §8.4]). In our opinion, computation often trumps cerebration.

In a 2010 computation using the BBP formula, Tse-Wo Zse of Yahoo! Cloud Computing calculated 256 binary digits of π starting at the *two quadrillionth bit*. He then checked his result using the following variant of the BBP formula due to Bellard:

$$\pi = \frac{1}{64} \sum_{k=0}^{\infty} \frac{(-1)^k}{1024^k} \left(\frac{256}{10k + 1} + \frac{1}{10k + 1} - \frac{64}{10k + 3} - \frac{4}{10k + 5} \right.$$

$$\left. - \frac{4}{10k + 7} - \frac{32}{4k + 1} - \frac{1}{4k + 3} \right). \tag{17}$$

In this case, both computations verified that the 24 hex digits beginning immediately after the 500 trillionth hex digit (i.e., after the two quadrillionth binary bit) are: E6C1294A ED40403F 56D2D764.

In 2012 Ed Karrel using the BBP formula on the *CUDA*[6] system with processing to determine that starting *after* the 10^{15} position the hex-bits are E353CB3F7F0C9ACCFA9AA215F2. This was done on four NVIDIA GTX 690 graphics cards (GPUs) installed in CUDA. *Yahoo!*'s run took 23 days; this took 37 days.[7]

---

[6] See http://en.wikipedia.org/wiki/CUDA.

[7] See www.karrels.org/pi/.

Some related BBP-type computations of digits of π^2 and Catalan's constant $G = \sum_{n\geq 1}(-1)^n/(2n+1)^2 = 0.9159655941\ldots$ are described in [17]. In each case, the results were computed two different ways for validation. These runs used 737 hours on a 16384-CPU IBM Blue Gene computer, or, in other words, a total of 1378 CPU-years.

2.10 Formal verification of proof

In 1611, Kepler described the stacking of equal-sized spheres into the familiar arrangement we see for oranges in the grocery store. He asserted that this packing is the tightest possible. This assertion is now known as the Kepler conjecture, and has persisted for centuries without rigorous proof. Hilbert implicitly included the irregular case of the Kepler conjecture in problem 18 of his famous list of unsolved problems in 1900: *whether there exist non-regular space-filling polyhedra?* the regular case having been disposed of by Gauss in 1831.

In 1994, Thomas Hales, now at the University of Pittsburgh, proposed a five-step program that would result in a proof: (a) treat maps that only have triangular faces; (b) show that the face-centered cubic and hexagonal-close packings are local maxima in the strong sense that they have a higher score than any Delaunay star with the same graph; (c) treat maps that contain only triangular and quadrilateral faces (except the pentagonal prism); (d) treat maps that contain something other than a triangular or quadrilateral face; and (e) treat pentagonal prisms.

In 1998, Hales announced that the program was now complete, with Samuel Ferguson (son of mathematician-sculptor Helaman Ferguson) completing the crucial fifth step. This project involved extensive computation, using an interval arithmetic package, a graph generator, and *Mathematica*. The computer files containing the source code and computational results occupy more than three Gbytes of disk space. Additional details, including papers, are available at http://www.math.pitt.edu/~thales/kepler98. For a mixture of reasons—some more defensible than others—the *Annals of Mathematics* initially decided to publish Hales' paper with a cautionary note, but this disclaimer was deleted before final publication.

Hales [36] has now embarked on a multi-year program to certify the proof by means of computer-based formal methods, a project he has named the "Flyspeck" project.[8] As these techniques become better understood, we can envision a large number of mathematical results eventually being confirmed by computer, as instanced by other articles in the same issue of the *Annals* as Hales' article. But this will take decades.

---

[8]He reported in December 2012 at an ICERM workshop that this was nearing completion.

3 The experimental paradigm in applied mathematics

The field of applied mathematics is an enormous edifice, and we cannot possibly hope, in this short essay, to provide a comprehensive survey of current computational developments. So we shall limit our discussion to what may be termed "experimental applied mathematics," namely employ methods akin to those mentioned above in pure mathematics to problems that had their origin in an applied setting. In particular, we will touch mainly on examples that employ either high-precision computation or integer relation detection, as these tools lead to issues similar to those already highlighted above.

First we will examine some historical examples of this paradigm in action.

3.1 Gravitational boosting or "slingshot magic"

One interesting space-age example is the unexpected discovery of *gravitational boosting* by Michael Minovitch, who at the time (1961) was a student working on a summer project at the Jet Propulsion Laboratory in Pasadena, California. Minovitch found that *Hohmann transfer ellipses* were not, as then believed, the minimum-energy way to reach the outer planets. Instead, he discovered, by a combination of clever analytical derivations and heavy-duty computational experiments on IBM 7090 computers (which were the world's most powerful systems at the time), that spacecraft orbits which pass close by other planets could gain a "slingshot effect" substantial boost in speed, compensated by an extremely small change in the orbital velocity of the planet, on their way to a distant location [44]. Some of his earlier computation was not supported enthusiastically by NASA. As Minovitz later wrote,

> Prior to the innovation of gravity-propelled trajectories, it was taken for granted that the rocket engine, operating on the well-known reaction principle of Newton's Third Law of Motion, represented the basic, and for all practical purposes, the only means for propelling an interplanetary space vehicle through the Solar System.[9]

Without such a boost from Jupiter, Saturn, and Uranus, the Voyager mission would have taken more than 30 years to reach Neptune; instead, Voyager reached Neptune in only ten years. Indeed, without gravitational boosting, we would still be waiting! We would have to wait much longer for Voyager to leave the solar system as it now apparently is.

---

[9]There are differing accounts of how this principle was discovered; we rely on the first-person account at http://www.gravityassist.com/IAF1/IAF1.pdf. Additional information on "slingshot magic" is given at http://www.gravityassist.com/ and http://www2.jpl.nasa.gov/basics/grav/primer.php.

3.2   Fractals and chaos

One premier example of 20th century applied experimental mathematics is the development of *fractal theory*, as exemplified by the works of Benoit Mandelbrot. Mandelbrot studied many examples of fractal sets, many of them with direct connections to nature. Applications include analyses of the shapes of coastlines, mountains, biological structures, blood vessels, galaxies, even music, art, and the stock market. For example, Mandelbrot found that the coast of Australia, the West Coast of Britain, and the land frontier of Portugal all satisfy shapes given by a fractal dimension of approximately 1.75.

In the 1960s and early 1970s, applied mathematicians began to computationally explore features of chaotic iterations that had previously been studied by analytic methods. May, Lorenz, Mandelbrot, Feigenbaum [43], Ruelle, York, and others led the way in utilizing computers and graphics to explore this realm, as chronicled, for example, in Gleick's book *Chaos: Making a New Science* [33]. By now fractals have found nigh countless uses. In our own research [12] we have been looking at expectations over self-similar fractals, motivated by modelling rodent brain-neuron geometry.

3.3   The uncertainty principle

Here we examine a principle that, while discovered early in the 20th century by conventional formal reasoning, could have been discovered much more easily with computational tools.

Most readers have heard of the *uncertainty principle* from quantum mechanics, which is often expressed as the fact that the position and momentum of a subatomic-scale particle cannot simultaneously be prescribed or measured to arbitrary accuracy. Others may be familiar with the uncertainty principle from signal-processing theory, which is often expressed as the fact that a signal cannot simultaneously be "time-limited" and "frequency-limited." Remarkably, the precise mathematical formulations of these two principles are identical (although the quantum mechanics version presumes the existence of de Broglie waves).

Consider a real, continuously differentiable, L^2 function $f(t)$, which further satisfies $|t|^{3/2+\varepsilon}f(t) \to 0$ as $|t| \to \infty$ for some $\varepsilon > 0$. (This assures convergence of the integrals below.) For convenience, we assume $f(-t) = f(t)$, so the Fourier transform $\hat{f}(x)$ of $f(t)$ is real, although this is not necessary. Define

$$E(f) = \int_{-\infty}^{\infty} f^2(t)\,dt \qquad V(f) = \int_{-\infty}^{\infty} t^2 f^2(t)\,dt$$

$$\hat{f}(x) = \int_{-\infty}^{\infty} f(t)e^{-itx}\,dt \qquad Q(f) = \frac{V(f)}{E(f)} \cdot \frac{V(\hat{f})}{E(\hat{f})}. \tag{18}$$

Table 1 Q values for various functions.

$f(t)$	Interval	$\hat{f}(x)$	$Q(f)$
$1 - t\,\mathrm{sgn}\,t$	$[-1, 1]$	$2(1 - \cos x)/x^2$	$3/10$
$1 - t^2$	$[-1, 1]$	$4(\sin x - x\cos x)/x^3$	$5/14$
$1/(1 + t^2)$	$[-\infty, \infty]$	$\pi \exp(-x\,\mathrm{sgn}\,x)$	$1/2$
$e^{-|t|}$	$[-\infty, \infty]$	$2/(1 + x^2)$	$1/2$
$1 + \cos t$	$[-\pi, \pi]$	$2\sin(\pi x)/(x - x^3)$	$(\pi^2 - 15/2)/9$

Then the uncertainty principle is the assertion that $Q(f) \geq 1/4$, with equality if and only if $f(t) = ae^{-(bt)^2/2}$ for real constants a and b. The proof of this fact is not terribly difficult but is hardly enlightening—see, for example, [24, pp. 183–188].

Let us approach this problem as an experimental mathematician might. As mentioned, it is natural when studying Fourier transforms (particularly in the context of signal processing) to consider the "dispersion" of a function and to compare this with the dispersion of its Fourier transform. Noting what appears to be an inverse relationship between these two quantities, we are led to consider $Q(f)$ in (18). With the assistance of *Maple* or *Mathematica*, one can explore examples, as shown in Table 1. Note that each of the entries in the last column is in the range $(1/4, 1/2)$. Can one get any lower?

To further study this problem experimentally, note that the Fourier transform $\hat{f}$ of $f(t)$ can be closely approximated with a *fast Fourier transform*, after suitable discretization. The integrals V and E can be similarly evaluated numerically.

Then one can adopt a search strategy to minimize $Q(f)$, starting, say, with a "tent function," then perturbing it up or down by some ϵ on a regular grid with spacing δ, thus creating a continuous, piecewise linear function. *When for a given δ, a minimizing function $f(t)$ has been found, reduce ϵ and δ, and repeat. Terminate when δ is sufficiently small, say 10^{-6} or so.* (For details, see [24].)

The resulting function $f(t)$ is shown in Figure 3. Needless to say, its shape strongly suggests a *Gaussian* probability curve. Figure 3 actually shows both $f(t)$ and the function $e^{-(bt)^2/2}$, where $b = 0.45446177$: they are identical to the resolution of the plot!

In short, it is a relatively simple matter, using 21st-century computational tools, to numerically "discover" the signal-processing form of the uncertainty principle. Doubtless the same is true of many other historical principles of physics, chemistry, and other fields, thus illustrating items 1, 2, 3, and 4 at the start of Section 2.

3.4 Chimera states in oscillator arrays

It is fair to say that the computational-experimental approach in applied mathematics has greatly accelerated in the 21st century. We show here a few specific illustrative

Fig. 3 Q-minimizer and matching Gaussian (identical to the plot resolution).

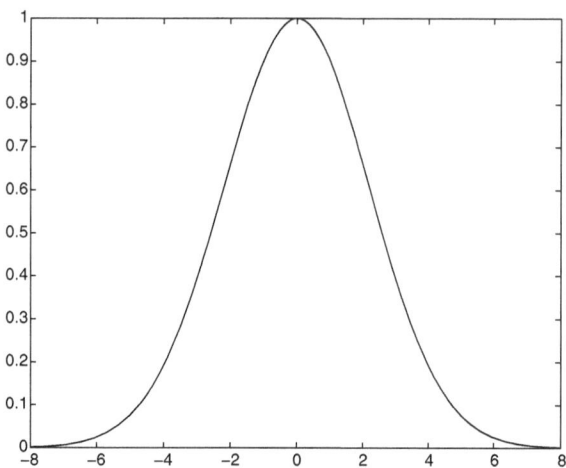

examples. These include several by the present authors, because we are familiar with them. There are doubtless many others that we are not aware of that are similarly exemplary of the experimental paradigm.

One interesting example was the 2002 discovery by Kuramoto, Battogtokh, and Sima of "chimera" states, which arise in arrays of identical oscillators, where individual oscillators are correlated with oscillators some distance away in the array. These systems can arise in a wide range of physical systems, including Josephson junction arrays, oscillating chemical systems, epidemiological models, neural networks underlying snail shell patterns, and "ocular dominance stripes" observed in the visual cortex of cats and monkeys. In chimera states, named for the mythological beast that incongruously combines features of lions, goats, and serpents, the oscillator array bifurcates into two relatively stable groups, the first composed of coherent, phased-locked oscillators, and the second composed of incoherent, drifting oscillators.

According to Abrams and Strogatz, who subsequently studied these states in detail [1], most arrays of oscillators quickly converge into one of four typical patterns: (a) synchrony, with all oscillators moving in unison; (b) solitary waves in one dimension or spiral waves in two dimensions, with all oscillators locked in frequency; (c) incoherence, where phases of the oscillators vary quasi-periodically, with no global spatial structure; and (d) more complex patterns, such as spatiotemporal chaos and intermittency. But in chimera states, phase locking and incoherence are simultaneously present in the same system.

The simplest governing equation for a continuous one-dimensional chimera array is

$$\frac{\partial \phi}{\partial t} = \omega - \int_0^1 G(x - x') \sin\left[\phi(x, t) - \phi(x', t) + \alpha\right] \, dx', \qquad (19)$$

Fig. 4 Phase of oscillations for a chimera system. The x-axis runs from 0 to 1 with periodic boundaries.

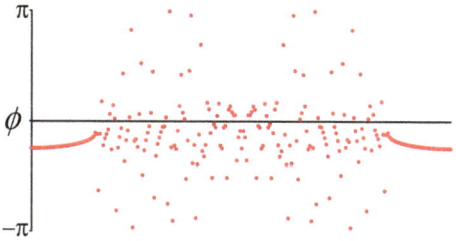

where $\phi(x, t)$ specifies the phase of the oscillator given by $x \in [0, 1)$ at time t, and $G(x - x')$ specifies the degree of nonlocal coupling between the oscillators x and x'. A discrete, computable version of (19) can be obtained by replacing the integral with a sum over a 1-D array $(x_k, 0 \le k < N)$, where $x_k = k/N$. Kuramoto and Battogtokh took $G(x - x') = C \exp(-\kappa |x - x'|)$ for constant C and parameter κ.

Specifying $\kappa = 4$, $\alpha = 1.457$, array size $N = 256$ and time step size $\Delta t = 0.025$, and starting from $\phi(x) = 6 \exp\left[-30(x - 1/2)^2\right] r(x)$, where r is a uniform random variable on $[-1/2, 1/2)$, gives rise to the phase patterns shown in Figure 4. Note that the oscillators near $x = 0$ and $x = 1$ appear to be phase-locked, moving in near-perfect synchrony with their neighbors, but those oscillators in the center drift wildly in phase, both with respect to their neighbors and to the locked oscillators.

Numerous researchers have studied this phenomenon since its initial numerical discovery. Abrams and Strogatz studied the coupling function is given $G(x) := (1 + A \cos x)/(2\pi)$, where $0 \le A \le 1$, for which they were able to solve the system analytically, and then extended their methods to more general systems. They found that chimera systems have a characteristic life cycle: a uniform phase-locked state, followed by a spatially uniform drift state, then a modulated drift state, then the birth of a chimera state, followed a period of stable chimera, then a saddle-node bifurcation, and finally an unstable chimera.[10]

3.5 Winfree oscillators

One closely related development is the resolution of the Quinn-Rand-Strogatz (QRS) constant. Quinn, Rand, and Strogatz had studied the *Winfree model* of coupled nonlinear oscillators, namely

$$\dot{\theta}_i = \omega_i + \frac{\kappa}{N} \sum_{j=1}^{N} -(1 + \cos \theta_j) \sin \theta_i \qquad (20)$$

---

[10]Various movies can be found on line. For example, http://dmabrams.esam.northwestern.edu/pubs/ngeo-video1.mov and http://dmabrams.esam.northwestern.edu/pubs/ngeo-video2.mov show two for groundwater flow.

for $1 \leq i \leq N$, where $\theta_i(t)$ is the phase of oscillator i at time t, the parameter κ is the coupling strength, and the frequencies ω_i are drawn from a symmetric unimodal density $g(w)$. In their analyses, they were led to the formula

$$0 = \sum_{i=1}^{N} \left(2\sqrt{1 - s^2(1 - 2(i-1)/(N-1))^2} \right.$$

$$\left. - \frac{1}{\sqrt{1 - s^2(1 - 2(i-1)/(N-1))^2}} \right),$$

implicitly defining a phase offset angle $\phi = \sin^{-1} s$ due to bifurcation. The authors conjectured, on the basis of numerical evidence, the asymptotic behavior of the N-dependent solution s to be

$$1 - s_N \sim \frac{c_1}{N} + \frac{c_2}{N^2} + \frac{c_3}{N^3} + \cdots,$$

where $c_1 = 0.60544365\ldots$ is now known as the QRS constant.

In 2008, the present authors together with Richard Crandall computed the numerical value of this constant to 42 decimal digits, obtaining

$$c_1 \approx 0.605443657196732749478922842244\ldots.$$

With this numerical value in hand, they were able to demonstrate that c_1 is the unique zero of the *Hurwitz zeta* function $\zeta(1/2, z/2)$ on the interval $0 \leq z \leq 2$. What's more, they found that $c_2 = -0.104685459\ldots$ is given analytically by

$$c_2 = c_1 - c_1^2 - 30 \frac{\zeta(-1/2, c_1/2)}{\zeta(3/2, c_1/2)}.$$

In this case experimental computation led to a closed form which could then be used to establish the existence and form of the critical point, thus engaging at the very least each of items #1 through #6 of our 'mathodology,' and highlighting the interplay between computational discovery and theoretical understanding.

3.6 High-precision dynamics

Periodic orbits form the "skeleton" of a dynamical system and provide much useful information, but when the orbits are unstable, high-precision numerical integrators are often required to obtain numerically meaningful results.

For instance, Figure 5 shows computed symmetric periodic orbits for the $(7+2)$-Ring problem using double and quadruple precision. The $(n+2)$-*body Ring problem* describes the motion of an infinitesimal particle attracted by the gravitational field

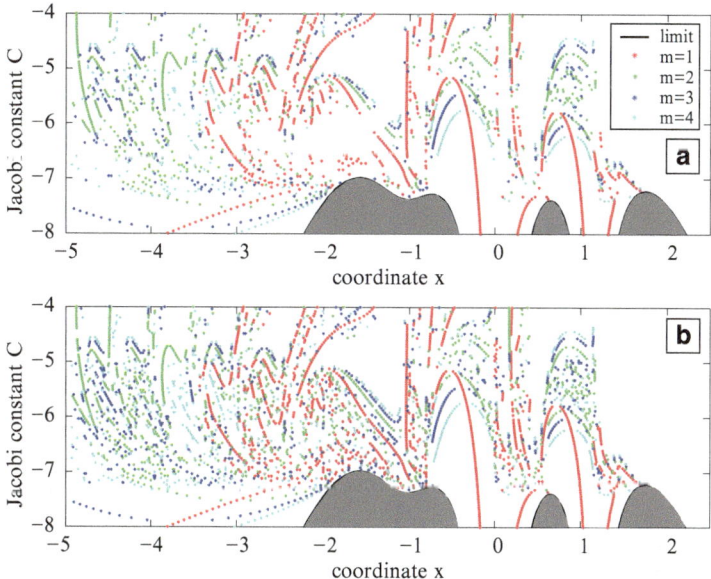

Fig. 5 Symmetric periodic orbits (*m* denotes multiplicity of the periodic orbit) in the most chaotic zone of the $(7 + 2)$-Ring problem using double (A) and quadruple (B) precision. Note "gaps" in the double precision plot. (Reproduced by permission.)

of $n + 1$ primary bodies, n of them at the vertices of a regular n-gon rotating in its own plane about the central body with constant angular velocity. Each point corresponds to the initial conditions of one symmetric periodic orbit, and the grey area corresponds to regions of forbidden motion (delimited by the limit curve). To avoid "false" initial conditions it is useful to check if the initial conditions generate a periodic orbit up to a given tolerance level; but for highly unstable periodic orbits double precision is not enough, resulting in gaps in the figure that are not present in the more accurate quad precision run.

Hundred-digit precision arithmetic plays a fundamental role in a 2010 study of the fractal properties of the LORENZ ATTRACTOR [3.XY] (see Figure 6). The first plot shows the intersection of an arbitrary trajectory on the Lorenz attractor with the section $z = 27$, in a rectangle in the $x - y$ plane. All later plots zoom in on a tiny region (too small to be seen by the unaided eye) at the center of the red rectangle of the preceding plot to show that what appears to be a line is in fact many lines.

The Lindstedt-Poincaré method for computing periodic orbits is based on the Lindstedt-Poincaré perturbation theory, Newton's method for solving nonlinear systems, and Fourier interpolation. Viswanath has used this in combination with high-precision libraries to obtain periodic orbits for the Lorenz model at the classical Saltzman's parameter values. This procedure permits one to compute, to high accuracy, highly unstable periodic orbits more efficiently than with conventional schemes, in spite of the additional cost of high-precision arithmetic. For these

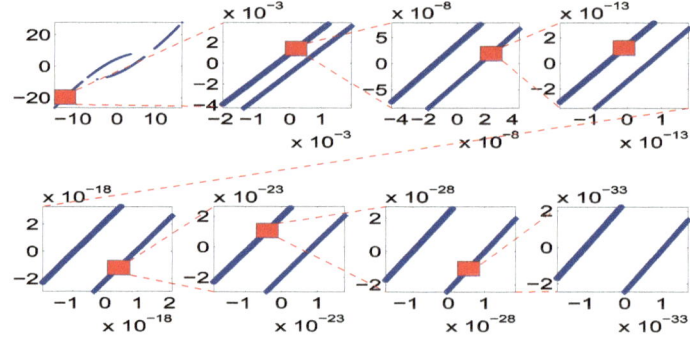

Fig. 6 Fractal property of the Lorenz attractor. (Reproduced by permission.)

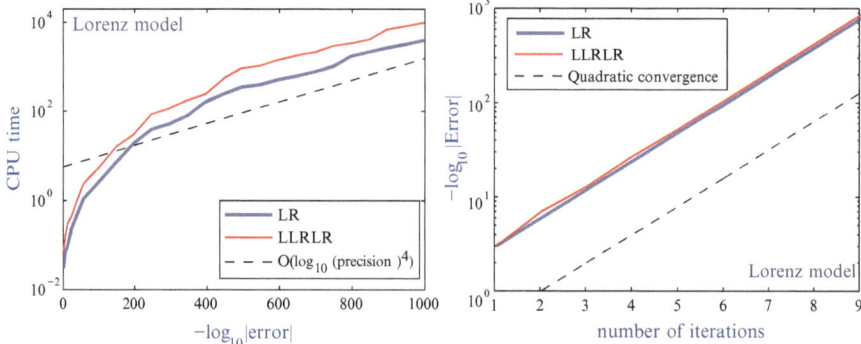

Fig. 7 Computational relative error vs. CPU time and number of iterations in a 1000-digit computation of the periodic orbits LR and LLRLR of the Lorenz model. (Reproduced by permission.)

reasons, high-precision arithmetic plays a fundamental role in the study of the fractal properties of the Lorenz attractor (see Figures 6 and 7), and in the consistent formal development of complex singularities of the Lorenz system using infinite series. For additional details and references, see [5].

3.7   Snow crystals

Computational experimentation has even been useful in the study of snowflakes. In a 2007 study, Janko Gravner and David Griffeath used a sophisticated computer-based simulator to study the process of formation of these structures, known in the literature as snow crystals and informally as *snofakes*. Their model simulated each of the key steps, including diffusion, freezing, and attachment, and thus enabled researchers to study, dependence on melting parameters. Snow crystals produced by their simulator vary from simple stars, to six-sided crystals with plate-ends, to crystals with dendritic ends, and look remarkably similar to natural snow crystals. Among the findings uncovered by their simulator is the fact that these crystals exhibit remarkable overall symmetry, even in the process of dynamically changing parameters. Their simulator is publicly available at http://psoup.math.wisc.edu/Snofakes.html.

> The latest developments in computer and video technology have provided a multiplicity of computational and symbolic tools that have rejuvenated mathematics and mathematics education. Two important examples of this revitalization are experimental mathematics and visual theorems. [38]

3.8 Visual mathematics

J. E. Littlewood (1885–1977) wrote [41]

> A heavy warning used to be given [by lecturers] that pictures are not rigorous; this has never had its bluff called and has permanently frightened its victims into playing for safety. Some pictures, of course, are not rigorous, but I should say most are (and I use them whenever possible myself).

This was written in 1953, long before the current graphics, visualization and dynamic geometric tools (such as *Cinderella* or *Geometer's Sketchpad*) were available.

The ability to "see" mathematics and mathematical algorithms as images, movies, or simulations is more-and-more widely used (as indeed we have illustrated in passing), but is still under appreciated. This is true for algorithm design and improvement and even more fundamentally as a tool to improve intuition, and as a resource when preparing or giving lectures (Fig. 8).

Fig. 8 Seeing often can and should be believing.

"Sometimes it is easier to see than to say."

3.9 Iterative methods for protein conformation

In [2] we applied continuous convex optimization tools (alternating projection and alternating reflection algorithms) to a large class of non-convex and often highly combinatorial image or signal reconstruction problems. The underlying idea is to consider a *feasibility* problem that asks for a point in the intersection $C := C_1 \cap C_2$ of two sets C_1 and C_2 in a Hilbert space.

The *method of alternating projections* (MAP) is to iterate

$$x_n \mapsto y_n = P_{C_1}(x_n) \mapsto P_{C_2}(y_n) =: x_{n+1}.$$

Here $P_A(x) := \{y \in A \colon \|x - y\| = \inf_{a \in A} \|x - a\| := d_A(x)\}$ is the *nearest point (metric) projection*. The *reflection* is given by $R_A(x) := 2x - P_A(x), (P_A(x) = \frac{x + R_A(x)}{2})$, and the most studied *Douglas-Rachford reflection method* (DR) can be described by

$$x_n \mapsto y_n = R_{C_1}(x_n) \mapsto z_n = R_{C_2}(y_n) \mapsto \frac{z_n + x_n}{2} =: x_{n+1}$$

This is nicely captured as 'reflect-reflect-average.'

In practice, for any such method, one set will be convex and the other sequesters non-convex information.[11] It is most useful when projections on the two sets are relatively easy to estimate but the projection on the intersection C is inaccessible. The methods often work unreasonably well but there is very little theory unless both sets are convex. For example, the optical aberration problem in the original Hubble telescope was 'fixed' by an alternating projection phase retrieval method (by Fienup and others), before astronauts could actually physically replace the mis-ground lens.

3.9.1 Protein conformation

We illustrate the situation with reconstruction of protein structure using only the short distances below about six Angstroms[12] between atoms that can be measured by nondestructive MRI techniques. This can viewed as a *Euclidean distance matrix completion problem* [2], so that only geometry (and no chemistry) is directly involved. That is, we ask the algorithm to predict a configuration in 3-space that is consistent with the given short distances.

---

[11] When the set A is non-convex the projection $P_A(x)$ may be a set and we must instead select some $y \in P_A(x)$.

[12] Interatomic distances below 6Å typically constitute less than 8% of the total distances between atoms in a protein.

Average (maximum) errors from five replications with reflection methods of six proteins taken from a standard database.

Protein	# Atoms	Rel. Error (dB)	RMSE	Max Error
1PTQ	404	−83.6 (−83.7)	0.0200 (0.0219)	0.0802 (0.0923)
1HOE	581	−72.7 (−69.3)	0.191 (0.257)	2.88 (5.49)
1LFB	641	−47.6 (−45.3)	3.24 (3.53)	21.7 (24.0)
1PHT	988	−60.5 (−58.1)	1.03 (1.18)	12.7 (13.8)
1POA	1067	−49.3 (−48.1)	34.1 (34.3)	81.9 (87.6)
1AX8	1074	−46.7 (−43.5)	9.69 (10.36)	58.6 (62.6)

What do the reconstructions look like? We turn to graphic information for 1PTQ and 1POA which were respectively our most and least successful cases.

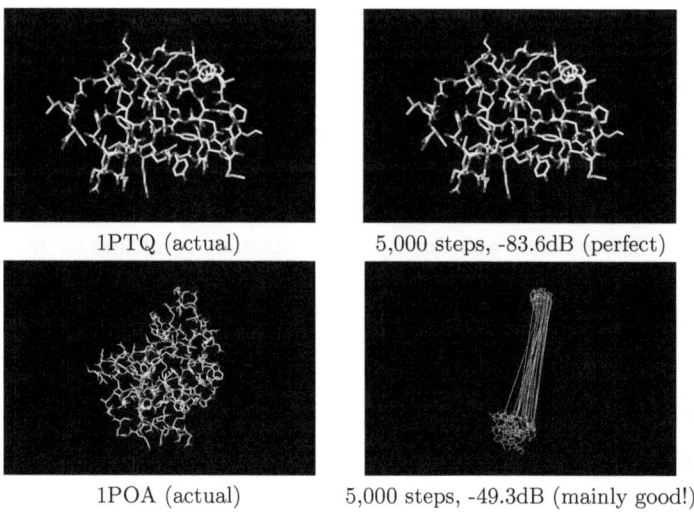

1PTQ (actual)   5,000 steps, -83.6dB (perfect)

1POA (actual)   5,000 steps, -49.3dB (mainly good!)

Note that the failure (and large mean or max error) is caused by a very few very large spurious distances[13]. The remainder is near perfect. Below we show the radical visual difference in the behavior of reflection and projection methods on IPTQ.

While traditional numerical measures (relative error in decibels, root mean square error, and maximum error) of success held some information, graphics-based tools have been dramatically more helpful. It is visually obvious that this method has successfully reconstructed the protein whereas the MAP reconstruction method, shown below, has not. This difference is not evident if one compares the two methods in terms of decibel measurement (beloved of engineers).

---

[13] After speeding up the computation by a factor of ten, and terminating when the decibel error was less than −100, this anomaly disappeared.

Douglas–Rachford reflection (DR) reconstruction: (of IPTQ[14])

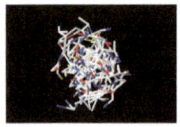
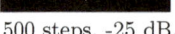
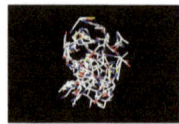
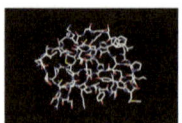
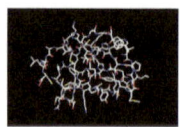

500 steps, -25 dB.   1,000 steps, -30 dB.   2,000 steps, -51 dB.   5,000 steps, -84 dB.

After 1000 steps or so, the protein shape is becoming apparent. After 2000 steps only minor detail is being fixed. Decibel measurement really does not discriminate this from the failure of the MAP method below which after 5000 steps has made less progress than DR after 1000.

Alternating projection (MAP) reconstruction: (of IPTQ)

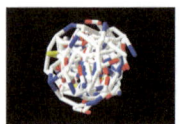

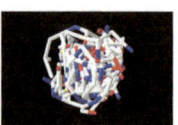

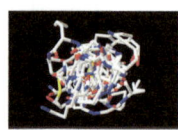

500 steps, -22 dB.   1,000 steps, -24 dB.   2,000 steps, -25 dB.   5,000 steps, -28 dB.

Yet MAP works very well for optical abberation correction of the Hubble telescope and the method is now built in to amateur telescope software. This problem-specific variation in behavior is well observed but poorly understood; it is the heart of our current studies.

3.10 Mathematical finance

Coincident with the rise of powerful computational hardware, sophisticated mathematical techniques are now being employed to analyze market data in real-time and generate profitable investment strategies. This approach, commonly known as "quantitative" or "mathematical" finance, often involves computationally exploring a wide range of portfolio options or investment strategies [30, 40].

One interesting aspect of these studies is the increasing realization of how easy it is to "over-compute" an investment strategy, in a statistical sense. For example, one common approach to finding an effective quantitative investment strategy is to tune the strategy on historical data (a "backtest"). Unfortunately, financial mathematicians are finding that beyond a certain point, examining thousands or millions of variations of an investment strategy (which is certainly possible with

---

[14]The first 3,000 steps of the 1PTQ reconstruction are available as a movie at http://carma. newcastle.edu.au/DRmethods/1PTQ.html.

today's computer technology) to find the optimal strategy may backfire, because the resulting scheme may "overfit" the backtest data—the optimal strategy will work well only with a particular set of securities or over a particular time frame (in the past!). Indeed, backtest overfitting is now thought to be one of the primary reasons that an investment strategy which looks promising on paper often falls flat in real-world practice [15, 16].

3.11 Digital integrity III

Difficulties with statistical overfitting in financial mathematics can be seen as just one instance of the larger challenge of ensuring that results of computational experiments are truly valid and reproducible, which, after all, is the bedrock of all scientific research. We discussed these issues in the context of pure mathematics in Section 2.9, but there are numerous analogous concerns in applied mathematics:

- Whether the calculation is numerically reproducible, i.e., whether or not the computation produces results acceptably close to those of an equivalent calculation performed using different hardware or software. In some cases, more stable numerical algorithms or higher-precision arithmetic may be required for certain portions of the computation to ameliorate such difficulties.
- Whether the calculation has been adequately validated with independently written programs or distinct software tools.
- Whether the calculation has been adequately validated by comparison with empirical data (where possible).
- Whether the algorithms and computational techniques used in the calculation have been documented sufficiently well in journal articles or publicly accessible repositories, so that an independent researcher can reproduce the stated results.
- Whether the code and/or software tool itself has been secured in a permanent repository.

These issues were addressed in a 2012 workshop held at the Institute for Computational and Experimental Research in Mathematics (ICERM). See [50] for details.

4 Additional examples of the experimental paradigm in action

4.1 Giuga's conjecture

As another measure of what changes over time and what doesn't, consider Giuga's conjecture:

Giuga's conjecture (1950): *An integer $n > 1$, is a prime if and only if $\mathcal{G}_n :=$ $\sum_{k=1}^{n-1} k^{n-1} \equiv n - 1 \mod n$.*

This conjecture is not yet proven. But it is known that any counterexamples are necessarily *Carmichael numbers*—square free 'pseudo-prime numbers'—and much more. These rare birds were only proven infinite in number in 1994. In [25, pp. 227], we exploited the fact that if a number $n = p_1 \cdots p_m$ with $m > 1$ prime factors p_i is a counterexample to Giuga's conjecture (that is, satisfies $\mathcal{G}_n \equiv n - 1 \bmod n$), then for $i \neq j$ we have that $p_i \neq p_j$, that

$$\sum_{i=1}^{m} \frac{1}{p_i} > 1,$$

and that the p_i form a *normal sequence*: $p_i \not\equiv 1 \mod p_j$ for $i \neq j$. Thus, the presence of '3' excludes $7, 13, 19, 31, 37, \ldots$, and of '5' excludes $11, 31, 41, \ldots$.

This theorem yielded enough structure, using some predictive experimentally discovered heuristics, to build an efficient algorithm to show—over several months in 1995—that any counterexample had at least 3459 prime factors and so exceeded 10^{13886}, extended a few years later to 10^{14164} in a five-day desktop computation. The heuristic is self-validating every time that the programme runs successfully. But this method necessarily fails after 8135 primes; at that time we hoped to someday exhaust its use.

In 2010, one of us was able to obtain almost as good a bound of 3050 primes in under 110 minutes on a laptop computer, and a bound of 3486 primes and 14,000 digits in less than 14 hours; this was extended to 3,678 primes and 17,168 digits in 93 CPU-hours on a Macintosh Pro, using *Maple* rather than C++, which is often orders-of-magnitude faster but requires much more arduous coding. In 2013, the same one of us with his students revisited the computation and the time and space requirements for further progress [27]. Advances in multi-threading tools and good Python tool kits, along with Moore's law, made the programming much easier and allowed the bound to be increased to **19,908** digits. This study also indicated that we are unlikely to exhaust the method in our lifetime.

4.2 Lehmer's conjecture

An equally hard number-theory related conjecture, for which much less progress can be recorded, is the following. Here $\phi(n)$ is Euler's *totient function*, namely the number of positive integers less than or equal to n that are relatively prime to n:

Lehmer's conjecture (1932). $\phi(n) \big| (n - 1)$ *if and only if n is prime.* Lehmer called this "as hard as the existence of odd perfect numbers."

Again, no proof is known of this conjecture, but it has been known for some time that the prime factors of any counterexample must form a normal sequence. Now there is little extra structure. In a 1997 Simon Fraser M.Sc. thesis, Erick Wong verified the conjecture for 14 primes, using normality and a mix of PARI, C++

and *Maple* to press the bounds of the "curse of exponentiality." This very clever computation subsumed the entire scattered literature in one computation, but could only extend the prior bound from 13 primes to 14.

For Lehmer's related 1932 question: *when does* $\phi(n) \mid (n + 1)$*?*, Wong showed there are eight solutions with no more than seven factors (six-factor solutions are due to Lehmer). Let

$$\mathcal{L}_m := \prod_{k=0}^{m-1} F_k$$

with $F_n := 2^{2^n} + 1$ denoting the *Fermat primes*. The solutions are

$$2, \mathcal{L}_1, \mathcal{L}_2, \ldots, \mathcal{L}_5,$$

and the rogue pair 4919055 and 6992962672132095, but analyzing just eight factors seems out of sight. Thus, in 70 years the computer only allowed the exclusion bound to grow by one prime.

In 1932 Lehmer couldn't factor 6992962672132097. If it had been prime, a ninth solution would exist: since $\phi(n)|(n + 1)$ *with* $n + 2$ *prime implies that* $N := n(n + 2)$ *satisfies* $\phi(N)|(N + 1)$. We say *couldn't* because the number is divisible by 73; which Lehmer—a father of much factorization literature–could certainly have discovered had he *anticipated* a small factor. Today, discovering that

$$6992962672132097 = 73 \cdot 95794009207289$$

is nearly instantaneous, while fully resolving Lehmer's original question remains as hard as ever.

4.3 Inverse computation and Apéry-like series

Three intriguing formulae for the Riemann zeta function are

$$(a)\ \zeta(2) = 3 \sum_{k=1}^{\infty} \frac{1}{k^2 \binom{2k}{k}}, \quad (b)\ \zeta(3) = \frac{5}{2} \sum_{k=1}^{\infty} \frac{(-1)^{k+1}}{k^3 \binom{2k}{k}}, \tag{21}$$

$$(c)\ \zeta(4) = \frac{36}{17} \sum_{k=1}^{\infty} \frac{1}{k^4 \binom{2k}{k}}.$$

Binomial identity (21)(a) has been known for two centuries, while (b)—exploited by Apéry in his 1978 proof of the irrationality of $\zeta(3)$—was discovered as early as 1890 by Markov, and (c) was noted by Comtet [11].

Using integer relation algorithms, bootstrapping, and the "Pade" function (*Mathematica* and *Maple* both produce rational approximations well), in 1996 David Bradley and one of us [11, 25] found the following unanticipated generating function for $\zeta(4n + 3)$:

$$\sum_{k=0}^{\infty} \zeta(4k + 3)\, x^{4k} = \frac{5}{2} \sum_{k=1}^{\infty} \frac{(-1)^{k+1}}{k^3 \binom{2k}{k}(1 - x^4/k^4)} \prod_{m=1}^{k-1} \left(\frac{1 + 4x^4/m^4}{1 - x^4/m^4} \right). \tag{22}$$

Note that this formula permits one to read off an infinity of formulas for $\zeta(4n + 3)$, for $n = 0, 1, 2, \ldots$ beginning with (21)(b), by comparing coefficients of x^{4k} on both sides of the identity.

A decade later, following a quite analogous but much more deliberate experimental procedure, as detailed in [11], we were able to discover a similar general formula for $\zeta(2n + 2)$ that is pleasingly parallel to (22):

$$\sum_{k=0}^{\infty} \zeta(2k + 2)\, x^{2k} = 3 \sum_{k=1}^{\infty} \frac{1}{k^2 \binom{2k}{k}(1 - x^2/k^2)} \prod_{m=1}^{k-1} \left(\frac{1 - 4x^2/m^2}{1 - x^2/m^2} \right). \tag{23}$$

As with (22), one can now read off an infinity of formulas, beginning with (21)(a). In 1996, the authors could reduce (22) to a finite form (24) that they could not prove,

$$\sum_{k=1}^{n} \frac{2n^2}{k^2} \prod_{i=1}^{n-1} \frac{(4k^4 + i^4)}{\prod_{\substack{i=1 \\ i \neq k}}^{n}(k^4 - i^4)} = \binom{2n}{n}, \tag{24}$$

but Almqvist and Granville did find a proof a year later. Both *Maple* and *Mathematica* can now prove identities like (24). Indeed, one of the values of such research is that it pushes the boundaries of what a CAS can do.

Shortly before his death, Paul Erdős was shown the form (24) at a meeting in Kalamazoo. He was ill and lethargic, but he immediately perked up, rushed off, and returned 20 minutes later saying excitedly that he had no idea how to prove (24), but that if proven it would have implications for the irrationality of $\zeta(3)$ and $\zeta(7)$. Somehow in 20 minutes an unwell Erdős had intuited backwards our whole discovery process. Sadly, no one has yet seen any way to learn about the irrationality of $\zeta(4n + 3)$ from the identity, and Erdős's thought processes are presumably as dissimilar from computer-based inference engines as Kasparov's are from those of the best chess programs.

A decade later, the Wilf-Zeilberger algorithm [46, 52]—for which the inventors were awarded the Steele Prize—directly (as implemented in *Maple*) certified (23) [24, 25]. In other words, (23) was both discovered and proven by computer. This is the experimental mathematician's "holy grail" (item 6 in the list at the beginning of Section 2), though it would have been even nicer to be led subsequently to an elegant human proof. That preference may change as future generations begin to take it for granted that the computer is a collaborator [7].

We found a comparable generating function for $\zeta(2n+4)$, giving (21) (c) when $x = 0$, but one for $\zeta(4n+1)$ still eludes us.

4.4 Ising integrals

High precision has proven essential in studies with Richard Crandall (see [9, 24]) of the following integrals:

$$C_n = \frac{4}{n!} \int_0^\infty \cdots \int_0^\infty \frac{1}{\left(\sum_{j=1}^n (u_j + 1/u_j)\right)^2} \, dU,$$

$$D_n = \frac{4}{n!} \int_0^\infty \cdots \int_0^\infty \frac{\prod_{i<j} \left(\frac{u_i - u_j}{u_i + u_j}\right)^2}{\left(\sum_{j=1}^n (u_j + 1/u_j)\right)^2} \, dU,$$

$$E_n = 2 \int_0^1 \cdots \int_0^1 \left(\prod_{1 \le j < k \le n} \frac{u_k - u_j}{u_k + u_j}\right)^2 \, dT,$$

where $dU = \frac{du_1}{u_1} \cdots \frac{du_n}{u_n}$, $dT = dt_2 \cdots dt_n$, and $u_k = \prod_{i=1}^k t_i$. Note that $E_n \le D_n \le C_n$.

The D_n arise in the Ising theory of mathematical physics and the C_n in quantum field theory. As far as we know the E_n are an entirely mathematical construct introduced to study D_n. Direct computation of these integrals from their defining formulas is very difficult, but for C_n it can be shown that

$$C_n = \frac{2^n}{n!} \int_0^\infty p K_0^n(p) \, dp,$$

where K_0 is the *modified Bessel function*. Indeed, it is in this form that C_n arises in quantum field theory, as we subsequently learned from David Broadhurst. We had introduced it to study D_n. Again an uncanny parallel arises between mathematical inventions and physical discoveries.

Then 1000-digit numerical values so computed were used with PSLQ to deduce results such as $C_4 = 7\zeta(3)/12$, and furthermore to discover that

$$\lim_{n \to \infty} C_n = 0.63047350\ldots = 2e^{-2\gamma},$$

with additional higher-order terms in an asymptotic expansion. One intriguing experimental result (not then proven) is the following:

$$E_5 \stackrel{?}{=} 42 - 1984 \operatorname{Li}_4\left(\frac{1}{2}\right) + \frac{189\pi^4}{10} - 74\zeta(3)$$

$$- 1272\zeta(3)\log 2 + 40\pi^2 \log^2 2 - \frac{62\pi^2}{3}$$

$$+ \frac{40\pi^2 \log 2}{3} + 88\log^4 2 + 464\log^2 2 - 40\log 2, \tag{25}$$

found by a multi-hour computation on a highly parallel computer system, and confirmed to 250-digit precision. Here $\operatorname{Li}_4(z) = \sum_{k\geq 1} z^k/k^4$ is the standard order-4 polylogarithm. We also provided 250 digits of E_6.

In 2013 Erik Panzer was able to formally evaluate all the E_n in terms of the *alternating multi zeta* values [45]. In particular, the experimental result for E_5 was confirmed, and our high-precision computation of E_6 was used to confirm Panzer's evaluation

$$E_6 = 86 - 88 \log 2 + 768 \log^4 2 + \frac{704}{3} \log^3 2 + 1360 \log^2 2 - 13964 \zeta_2\zeta_3$$

$$- 348 \zeta_2 - 6048 \zeta_{1,-3} + 134 \zeta_3 + \frac{53775}{2} \zeta_5 + 27904 \zeta_{1,1,-3} + 830 \zeta_2{}^2$$

$$- 2632 \log 2 \zeta_3 - 272 \log 2\zeta_2 + 512 \log^2 2 \zeta_2 + \frac{1024}{3} \log^3 2 \zeta_2 + 384 \log^2 2 \zeta_3$$

$$- \frac{3216}{5} \log 2 \zeta_2^2 + 11520 \log 2\zeta_{1,-3} - \frac{4096}{15} \log^5 2, \tag{26}$$

where ζ_3 is a short hand for $\zeta(3)$ etc.

Here again we see true experimental applied mathematics, wherein our conjecture (25) and our numerical data for (26) (think of this as prepared biological sample) was used to confirm further discovery and proof by another researcher. Equation (26) was automatically converted to Latex from Panzer's *Maple* worksheet.

Maple does a lovely job of producing correct but inelegant TEX. In our experience many (largely predictable) errors creep in while trying to prettify such output. For example, as in (1), while 704/3 might have been 70/43, it is less likely that a basis element is entirely garbled (although a power might be wrong). Errors are more likely to occur in trying to directly handle the TEX for complex expressions produced by computer algebra systems.

4.5 Ramble integrals and short walks

Consider, for complex s, the n-dimensional *ramble integrals* [10]

$$W_n(s) = \int_{[0,1]^n} \left| \sum_{k=1}^{n} e^{2\pi x_k i} \right|^s dx, \tag{27}$$

which occur in the theory of uniform random walk integrals in the plane, where at each step a unit-step is taken in a random direction as first studied by Pearson (who did discuss 'rambles'), Rayleigh and others a hundred years ago. Integrals such as (27) are the s-th moment of the distance to the origin after n steps. As is well known, various types of random walks arise in fields as diverse as aviation, ecology, economics, psychology, computer science, physics, chemistry, and biology.

In 2010–2012 work (by J. Borwein, A. Straub , J. Wan and W. Zudilin), using a combination of analysis, probability, number theory and high-precision numerical computation, produced results such as

$$W_n'(0) = -n \int_0^\infty \log(x) J_0^{n-1}(x) J_1(x) dx,$$

were obtained, where $J_n(x)$ denotes the *Bessel function* of the first kind. These results, in turn, lead to various closed forms and have been used to confirm, to 600-digit precision, the following *Mahler measure* conjecture adapted from Villegas:

$$W_5'(0) \stackrel{?}{=} \left(\frac{15}{4\pi^2} \right)^{5/2} \int_0^\infty \{ \eta^3(e^{-3t}) \eta^3(e^{-5t})$$
$$+ \eta^3(e^{-t}) \eta^3(e^{-15t}) \} \, t^3 \, dt,$$

where the *Dedekind eta-function* can be computed from: $\eta(q) =$

$$q^{1/24} \prod_{n \geq 1} (1 - q^n) = q^{1/24} \sum_{n=-\infty}^{\infty} (-1)^n q^{n(3n+1)/2}.$$

There are remarkable and poorly understood connections between diverse parts of pure, applied and computational mathematics lying behind these results. As often there is a fine interplay between developing better computational tools—especially for special functions and polylogarithms—and discovering new structure.

4.6 Densities of short walks

One of the deepest 2012 discoveries is the following closed form for the *radial density* $p_4(\alpha)$ of a four step uniform random walk in the plane: For $2 \leq \alpha \leq 4$ one has the real hypergeometric form:

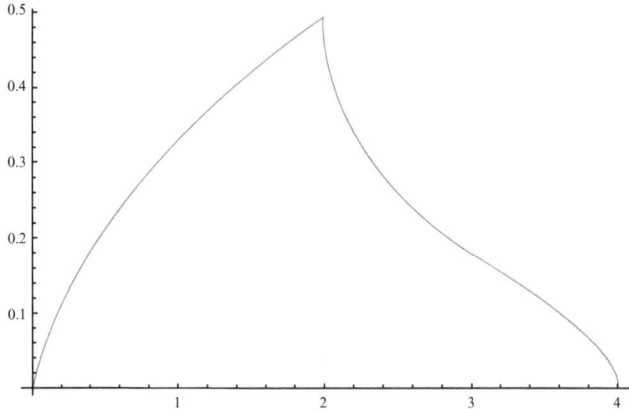

Fig. 9 The "shark-fin" density of a four step walk.

$$p_4(\alpha) = \frac{2}{\pi^2} \frac{\sqrt{16-\alpha^2}}{\alpha} {}_3F_2 \left(\begin{matrix} \frac{1}{2}, \frac{1}{2}, \frac{1}{2} \\ \frac{5}{6}, \frac{7}{6} \end{matrix} \middle| \frac{(16-\alpha^2)^3}{108\alpha^4} \right).$$

Remarkably, $p_4(\alpha)$ is equal to the real part of the right side of this identity everywhere on $[0, 4]$ (not just on $[2, 4]$), as plotted in Figure 9. This was an entirely experimental discovery—involving at least one fortunate error—but is now fully proven.

4.7 Moments of elliptic integrals

The previous study on ramble integrals also led to a comprehensive analysis of moments of elliptic integral functions of the form:

$$\int_0^1 x^{n_0} K^{n_1}(x) K'^{n_2}(x) E^{n_3}(x) E'^{n_4}(x) dx,$$

where the elliptic functions K, E and their complementary versions are

$$K(x) = \int_0^1 \frac{dt}{\sqrt{(1-t^2)(1-x^2 t^2)}},$$

$$E(x) = \int_0^1 \frac{\sqrt{1-x^2 t^2}}{\sqrt{1-t^2}} dt,$$

$$K'(x) = K(\sqrt{1-x^2}), \qquad E'(x) = E(\sqrt{1-x^2}).$$

Computations of these integrals to 3200-digit precision, combined with searches for relations using the PSLQ algorithm, yielded thousands of unexpected relations among these integrals (see [10]). The scale of the computation was required by the number of integrals under simultaneous investigation.

4.8 Computation and Constructivism

The fact that

$$\left(\sqrt{2}^{\sqrt{2}}\right)^{\sqrt{2}} = \sqrt{2}^2 = 2$$

provides a one line proof that *there exist irrational numbers α and β such that α^β is rational.* Indeed, either (a) $\sqrt{2}^{\sqrt{2}}$ is rational or (b) it is irrational; and can play the role of α. This lovely example of the *principle of the excluded middle* is rejected by constructivists and intuitionists alike. Indeed (b) holds but a constructive proof of such fact is quite elaborate. This is discussed in more detail in [24, §2.9] and [22].

Here we want to make the point that a computational sensibility leads one to want to know which case occurs, both for existential reasons, and because any algorithm with a conditional (if–then) step is probably worthless if one cannot determine which step to take.

4.9 Limits of computation

As we noted above, experimental studies have raised the question of whether one can truly trust—in the mathematical sense—the result of a computation, since there are many possible sources of errors: unreliable numerical algorithms; bug-ridden computer programs implementing these algorithms; system software or compiler errors; hardware errors, either in processing or storage; insufficient numerical precision; and obscure errors of hardware, software or programming that surface only in particularly large or difficult computations. Thus when applying experimental methods, it is essential to utilize rigorous validity checks, such as computing a key result by two completely different algorithms, and verifying that they produce the same result.

Along this line, it is important to keep in mind that even high-precision numerical evaluations can sometimes mislead. One remarkable example is the following:

$$\int_0^\infty \cos(2x) \prod_{n=1}^\infty \cos(x/n)\, \mathrm{d}x = \tag{28}$$

0.392699081698724154807830422909937860524645434187231595926...

The computation of this integral to high precision can be performed using a scheme described in [14]. When we first did this computation, we thought that the result was $\pi/8$, but upon careful checking with the numerical value

$$0.39269908169872415480783042290993786052464617492188227621\ldots,$$

it is clear that the two values disagree beginning with the 43rd digit!

The late Richard Crandall [29, §7.3] later explained this mystery. By employing a physically motivated analysis of *running out of fuel* random walks, he showed that $\pi/8$ is given by the following very rapidly convergent series expansion, of which formula (28) above is merely the first term:

$$\frac{\pi}{8} = \sum_{m=0}^{\infty} \int_0^{\infty} \cos[2(2m+1)x] \prod_{n=1}^{\infty} \cos(x/n) \, dx. \tag{29}$$

Two terms of the series above suffice for 500-digit agreement.

As a final sobering example in this section, we offer the following "sophomore's dream" identity

$$\sigma_{29} := \sum_{n=-\infty}^{\infty} \operatorname{sinc}(n) \operatorname{sinc}(n/3) \operatorname{sinc}(n/5) \cdots \operatorname{sinc}(n/23) \operatorname{sinc}(n/29)$$

$$= \int_{-\infty}^{\infty} \operatorname{sinc}(x) \operatorname{sinc}(x/3) \operatorname{sinc}(x/5) \cdots \operatorname{sinc}(x/23) \operatorname{sinc}(x/29) \, dx, \tag{30}$$

where the denominators range over the odd primes, which was first discovered empirically. More generally, consider

$$\sigma_p := \sum_{n=-\infty}^{\infty} \operatorname{sinc}(n) \operatorname{sinc}(n/3) \operatorname{sinc}(n/5) \operatorname{sinc}(n/7) \cdots \operatorname{sinc}(n/p)$$

$$\stackrel{?}{=} \int_{-\infty}^{\infty} \operatorname{sinc}(x) \operatorname{sinc}(x/3) \operatorname{sinc}(x/5) \operatorname{sinc}(x/7) \cdots \operatorname{sinc}(x/p) \, dx. \tag{31}$$

Provably, the following is true: The "sum equals integral" identity, for σ_p remains valid at least for p among the first 10^{176} primes; but stops holding after some larger prime, and thereafter the "sum less the integral" is strictly positive, but *they always differ by much less than one part in a googol* $= 10^{100}$. An even stronger estimate is possible assuming the Generalized Riemann Hypothesis (see [29, §7] and [19]). *What does it mean for an identity to be provably false but, assuming that the universe is finite, at a level that is intrinsically infinitesimal?*

As a single example of the sorts of difficulties that can arise even when one relies on well-maintained commercial software, the present authors found that neither *Maple* nor *Mathematica* was able to numerically evaluate constants of the form

$$\frac{1}{2\pi} \int_0^{2\pi} f(e^{i\theta})\, d\theta$$

where $f(\theta) = \text{Li}_1(\theta)^m \text{Li}_1^{(1)}(\theta)^p \text{Li}_1(\theta + \pi)^n \text{Li}_1^{(1)}(\theta - \pi)^q$ (for $m, n, p, q \geq 0$ integers) to high precision in reasonable run time. In part this was because of the challenge of computing polylog and polylog derivatives (with respect to order) at complex arguments. The version of *Mathematica* that we were using was able to numerically compute $\partial \text{Li}_s(z)/\partial s$ to high precision, which is required here, but such evaluations were not only many times slower than $\text{Li}_s(z)$ itself, but in some cases did not even return a tenth of the requested number of digits correctly [6].

For such reasons, experienced programmers of mathematical or scientific computations routinely insert validity checks in their code. Typically such checks take advantage of known high-level mathematical facts, such as the fact that the product of two matrices used in the calculation should always give the identity, or that the results of a convolution of integer data, done using a fast Fourier transform, should all be very close to integers.

5 Concluding Remarks

Let us start by reminding ourselves that in matters of practice what we do and what we say we do are not to be conflated. As Irving Kaplansky (1917–2006) remarked [37]:

> We [Kaplansky and Halmos] share a philosophy about linear algebra: we think basis-free, we write basis-free, but when the chips are down we close the office door and compute with matrices like fury.

The central issues of how to view experimentally discovered results have been discussed before. In 1993, Arthur Jaffe and Frank Quinn warned of the proliferation of not-fully-rigorous mathematical results and proposed a framework for a "healthy and positive" role for "speculative" mathematics [39]. Numerous well-known mathematicians responded [3]. Morris Hirsch, for instance, countered that even Gauss published incomplete proofs, and the 15,000 combined pages of the proof of the classification of finite groups raises questions as to when we should certify a result. He suggested that we attach a label to each proof—e.g., "computer-aided," "mass collaboration," "constructive," etc. Saunders Mac Lane quipped that "we are not saved by faith alone, but by faith and works," meaning that we need both intuitive work and precision.

At the same time, computational tools now offer remarkable facilities to confirm analytically established results, as in the tools in development to check identities in equation-rich manuscripts, and in Hales' project to establish the Kepler conjecture by formal methods.

The flood of information and tools in our information-soaked world is unlikely to abate. Thus in today's computer-rich world it is essential to learn and teach judgment when it comes to using what is possible digitally. This means mastering the sorts of techniques we have illustrated and having some idea why a software system does what it does. It requires knowing when a computation is or can—in principle or practice—be made into a rigorous proof and when it is only compelling evidence (or is, in fact, entirely misleading). For instance, even the best commercial linear programming packages of the sort used by Hales will not certify any solution, although the codes are almost assuredly correct.

An effective digital approach to mathematics also requires rearranging hierarchies of what we view as hard and as easy. Continued fractions, moderate size integer or polynomial factorization, matrix decompositions, semi-definite programs, etc. are very different objects *in vivo* and *in silico* (see [24, Ch. 8, ed. 2]). It also requires developing a curriculum that carefully teaches experimental computer-assisted mathematics, either as a stand-alone course or as part of a related course in number theory or numerical methods. Some efforts along this line are already underway, mostly at the undergraduate level, by Marc Chamberland at Grinnell (http://www.math.grin.edu/~chamberl/courses/MAT444/syllabus.html), Victor Moll at Tulane, Jan de Gier in Melbourne, and Ole Warnaar at University of Queensland.

Judith Grabner has noted that a large impetus for the development of modern rigor in mathematics came with the Napoleonic introduction of regular courses: lectures and textbooks force a precision and a codification that apprenticeship obviates. But it will never be the case that quasi-inductive mathematics supplants proof. We need to find a new equilibrium. That said, we are only beginning to tap new ways to enrich mathematics. As Jacques Hadamard said more eloquently in French [47]:

> The object of mathematical rigor is to sanction and legitimize the conquests of intuition, and there was never any other object for it.

Often in our experience, we follow a path empirically, computationally (numerically, graphically and symbolically) to a dead-end with a correct but unhelpful result. With non-traditional methods of rigor such as we have illustrated, in those cases we have frequently been spared proving intermediate results which while hard are ultimately sterile (i.e., item 5 in the list at the start of Section 2).

Never have we had such a cornucopia of ways to generate intuition. The challenge is to learn how to harness them, how to develop and how to transmit the necessary theory and practice. The Priority Research *Centre for Computer Assisted Research Mathematics and its Applications* (CARMA), http://www.newcastle.edu. au/research/centres/carmacentre.html, which one of us directs, hopes to play a lead role in this endeavor, an endeavor which in our view encompasses an exciting mix of exploratory experimentation and rigorous proof. And an exciting development showing the growing centrality of experimental mathematics is the opening of an NSF funded *Institute for Computational and Experimental Research in Mathematics* (ICERM), http://icerm.brown.edu/:

> The mission of the Institute for Computational and Experimental Research in Mathematics (ICERM) is to support and broaden the relationship between mathematics and computation:

specifically, to expand the use of computational and experimental methods in mathematics, to support theoretical advances related to computation, and address problems posed by the existence and use of the computer through mathematical tools, research and innovation.

Finally, it seems clear to us that experimental computation can end up reinforcing either an idealist or a constructivist approach, largely depending on one's prior inclinations [42]. It is certainly both a supplement and a complement to traditional post-Cauchy modes of rigor. Experimental computation is also an interrogable *lingua franca*, allowing quite disparate parts of mathematics to more fruitfully entangle. For this we should all be very grateful.

References

1. D. M. Abrams and S. H. Strogatz, "Chimera states in a ring of nonlocally coupled oscillators," *Intl. J. of Bifur. and Chaos*, vol. 16 (2006), 21–37
2. F. Aragon, J. M. Borwein and M. Tam, "Douglas-Rachford feasibility methods for matrix completion problems." ANZIAM Journal, July 2014, http://arxiv.org/abs/1308.4243.
3. M. Atiyah, et al, "Responses to 'Theoretical Mathematics: Toward a Cultural Synthesis of Mathematics and Theoretical Physics,' by A. Jaffe and F. Quinn," *Bulletin of the American Mathematical Society*, vol. 30, no. 2 (Apr 1994), 178–207.
4. J. Avigad, "Computers in mathematical inquiry," in *The Philosophy of Mathematical Practice*, P. Mancuso ed., Oxford University Press, 2008, 302–316.
5. D. H. Bailey, R. Barrio and J. M. Borwein, "High-precision computation: Mathematical physics and dynamics," *Appl. Math. and Comp.*, vol. 218 (2012), 10106–10121.
6. D. H. Bailey and J. M. Borwein, "Computation and theory of extended Mordell-Tornheim-Witten sums. Part II," Journal of Approximation Theory (special issue in honour of Dick Askey turning 80), vol. 189 (2015), 115–140. DOI: http://dx.doi.org/10.1016/j.jat.2014.10.004.
7. D. H. Bailey and J. M. Borwein, "Computer-assisted discovery and proof." *Tapas in Experimental Mathematics*, 21–52, in *Contemporary Mathematics*, vol. 457, American Mathematical Society, Providence, RI, 2008.
8. D. H. Bailey and J. M. Borwein, "Experimental Applied Mathematics." As a *Final Perspective* in the *Princeton Companion to Applied Mathematics*. Entry VIII.6, 925–932, 2015.
9. D. H. Bailey and J. M. Borwein, "Exploratory experimentation and computation," *Notices of the AMS*, vol. 58 (Nov 2011), 1410–19.
10. D. H. Bailey and J. M. Borwein, "Hand-to-hand combat with thousand-digit integrals," *J. Comp. Science*, vol. 3 (2012), 77–86.
11. D. Bailey, J. Borwein, N. Calkin, R. Girgensohn, R. Luke and V. Moll, *Experimental Mathematics in Action*, A K Peters, Natick, MA, 2007.
12. D. Bailey, J. Borwein, R. Crandall and M. Rose, "Expectations on Fractal Sets." *Applied Mathematics and Computation*. vol. 220 (Sept), 695–721. DOI: http://dx.doi.org/10.1016/j.amc.2013.06.078.
13. D. H. Bailey, J. M. Borwein and A. D. Kaiser, "Automated simplification of large symbolic expressions," Journal of Symbolic Computation, vol. 60 (Jan 2014), 120–136, http://www.sciencedirect.com/science/article/pii/S074771711300117X.
14. D. H. Bailey, J. M. Borwein, V. Kapoor and E. Weisstein,"Ten Problems in Experimental Mathematics," *American Mathematical Monthly*, vol. 113, no. 6 (Jun 2006), 481–409.
15. D. H. Bailey, J. M. Borwein, M. Lopez de Prado and Q. J. Zhu, "Pseudo-mathematics and financial charlatanism: The effects of backtest over fitting on out-of-sample performance," *Notices of the American Mathematical Society*, May 2014, http://ssrn.com/abstract=2308659.

16. D. H. Bailey, J. M. Borwein, M. Lopez de Prado and Q. J. Zhu, "The probability of backtest overfitting," Financial Mathematics, to appear, March 2015. http://ssrn.com/abstract=2326253.
17. D. H. Bailey, J. M. Borwein, A. Mattingly and G. Wightwick, "The Computation of Previously Inaccessible Digits of π^2 and Catalan's Constant," *Notices of the American Mathematical Society*, vol. 60 (2013), no. 7, 844–854.
18. D. H. Bailey, P. B. Borwein and S. Plouffe, "On the Rapid Computation of Various Polylogarithmic Constants," *Mathematics of Computation*, vol. 66, no. 218 (Apr 1997), 903–913.
19. R. Baillie, D. Borwein, and J. Borwein, "Some sinc sums and integrals," *American Math. Monthly*, vol. 115 (2008), no. 10, 888–901.
20. J. M. Borwein, "The Experimental Mathematician: The Pleasure of Discovery and the Role of Proof," *International Journal of Computers for Mathematical Learning*, 10 (2005), 75–108. CECM Preprint 02:178; 264, http://docserver.carma.newcastle.edu.au/264.
21. J. M. Borwein, "Exploratory Experimentation: Digitally-assisted Discovery and Proof," in *ICMI Study 19: On Proof and Proving in Mathematics Education*, G. Hanna and M. de Villiers, eds., New ICMI Study Series, 15, Springer-Verlag, 2012, http://www.carma.newcastle.edu.au/jon/icmi19-full.pdf.
22. J. M. Borwein, "Implications of Experimental Mathematics for the Philosophy of Mathematics," Ch. 2 (33–61) and cover image in *Proof and Other Dilemmas: Mathematics and Philosophy*, Bonnie Gold and Roger Simons, eds., MAA Spectrum Series, July 2008. D-drive Preprint 280, http://www.carma.newcastle.edu.au/jon/implications.pdf.
23. J. M. Borwein, "The SIAM 100 Digits Challenge," Extended review in the *Mathematical Intelligencer*, vol. 27 (2005), 40–48.
24. J. M. Borwein and D. H. Bailey, *Mathematics by Experiment: Plausible Reasoning in the 21st Century*, A K Peters, second edition, 2008.
25. J. M. Borwein, D. H. Bailey and R. Girgensohn, *Experimentation in Mathematics: Computational Roads to Discovery*, A K Peters, Natick, MA, 2004.
26. J. M. Borwein, O-Y. Chan and R. E. Crandall, "Higher-dimensional box integrals," *Experimental Mathematics*, vol. 19 (2010), 431–435.
27. J. M. Borwein, C. Maitland and M. Skerritt, "Computation of an improved lower bound to Giuga's primality conjecture," *Integers* 13 (2013) #A67, http://www.westga.edu/~integers/cgi-bin/get.cgi.
28. J. M. Borwein, I. J. Zucker and J. Boersma, "The evaluation of character Euler double sums," *Ramanujan Journal*, vol. 15 (2008), 377–405.
29. R. E. Crandall, "Theory of ROOF Walks," 2007, http://www.perfscipress.com/papers/ROOF11_psipress.pdf.
30. D. Easley, M. Lopez de Prado and M. O'Hara, *High-Frequency Trading*, Risk Books, London, 2013.
31. L. R. Franklin, "Exploratory Experiments," *Philosophy of Science*, vol. 72 (2005), 888–899.
32. M. Giaquinto, *Visual Thinking in Mathematics. An Epistemological Study*, Oxford University Press, New York, 2007.
33. J. Gleick, *Chaos: The Making of a New Science*, Viking Penguin, New York, 1987.
34. J. Gravner and D. Griffeath, "Modeling snow crystal growth: A three-dimensional mesoscopic approach," *Phys. Rev. E*, vol. 79 (2009), 011601.
35. J. Guillera, "Hypergeometric identities for 10 extended Ramanujan-type series," *Ramanujan Journal*, vol. 15 (2008), 219–234.
36. T. C. Hales, "Formal Proof," *Notices of the American Mathematical Society*, vol. 55, no. 11 (Dec. 2008), 1370–1380.
37. Paul Halmos, *Celebrating 50 Years of Mathematics*, Springer, 1991.
38. G. Hanna and M. de Villiers (Eds.), *On Proof and Proving in Mathematics Education*, the 19th ICMI Study, New ICMI Study Series, 15, Springer-Verlag, 2012.
39. A. Jaffe and F. Quinn, "'Theoretical Mathematics': Toward a Cultural synthesis of Mathematics and Theoretical Physics," *Bulletin of the American Mathematical Society*, vol. 29, no. 1 (Jul 1993), 1–13.

40. Mark Joshi, *The Concepts and Practice of Mathematical Finance*, Cambridge University Press, 2008.
41. J. E. Littlewood, *A Mathematician's Miscellany*, Methuen, London, 1953, reprinted by Cambridge University Press, 1986.
42. M. Livio, *Is God a Mathematician?*, Simon and Schuster, New York, 2009.
43. B. B. Mandelbrot, *The Fractal Geometry of Nature*, W. H. Freeman, New York, 1982.
44. M. Minovitch, "A method for determining interplanetary free-fall reconnaissance trajectories," *JPL Tech. Memo TM-312-130*, (23 Aug 1961), 38–44.
45. Erik Panzer, "Algorithms for the symbolic integration of hyperlogarithms with applications to Feynman integrals," http://arxiv.org/abs/1403.3385.
46. M. Petkovsek, H. S. Wilf, D. Zeilberger, $A = B$, A K Peters, Natick, MA, 1996.
47. G. Pólya, *Mathematical discovery: On understanding, learning, and teaching problem solving*, (Combined Edition), New York, John Wiley and Sons, New York, 1981.
48. R. Preston, "The Mountains of Pi," *New Yorker*, 2 Mar 1992, http://www.newyorker.com/archive/content/articles/050411fr_archive01.
49. H. K. Sørenson, "Exploratory experimentation in experimental mathematics: A glimpse at the PSLQ algorithm," in B. Lowe, ed., *Philosophy of Mathematics: Sociological Aspects and Mathematical Practice*, Thomas Muller, London, 20–10, 341–360.
50. V. Stodden, D. H. Bailey, J. M. Borwein, R. J. LeVeque, W. Rider and W. Stein, "Setting the default to reproducible: Reproducibility in computational and experimental mathematics," manuscript, 2 Feb 2013, http://www.davidhbailey.com/dhbpapers/icerm-report.pdf.
51. H. S. Wilf, "Mathematics: An experimental science," *The Princeton Companion to Mathematics*, Princeton University Press, 2008.
52. H. S. Wilf and D. Zeilberger, "Rational Functions Certify Combinatorial Identities," *Journal of the American Mathematical Society*, vol. 3 (1990), 147–158.
53. S. Wolfram, *A New Kind of Science*, Wolfram Media, Champaign, IL, 2002.
54. W. Zudilin, "Ramanujan-type formulae for $1/\pi$: A second wind," 19 May 2008, http://arxiv.org/abs/0712.

Mathematical products

Philip J. Davis

A prominent mathematician recently sent me an article he had written and asked me for my reaction. After studying it, I said that he was proposing a mathematical "product" and that as such it stood in the scientific marketplace in competition with nearby products. He bridled and was incensed by my use of the word "product" to describe his work. Our correspondence terminated. What follows is an elaboration of what I mean by mathematical products and how I situate them within the mathematical enterprise.

Civilization has always had a mathematical underlay, often informal, and not always overt. I would say that mathematics often lies deep in formulaic material, procedures, conceptualizations, attitudes, and now in chips and accompanying hardware. In recent years the mathematization of our lives has grown by leaps and bounds. A useful point of view is to think of this growth in terms of products. Mathematical products serve a purpose; they can be targeted to define, facilitate, enhance, supply, explain, interpret, invade, complicate, confuse, and create new requirements or environments for life.

What? Mathematical "products"? Products in an intellectual area that is reputed to contain the finest result of pure reason and logic: a body of material that in its early years was in the classical quadrivium along with astronomy and music? How gross of me to bring in the language of materialistic commerce and in this way sully or besmirch the reputation of what are clean, crisp idealistic constructions! Products are the routine output of factories, not of skilled craftsmen whose sharp minds frequently reside far above the usual rewards of life. The notion that mathematics has products or that its content is merchandise, might tarnish both its image and the self image of the creators of this noble material.

P.J. Davis (✉)
Department of Applied Mathematics, Brown University,
182 George Street, Providence, RI 02912, USA
e-mail: Philip_Davis@brown.edu

© Springer International Publishing Switzerland 2015
E. Davis, P.J. Davis (eds.), *Mathematics, Substance and Surmise*,
DOI 10.1007/978-3-319-21473-3_4

And yet … The world is full of mathematical products—mathematics *produces* knowledge hence we have *mathematical products*—many of them. As O'Halloran [1, 2] claims, mathematics is functional; it permits us to construe and reason about the world in new ways that extend beyond our linguistic formulations. The world of today embraces the product of that knowledge.

1 Examples of Mathematical Products

Yes, the world is full of mathematical products of all sorts. I will name a few. A slide rule is a product. A French curve is a product. An algorithm (recipe) for solving linear equations is a product. A theorem is a product and stands among hundreds of thousands of theorems, ready to be interpreted, appreciated, used, updated, reworked or neglected. A text book on linear algebra is a product. A polling system is a product. The statutory rule for allocating representatives after a new census is a product. A tax or a lottery or an insurance scheme or even a Ponzi scheme is a product. Telephone numbers are a product. A professional mathematical society is a product. A medical decision that depends in a routine manner on some sort of quantification is a product. A computer language is a product. A supermarket cash register and the Julian calendar are products. The act of taking a number at a delicatessen or a bakeshop counter to facilitate ones' "next" is a product. Matlab is a product. Google is a product. Encryption schemes are products. Sometimes a mathematical product is designed for very specialized usage; it may then be called a package or a tool box.

Admittedly these examples might seem to indicate that in my mind anything at all that has to do with mathematics can be considered a product. Is Cantor's diagonalization process a product? Is a T-shirt imprinted with the face of Kurt Gödel a mathematical product? Well, I would find it exceedingly difficult to propose a formal definition. In any case, let us see the extent to which one might describe and discuss the mathematical enterprise from the point of view of its products that I have or will cite.

What is the clientele for mathematical products? While mathematical products are the brain children of inspired individuals or groups, the targeted users of the products may vary from a few individuals to entire populations. Those targeted may be aware of the availability of a product that has been claimed to be of use; they may either use it or reject it. In many cases the product is built into a whole social system and one cannot easily opt out of its use. Examples: phone numbers and area codes; the US Census; more locally, passwords at the ATM around the corner.

1.1 Scientific/Technological aspects of Mathematical Products

Mathematics was called by Gauss "the Queen of the Sciences" and a good fraction of its products relate to science/technology: e.g., packages for the factorization of large integers; for the analysis of architectural structures or packages marketed; for on-site DNA analysis. A scheme for constructing and interpreting a horoscope can be a mathematical product of considerable sophistication and complexity. The "wise" may reject its conclusions, yet the product flourishes.

Without in any way dethroning the Queen, it should be pointed out that the employment of mathematics has always gone far beyond what are now called "the sciences." Mathematics has made an impact on commerce, trade, medicine, biology, mysticism, theology, entertainment, warfare, etc.

1.2 The Transmission or Communication of Mathematical Products

Transmission is done by a wide variety of "signs" or "semiotic products." Short texts, books, pictures, programs, flash drives, chips, formal classroom teaching or the informal master/apprentice relationship, word of mouth, etc. The international or intercultural transmission and absorption of mathematical products (e.g., the adoption by the West of Arabic numerals) has been and still is the object of scholarly studies.

1.3 Commercial aspects of Mathematical Products

The commercialization of mathematical products has grown by leaps and bounds since World War II. A mathematical product can be promoted in many of the same ways that a brand of breakfast food is promoted: by ads, by the praises of well-known personalities or groups, etc, (plugs). On MATLAB's website, you can find a list of MATLAB's available products, listed openly and labeled clearly as "products." Investment and insurance schemes are called "products."

A product can be sold, e.g., a hand held computer or the *Handbook of Mathematical Functions*. A product can be licensed for usage, or it can be made available as a freebie. In the case of taxes (*qua* mathematical product), it is "promoted" by laws and threats of punishment. Rubik's Cube, a mathematical product, caught the imagination and challenged the wits of millions of people and has earned fortunes. Sudoku, a mathematical puzzle, is sold in numerous formats. If a product is income producing, its sellers can be taxed. A product can be copyrighted or patented; the owners of such can be contested, sued for infringement.

1.4 Competitive Aspects of Mathematical Products

A mathematical product is often subject to competition from nearby products. Think of the innumerable ways of solving a set of linear equations. Textbooks, a source of considerable income, compete in a mathematical market place that involves educationists, testing theorists and outfits, unions, publishers, parents' groups, local state and national governments.

1.5   Social aspects of Mathematical Products

If a mathematical product finds widespread usage, it may have social, economic, ethical, legal or political implications or consequences. The repugnant Nuremberg Racial Laws (Germany, 1935) with their numerical criteria caused incredible suffering. DNA sequencing and its interpretations is a relatively new branch of applied mathematics, resulting in a host of new products. In a number of States, the level of mathematical tests for the lower school grades has been questioned. The social consequences of mathematical products, benign or otherwise, may not emerge for many years.

1.6   Legal aspects of Mathematical Products

There are innumerable examples of this. The US Constitution is full of number processes. Consider

> "Representatives and direct Taxes shall be apportioned among the several States which may be included within this Union, according to their respective Numbers, which shall be determined by adding to the whole Number of free Persons, including those bound to Service for a Term of Years, and excluding Indians not taxed, three fifths of all other Persons."
> (Later: Amended!)

Some mathematical products have been subject to judicial review. As an example, the mathematical scheme for the 2010 Census was vetted and restricted by the US Supreme Court.

An example of a statutory product is the method of least proportions used to allocate representatives in Congress. It was approved by the Supreme Court in Dept. of Commerce v. Montana, 503 U.S. 442 (1992). Another example: A multiple regression model used in an employment discrimination class action is another such example; it was approved by the Supreme Court in Bazemore v. Friday, 478 U.S. 385 (1986).

1.7   Logical or Philosophical aspects of Mathematical Products

A mathematical product, considered as such, is neither true nor false. Of course, it may embody certain principles of deductive logic, but these do not automatically make the employment of the product plausible or advisable. A product can be made plausible, moot or useless on the basis of certain internal or external considerations. An interesting historical example of this is the dethroning of Euclidean geometry as the unique geometry by the discovery of non-Euclidean geometries.

A product may raise or imply philosophical questions such as the distinction between the subjective and the objective or between the qualitative and the quantitative, between the deterministic and the probabilistic, the tangible and the intangible, the hidden and the overt.

Numerical indexes of this thing and that thing abound. Cases of subjectivity occur when a product asks a person or a group of people to pass judgment on some issue: "On a scale of zero to ten, how much do you like tofu?" The well-known *Index of Economic Freedom* embodies a number of items, expressed numerically:

> "We measure ten components of economic freedom, assigning a grade in each using a scale from 0 to 100, where 100 represents the maximum freedom. The ten component scores are then averaged to give an overall economic freedom score for each country. The ten components of economic freedom are: Business Freedom | Trade Freedom | Fiscal Freedom | Government Size | Monetary Freedom | Investment Freedom | Financial Freedom | Property rights | Freedom from Corruption | Labor Freedom"

1.8   Moral aspects of Mathematical Products

Society asks many questions. Does the manner of taking the US Census account properly for the homeless? Are tests in Algebra slanted towards certain subcultures? Does the tremendous role that mathematics plays in war raise questions or angst in the minds of those who are responsible for its application? Are results of IQ testing being misused?

2   Judgments of Mathematical Products

As mentioned, mathematical products serve a purpose; they can be targeted to define, facilitate, enhance, invade, any of the requirements or aspects of life. Ultimately, a product can be judged in the same way that any product can be judged: by the response of its targeted users or purchasers. In the case of a mathematical product what criteria are in play? The cheapest? The most convenient? The most useful? The most comprehensive? The most accurate? The most original? The most seminal? The most reassuring? The safest or least vulnerable? The most esthetic? The most moral? Is the product unique? Are there pressures from investors or the various foundations that support their production?

Is "survival of the fittest" a good description of the judgment process? Probably not. There are fashions in the product world attracting both excited consumers and producers. Time, chance and what the larger world requires, appreciates, or suffers from mathematizations that are always in play.

References

1. O'Halloran, K. L. (2005). *Mathematical Discourse: Language, Symbolism and Visual Images.* London and New York: Continuum.
2. O'Halloran, K. L. (2015). The Language of Learning Mathematics: A Multimodal Perspective. *The Journal of Mathematical Behaviour.* http://dx.doi.org/10.1016/j.jmathb.2014.09.002

How should robots think about space?

Ernest Davis

Abstract A robot's action are carried out over time through space. The computational mechanisms that the robot uses to choose and guide actions must therefore at some level reflect the structure of time and space. This essay discusses what ontology for space and time should be built into a robot. I argue that the designers of a robot generally need to have in mind explicit otologies of time and space, and that these will generally be the real line and three-dimensional Euclidean space, though it is often advantageous to project down to a lower-dimensional space, when possible. I then discuss which spatial or spatio-temporal regions should be considered "well-behaved" and which should be excluded as "monsters." I propose some principles that the class of well-behaved regions should observe, and discuss some examples.

1 Artificial Intelligence and Ontology

Artificial intelligence (AI), like computer science generally, is a branch of engineering, not science or philosophy or scholarship. The goal of the field, in ambitious moods, is to construct programs and robots that can behave as intelligently as people across a wide range of activities; in more modest moods, to build programs that can do fairly well at some activity that requires intelligence; but in any case to build something. Therefore, from the standpoint of AI research the question of whether *X actually* exists is not particularly important; the question is, will it be helpful to a robot[1] to think about *X*. The choice or design of an ontology is part of the software engineering process; and if an ontology is helpful in building a robot, that is all we

---

[1]For purposes of both brevity and concreteness, I will largely frame the discussion in this essay in terms of robots. However, most AI systems are not connected to physical robots; and much of the discussion in this paper applies equally to AI systems of many other kinds that deal with spatial information; e.g. to natural language processing systems that have to interpret spatial terms in text.

E. Davis (✉)
Department of Computer Science, Courant Institute of Mathematical Sciences, New York University, 251 Mercer Street, Room 329, New York, NY 10012, USA
e-mail: davise@cs.nyu.edu

© Springer International Publishing Switzerland 2015
E. Davis, P.J. Davis (eds.), *Mathematics, Substance and Surmise*,
DOI 10.1007/978-3-319-21473-3_5

75

ask of it. Of course, since robots work in the real world, there are often advantages to using ontologies that bear some relation to reality; however, verisimilitude is definitely a secondary consideration as compared to usefulness. Likewise, since human beings are incomparably more intelligent than any AI program that exists, the AI programmer may well want to pay attention to what psychologists, linguists, and anthropologists tell her about how the human mind works or how natural language works; however, she is not bound by those findings if she decides that they don't fit into her program design.

What ontologies a robot needs depends, of course, on what tasks it is supposed to be doing; and since human intelligence is open-ended and there is no end to the things that people can think about, there is likewise no end to AI ontologies. If the robot you are trying to build is supposed to answer questions about cell biology, then it will need an ontology that includes ribosomes and endoplasmic reticulums; if it is supposed to prove theorems in category theory, it will need an ontology that includes categories and morphisms; if it is supposed to read and understand the *Harry Potter* books, it will need to suspend disbelief and use an ontology that includes horcruxes, magic wands, and potions. Ultimately, a robot with full-fledged intelligence should be develop these ontologies by itself on the fly as it learns biology or category theory or reads *Harry Potter,* the way people do. How that is to be done is a deep problem for AI research, about which not very much is known; I am not going to address that in this essay.

Rather, this essay focuses on one particular type of core competency: how a robot could reason about spatial relations and about motion for a variety of everyday settings, and what ontology of space it should use for that. I will begin by making this concrete by giving a few examples of the kinds of everyday settings that involve spatial reasoning that I have in mind. I will then discuss the role of an ontology in building a program; the need for ontologies; the need for a distinctive spatial ontology; the argument for using multiple spatial ontologies; and the advantages and difficulties of using standard Euclidean geometry ($\mathbb{R}^3$) as a spatial ontology.

2 Examples

Let me begin by discussing a few examples of the kinds of reasoning I have in mind. These have been chosen to be natural and simple examples that illustrate a variety of geometric features, geometric entities, and forms of intelligent spatial reasoning.

2.1 Exhibit A: A cheese grater

Figure 1 shows a common make of cheese grater. (Readers who own this object and happen to be near their kitchen might find it useful to take it out and contemplate it directly.)

What do we want a household robot to understand about a cheese grater? Most obviously, the robot should know how to use the grater: Place the grater vertically with the base flush on a table top; hold the handle in one hand; hold the cheese flush against the appropriate side of the grater at the top; slide the cheese downward against the face of the grater, pushing inward; repeat until enough cheese has been grated; collect the grated cheese from the table inside the grater.

If all we want the robot to do is to carry out this procedure by rote, and we are willing to preprogram it for this particular procedure, then we may well be able to get away without any kind of general reasoning ability or ontology; all we would need is a set of rules mapping percepts to the corresponding action. We will discuss this further in section 4.

But a truly intelligent robot should be able to go beyond just rote execution. A truly intelligent robot would be able to adapt this plan to varying circumstances: If the table surface is slanted, then the base should be placed flush against it, leaving the grater off the vertical; in this case, the grater should generally be placed so that the surface being used is uppermost. If the grater would scratch the table and so cannot be placed directly on it, then either another object should be placed between the grater and the table, or the grater can be held the air with the other hand; this will require a tighter grip than grating in the usual way. If the holes are blocked on the inside, then they must be cleared before the grater can be used. The robot should be able to predict what will happen if the grater is used in some unusual way; for instance, if the food is pulled up the side of the grater rather than down, that will work for the small holes, but not for the circular "cheddar cheese" holes (in the foreground in figure 1). The robot should be able to understand verbal instructions for using the grater or to give verbal instructions. The robot should be able to apply its knowledge of this particular grater to other graters of related but different design. These kinds of tasks presumably require the robot to understand the geometrical features of the grater and their relation to the physical interactions of grater, manipulators, and food being grated.

The geometry of the grater is rather complex: a complete description needs to specify the overall frustum shape plus handle, the shapes of the individual holes, and the patterns in which the holes are laid out. Ontologically, however, the spatial entity associated with the grater is fairly simple. The grater is, for virtually all intents and purposes, a rigid object. Therefore the regions that any particular grater occupies at two different times are congruent to one another, and the grater moves continuously as a function of time. Two graters of the same make occupy nearly congruent spatial regions; two different graters of this general category occupy regions that are not congruent but have a number of geometric features in common.

The geometry of the cheese as it is being grated is harder to characterize. For concreteness, let us consider a cheddar cheese being grated on the circular holes. There is the block of cheese held in the hand, the pile of strips that have been grated lying in a pile, and transient states when a strip of cheese has been partially cut off from the main block. It would be laborious, and certainly beyond the scope of this paper, to give a detailed characterization of the geometric properties of the cheese. However, it is clear that a truly intelligent robot of the kind we are contemplating, or a human using the grater, has some understanding of the geometric features of the cheese. For instance, the robot should know that the strips come off the part of the cheese that has been in contact with the grater, and that they leave semi-cylindrical indentations in that end of the cheese; it should be astonished if the block of cheese started wearing away at the opposite end or in the middle or if, after grating, the block of cheese had deep conical pits going inward into the block.

2.2   Exhibit B: A string bag

Figure 2 shows a string bag with vegetables. For our purposes, the issues are not very different in kind from those involved in the cheese graters; but we bring the example to illustrate what a broad range of spatial features are involved in even simple tools of this kind. Again, we would expect a truly intelligent robot, not merely to be able to use a particular bag in a standard way but to be able to reason about how the bag would work in unusual circumstances or could be used for unusual purposes.

As with the grater, the spatial ontology of any particular string bag can be characterized in terms of the space of regions that the bag could be made to occupy. If it is acceptable to approximate the individual strings as one-dimensional curves, then a feasible region for the bag is a mapping of the strings into space that preserves the arc length of strings and the connections between them. If it is necessary to deal with the three-dimensionality of the strings, then the definition becomes more complex. Still more complex is the geometric characterization of string bags in general (i.e., networks of strings that will successfully function as a string bag)

Fig. 2 String Bag with Vegetables. Downloaded from http://commons.wikimedia.org/wiki/File: String_bag.jpg12/28/13 and reproduced under a Creative Commons licence

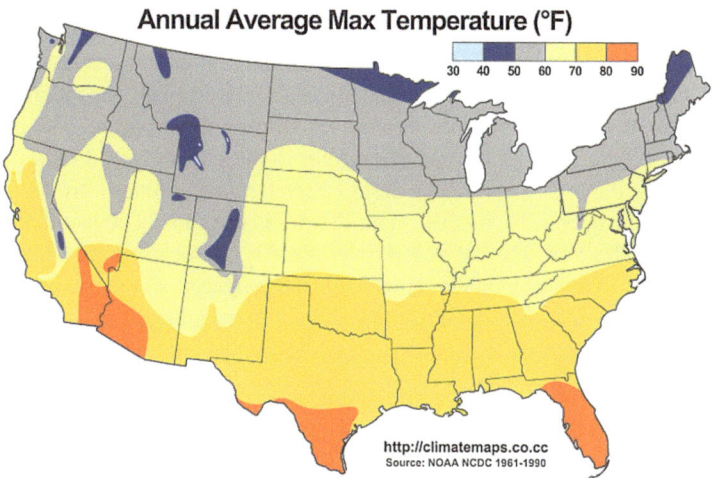

Fig. 3 Annual Average Max Temperature. Downloaded from http://commons.wikimedia. org/wiki/File:Average_Annual_High_Temperature_of_the_United_States.jpg12/28/13 and reproduced under a Creative Commons licence

2.3 Exhibit C: A Temperature Map

Our third example is the ground air temperature distribution over a large region, illustrated in figure 3. This illustrates that spatial knowledge is not always knowledge about a particular thing at a particular place. The air temperature is a continuous field over the cross-product of the surface of the earth times the time line.

Unless the robot is working as a meteorologist, it probably does not need a detailed physical model of the weather, but even the robot on the street should have some basic spatial information available, much of it in the form of general rules rather than absolute rules: At any given time, temperatures are lower further north and in the mountains. At any place, date of the year, and time of day the temperature is predictable with reasonably high confidence in a range of about plus or minus 20 degrees Fahrenheit, though outliers are not very uncommon. The temperature difference between places ten miles apart is generally not very large, though again outliers can occur, particularly when there are strong differences in topography or a topographical boundary. The body of knowledge for the average person, other than purely chronological ("When I went to Chicago in January, it was $-10°$ with a 30 mph wind"), and the range of types of reasoning are both probably much smaller than with a grater; but some significant reasoning must be done, and it involves knowledge of a form that is quite different from a grater.

3 What is the role of an ontology in a program?

The word "ontology" has become a buzz-word in AI, with a number of different meanings, some quite banal (e.g., what kinds of buttons are on a web-based order form). In this essay, what I mean by the word is the underlying conceptualization of the world that the programmer has in mind in building the program. The meanings of some of the data structures—those that refer to the outside world rather than the internal state of the program—can be described in terms of entities or relations in the ontology, and the operations on these data structures can mostly be justified in terms of the ontology. Importantly, the underlying ontology can be different from the class of entities that can be individually denoted by the program's data structures.

Let me give a couple of examples. Almost all numerical calculations on computers that require non-integer values are done using floating point numbers. There are fairly complex rules for doing various kinds of mathematical computations over these, and still more complex rules for estimating the accuracy of your answer when you're done. However, the underlying ontology, as I am using the term, is the traditional real line $\mathbb{R}$, in the sense that the real line is the "gold standard" to which floating point operations are essentially always referred. When one speaks of the accuracy of a floating point calculation, one means the accuracy as compared to the true value given in the theory of the reals; and a floating point algorithm is acceptable to the degree that it gives answers that correspond to the true real value. For example, if I give MATLAB the problem

$$(\sqrt{153} + \sqrt{147}) \cdot (\sqrt{153} - \sqrt{147})$$

I get the answer $6 - 1.243 \cdot 10^{-14}$. The reason that this is a good answer is that (a) it is very close to 6, which is the true answer in the theory of the reals; (b) the people who know about this kind of thing assure me that it would be difficult or impossible to do systematically better than MATLAB if one is committed to computing intermediate results in floating point form, rather than manipulating the symbolic expression.

There are alternative methods for representing real numbers. Standard double-precision floating point notation represents the number "two and two-thirds" as (essentially) "$0.101 \ldots$ (52 bits) $10 \cdot 2^2$". One could use decimal notation and represent it as "$0.26666 \ldots 67 \cdot 10^1$". One could use an exact fraction: "8/3". One can go further and represent an arbitrary algebraic number x exactly[2] as a pair of a polynomial p for which x is a zero and an interval range that contains x and no other zero of p; for instance, "$\langle x^5 - 7x + 1, [0, 1] \rangle$" represents the root of $x^5 - 7x + 1$ that lies between 0 and 1 (≈ 0.14287). The set of numbers that are exactly representable in each case is different, and so are the algorithms that carry out a given calculation. But the underlying ontology is the same: namely, the real line.

Another, similar example: If a programmer is writing a "physics engine" to simulate interactions among rigid solid objects, she will probably begin by choose some specific idealized model of solid object physics: e.g., perfect rigidity, uniform density, Newtonian mechanics, Coulomb friction, inelastic collisions. This, then, is the ontology. The program will almost certainly not calculate the results exactly— for example, calculating the exact time of collisions between two rotating objects is in general very difficult—rather, the calculations are an approximation of this model, and the algorithms are tuned to achieve the best possible approximation of the model. If the programmer later needs to incorporate some phenomenon that is outside the scope of this idealization, she will probably first consider how to modify the model, or create a separate model for the phenomenon, rather than simply diving directly into the code and tweaking it to get the phenomenon. Certainly, in most cases, she is well advised to do so.

This definition of ontology gives a partial answer to what is known as the "symbol grounding" problem: What determines the meaning of a symbol in a computer program? On this view of ontology, a symbol means what the programmer intends that it should mean. That does not solve the general symbol grounding problem because it does not give any account of symbols that the program generates on the fly but, as far as it goes, it seems reasonable.

4   Is it necessary to have an ontology of space?

Is it useful for a robot to have an ontology of space? That is, is it a good idea for someone programming a robot to think through what kinds of spatial entities and spatial relations she intends to use, and how the data structures express those?

---

[2]This data structure *by definition* denotes this exact algebraic number, because that is what the programmer means by the data structure. The programmer presumably provides functions that carry out exact calculations over these data structures, e.g. an addition function that takes two addends represented in this form and returns the sum, also represented in this form.

Perhaps not. Recently, remarkable successes in AI tasks have been achieved by systems in which the AI engineer presents a powerful, general-purpose machine learning program with a large collection of sample inputs and outputs, and leaves it to the program to work out all the intermediate complexities. For example, Cho et al. [1] have built a machine translation system, of state-of-the-art quality, which works purely by being shown parallel texts in English and French, with no built-in ideas of syntax, semantics, or meaning in either language.

In some cases, a machine learning algorithm of this kind will spontaneously develop internal structures that a human observer can see correspond to "natural" domain features or concepts. However, even when that occurs, that is only an interesting side effect; it is not an inherent or an intended part of the working of the algorithm. In many cases, when the learning algorithm is done, human observers really have no idea why it works or what its internal structures might "mean"; all they can say is that, experimentally, it carries out the task successfully.

As of the date of writing, there have been few applications of this technology to robotic control and manipulation, but no doubt those are coming as well. Many enthusiasts for this kind of approach believe that it will suffice to achieve all of AI. If so, then the answer to the question in the chapter title would be "There is no need for a robot to think about space," or at least, "There is no need for us to think about how a robot should think about space," or indeed any other abstract concept.

This approach has a number of points in its favor. First, it passes the buck; rather than worry about what to do with space, we can simply rely on a machine learning facility which we will need anyway for other purposes. Second, it will presumably by design tend to home in on spatial properties that are important to the robot and ignore those that are not, which can be a difficult problem for hand-designed ontologies.

Third, there is a strong argument to be made that this must in principle be possible. After all, any concept that a person knows was acquired in one of three ways: he learned it himself from experience, or he was, explicitly or implicitly, taught it from a store of cultural capital, or it was innate. In each of these cases, the concept was learned from experience at some point at some level; either the individual learned it, or the culture learned it, or the species or one of its ancestors learned it through evolution. In any case, the concept must be learnable.

However, I personally suspect that there are limits to how far this kind of approach can be taken, with anything like the kinds of learning techniques and data sources currently under consideration. These techniques shine at narrowly defined tasks and achieve a certain level of success at broadly defined tasks; but achieving human levels of performance on complex tasks would seem to require sophisticated representations of knowledge and reasoning abilities that are substantially beyond the machine-learning technology we currently have. For example, this kind of technology might well be able to learn to grate cheese, but it is harder to see how it could achieve the kind of flexible planning and other kinds of reasoning described in section 2 without being separately trained on each instance.

Another alternative approach uses a programmer, rather than a machine learning algorithm, to deal with the spatial issues, but releases her from the responsibility of thinking through a coherent theory of space. Rather she programs catch-as-catch-can; recording the spatial information that the robot will need in unsystematic and probably subtly inconsistent data structures, and writing subroutines to deal with specific tasks.

There is certainly a lot of this kind of software out there, and some of it works well enough. But it is not what we who teach computer science recommend as good software engineering; it tends to lead to code that is brittle, inextensible, and hard to maintain. Particularly in AI applications, where one wants to attain the maximum degree of flexibility, and to be able to use the same knowledge for a multitude of different purposes, it is important to have data structures with well-defined meanings and functions that carry out well-defined computations or actions. For these kinds of definitions, a clear ontology is an invaluable asset.

5   Should a robot have a separate ontology of space?

Should a robot have a clearly delineated ontology of space as a separate conceptual entity or should space be just an aspect of other theories? The second alternative might be something along the following lines. The robot knows about concrete real-world entities—e.g., the cheese, the grater, itself—and it knows about relations among these. Some of these relations are strongly spatial, e.g. "The block of cheese is touching the face of the grater"; some are non-spatial, e.g. "This is mild, low-fat cheddar cheese"; and some combine spatial and non-spatial elements, e.g. "I am pushing the cheese downward against the face of the grater." But there are no abstract spatial entities—there are triangular objects, but no triangles—and reasoning that is based purely on spatial relations does not form a separate or particularly distinct part of the overall reasoning apparatus. This is a somewhat holistic view or a Machian view.

I do not know of any attempts to develop an AI system or robot along these lines. From the standpoint of software engineering, there are good reasons for having a separate spatial reasoning component with abstract spatial objects; these are, in fact, much the same reasons that geometry, abstracted from its specific applications, became a separate discipline 2500 years ago.

- There is generally a clear distinction between purely geometric reasoning and reasoning that involves other issues. (As a point of contrast, it can be very difficult to draw a line between folk psychology and folk sociology.)
- The validity of a geometric inference is essentially independent of the application, though the importance of one or another type of geometric inference depends strongly on the application.
- There is a largely self-contained body of techniques used for geometric calculation and geometric inference which are not applicable to other subject matters.

Therefore placing geometric reasoning in a separate component with abstract geometric entities serves the software engineering goal of modularization; this significantly reduces the risk of unnecessary repetition of code, of inconsistencies, and of errors.

6 Should a robot have a single ontology of space or multiple ontologies?

Perhaps a robot should be able to think about space in a number of different ways corresponding to different tasks or circumstances. For instance, it could model space as three-dimensional when grating cheese; as the two-dimensional surface of a sphere when thinking about geography; and as a network of points and edges when planning a route through the subway. There is, in fact considerable psychological evidence that suggests that people use multiple spatial models in this way. In particular, people are often extremely poor at geometric reasoning in three dimensions—for instance, at visualizing the three dimensional rotation of an object of any complexity—and prefer to project down to two dimensions, if that is in any way an option.

From an engineering standpoint the chief advantages of using multiple models is that a specialized model can be tuned to the task and can support more efficient algorithms. There are two drawbacks. The first drawback of multiple models are that when two tasks that use different models interact, information has to be translated from one model to the other. The second drawback is that if the wrong model is chosen for a task, incorrect or suboptimal results may be obtained.

In some cases these drawbacks hardly apply. For instance, the low levels of vision proceed with little or no input from higher-level reasoning; the representation used here can therefore be entirely optimized to this particular process. The only constraint is that, at the end of the vision process, the output has to be presented in a form that other processes (e.g., manipulation) can use. However, aside from a few autonomous, specialized processes like low-level vision, intelligent spatial reasoning would seem to require that information be quite freely exchangeable between all kinds of cognitive tasks: high-level vision, physical reasoning, navigation, cognitive mapping, planning, natural language understanding, and so on.

The general opinion here, with which I concur, is that the simplifications offered by working in specialized ontologies outweigh the costs of moving back and forth between them. In particular, the ability to reduce problems from three to one or two dimensions seems to be extremely powerful, when it is possible. However, we do not actually know what difficulties may lie here; the technical issues of transferring information between spatial ontologies have not been extensively studied, and may be problematic.

7 What is the model of space?

What ontology of space is appropriate for tasks that substantively involve fully three-dimensional space, like grating cheese? The traditional answer, of course, is $\mathbb{R}^3$, three-dimensional Euclidean space. Indeed this is so universally accepted that many mathematicians would no doubt be surprised to learn that it is at all controversial. But it is. Many people are quite uncomfortable with $\mathbb{R}^3$. And it certainly has features that give one pause: it is full of monsters such as Zeno's paradoxes, the space-filling curve, and the Banach-Tarski paradox;[3] it raises all kinds of unreal distinctions such as the difference between a topologically open region and its closure; and most of its elements cannot be finitely named.

It seems to me, however, that the conventional wisdom has settled on $\mathbb{R}^3$ for a number of very compelling reasons that more than make up for these drawbacks.

Correspondence to reality: The physicists tell us that Euclidean geometry certainly breaks down on the cosmic scale and may break down on the smallest scales (10^{-43} meters or so).[4] However within the eleven orders of magnitude from about 0.2 mm to about 20,000 km that span the range of normal human interaction, it is spectacularly accurate to within the accuracy of measurement. Discrepancies can almost always be convincingly attributed to problems with the measurement technique rather than discrepancies between the world's geometry and $\mathbb{R}^3$.

Continuity. If an object goes from one side of a line to the other side, it must cross the line. If an object goes from outside a closed curve to inside, it must cross the curve.

Invariance under rotation: Rigid objects can rotate while preserving their shape, and the theory of space must have a way of describing this. This is not so easy to achieve.[5] An interesting point of comparison is rotating a pixel image. In computer graphics, if you have an image of a long narrow bar, and you use a naive algorithm to rotate it, the result may well have a visibly jagged edge. If you carry out a long sequence of rotations, computing each new shape from the previous one, without reference to the original shape, the geometric information can be entirely lost; depending on the algorithm, the bar may split or may fatten into a circle. The graphicists have tricks for getting around this; these are based, partially on the $\mathbb{R}^2$ theory of rotation, partially on an understanding of how the eye reads images, partly on empirical findings of what visually works. But the point, for our purposes, is that

---

[3]The space-filling curve, the Banach-Tarski paradox, and some other technical terms are defined in a glossary at the end of the chapter.

[4]It also breaks down in the neighborhood of a black hole; but that is the least of the problems facing a robot in the neighborhood of a black hole.

[5]The Pythagorean proof that the ratio of the diagonal of an isosceles right triangle to its leg is irrational fundamentally depends on rotation. Comparing the lengths of two non-parallel line segments depends on rotating one of the lines so that the two are collinear. In affine geometry, where rotation is undefined, the ratio of the lengths of two non-parallel lines is also undefined.

the geometric aspects of these corrections are based on a gold standard which is the $\mathbb{R}^2$ theory of rotation; so in the terms discussed in section 3, the ontology used here is in fact $\mathbb{R}^2$.

Moreover, a physical theory of solid objects should presumably allow them to rotate at a uniform rate, and to be in a well-defined spatial state at exact fractions of one complete rotation. In a real-based theory, this involves terms of the form $\cos(\pi/n)$ and $\sin(\pi/n)$ for integer n; an alternative ontology must provide something analogous. A physical theory that gives rise to non-uniform rotation, like a solid pendulum, similarly entails the existence of various kinds of non-elementary functions. What one can effectively compute or what one needs to compute about these in any given situation is a secondary question; the critical point is that the theory of space and motion had better allow these to exist. If an alternative theory of space is to be used, then one has to find an alternative formulation of the laws of motion that similarly guarantees that future states are geometrically meaningful.[6]

Other measures give other kinds of real values; for instance, the ratio of the volume of a sphere to the volume of the circumscribing cube is $\pi/6$. However, here it seems to me there is more ontological wiggle room to escape these commitments; one could possibly have an ontology in which there are no exact spheres or no exact cubes or no exact measurements of volumes. The invariance of rigid objects under rotation is less forgiving; you want to be sure that, if you pick up the cheese grater and carry it around, its geometric properties remain the same. Of course, solid objects are not perfectly rigid, but it should be the business of the physical theory to tell us about the slight deformations and wear, not the geometric theory. If your spatial ontology makes it *geometrically impossible* for the grater to maintain its shape, it seems to me that you are off to a bad start.

It is sometimes argued that, since programs cannot actually manipulate real numbers, they cannot actually be working in $\mathbb{R}^3$. However, on the view of ontology that I have taken in section 3, that argument does not hold water. The programs are using one representation or another, but what they are doing makes sense because the representations are approximations to $\mathbb{R}^3$.

I am not, of course, arguing that the man on the street explicitly knows the theory of the reals as formulated by Cauchy, Dedekind, etc. I am not even arguing that he knows it implicitly in the same sense that a native speaker of a language has implicit knowledge of the grammatical rules. What I am saying is that the man on the street has an understanding of spatial relations, much of it non-verbalized, that he uses for physical reasoning and other cognitive tasks, and that, in general, $\mathbb{R}^3$ is an extremely

---

[6]For instance, an ontology of space in which all points have rational coordinates would be consistent with a theory in which all rotations are rational rotations, i.e. (in 2D) rotations by angle θ such that $\cos(\theta) = a/c$ and $\sin(\theta) = b/c$ where a, b, c are a Pythagorean triple (see Glossary). But then the position at which a rotating object collides with a stationary object would in general be undefined; it would either have to overshoot (in which the two objects interpenetrate) or undershoot (in which the collision occurs while there is still a finite gap between the objects).

good model for that understanding, much better than any other model we know of. Likewise, if a programmer wants to build a robot to do this kind of reasoning, she will do better to ground her program in $\mathbb{R}^3$ than any other alternative.

8 Welcoming friends while keeping out monsters

If we accept Euclidean geometry as our ontology of space, and we view the spatial extent of an object or other physical entity as a set of points in $\mathbb{R}^3$, then we are faced with the problem of distinguishing which sets of points can be reasonably viewed as physically meaningful and which are purely mathematical constructs of no physical significance. Our general approach here follows the tradition of many generations of mathematicians;[7] whatever we are doing, we build a castle with a moat and a drawbridge. Mathematical entities that will be well behaved for what we are doing are welcomed in; monsters that will make trouble are kept out. Which mathematical entities are considered well behaved and which are monsters depends on what we are doing inside.[8]

However, deciding on a proper admissions policy is not always easy. A successful policy must exclude the monsters, but it must admit all the mathematical entities that the work inside the castle requires. Moreover, one has to be sure that the work inside the castle will not *breed* monsters inside the keep; it is often very difficult to be sure that cannot happen. I do not have a final answer to propose on this question for the problem of the reasoning about space and motion by a household robot. After some preliminary groundwork discussing physical idealization, we will first consider the policies for inviting friends and then consider a few examples of monsters and the issues they raise.

8.1 A preliminary comment on physical idealization

All of the examples that I will discuss below arise from the use of one or another idealized models of physical reality. The specific models in my discussion are, first, the general model of matter as continuous rather than atomic, and, second, the

---

[7]Physicists and engineers rarely if ever worry about this; they use their physical intuition to avoid worrying about physically meaningless problems. But AI programs only have the physical intuitions that we program into them. The purpose of our analysis here is precisely to determine what *are* the physical intuitions that will enable the AI program to avoid worrying about meaningless problems.

[8]One semester as an undergraduate, I took both Analysis II, which was mostly the theory of distributions, and Functions of a Complex Variable. In Analysis, the *ne plus ultra* of well-behaved functions were C^∞ functions with compact support; distributions could be defined as linear functionals over that space. In Complex Variables, these were all monsters because they are not analytic.

specific model of rigid solid objects as perfectly rigid; additional examples would involve other kinds of idealizations (e.g., liquids as perfectly incompressible). All of the particular mathematical anomalies that I discuss go away if a more sophisticated model is used. Thus each of our castles and moats is designed for the purpose of properly formulating a theory that in any case is a false idealization. One can certainly reasonably ask whether this is worth the trouble; might we not simply be better off working once and for all with the true physical theory?

I don't think so. First, quantum theory is, of course, by no means clear sailing logically, and even the reduction of thermodynamics to statistical mechanics has its slippery patches. More importantly for our purposes, the atomic/quantum model of matter is extremely difficult to use for large-scale problems of mechanics, and it is therefore rarely used for these. There is a huge scientific and engineering literature that uses only a continuous model of matter, and then there is another huge literature that integrates continuous and atomic models. Therefore it is important to work out how continuous models can be reasoned about. Idealized models of physics such as continuum mechanics are necessarily somewhat wrong when one measures sufficiently accurately, and completely wrong when applied far outside their limits (e.g., on the very small scale), but they should at least be consistent, and, within their limits, they should not give answers that are nonsensical. We will return to this point in section 9.

There is also a historical/epistemological issue. The atomic theory was fiercely rejected by many physicists throughout the nineteenth century, and there were holdouts as late as the 1910s; the issue was not absolutely settled until experiments in the 1910s by Perrin and others that reliably determined the size of the atom.[9] To the best of my knowledge, no one ever formulated a precise formal axiomatization of the continuous theory of matter, but presumably the theory is sufficiently logically coherent that this could have been done; and a precise statement of this kind would have had to address the issues that we raise below. If no such axiomatization could have been formulated, then one is left with the strange conclusion that the continuous theory of matter could have been excluded as impossible purely on the basis of observations at the medium scale with no experiments at the small scale. That seems counterintuitive.

8.2   What spatial regions do we need?

We turn now to the following question. If we accept that the spatial extent of a solid object is a set of points in $\mathbb{R}^3$, what kinds of sets can meaningfully be considered as the extent of a single solid object? For instance: A sphere seems reasonable, a single

---

[9]Both Dalton's chemical theory and the Maxwell/Boltzmann kinetic theory of heat have the unsettling feature that, though they are based on particles, they give no indication of how large the particles should be.

point seems doubtful, a disconnected set seems very doubtful, as does the set of all points with rational coordinates. For convenience of reference we use the phrase "object-region" to mean a region that could plausibly be the extent of a solid object.

It turns out, in fact, that what needs to be done is not so much to identify a particular hard-and-fast rule for distinguishing object-regions, but rather to choose a collection of object-regions that will work together well.

Ultimately, a full specification of the ontology should precisely define the collection of object-regions to be used; but some aspects of this definition are rather arbitrary and we will not try to resolve them here. For instance: do we want to assume that all regions are polyhedra, or that all regions are smooth, or do we want an ontology that includes both smooth regions and polyhedra? The advantage of including polyhedra is that rectangular boxes and so on are useful entities with nice physical properties; and artifacts that are actually polyhedral to a very high precision are quite common. The drawbacks of including polyhedral regions are that some aspects of physics are defined in terms of the normal to the surface, and for non-smooth objects, since there is no unique normal, one has to work around this (which can be done, but it takes work); and that most though not all ostensibly polyhedral objects actually have visibly curved edges if you look closely enough. The advantages of restricting the class of regions to polyhedral objects are, first, that polyhedra support comparatively efficient algorithms when there are exact specifications; and, second, that the class of polyhedra has nice closure properties, e.g. it is closed under union. The advantage of allowing non-polyhedral objects is that sphere, ovoids, helical screws, and so on are also useful physical entities; they can be manufactured or arise in nature to very high precision; and it is a substantial nuisance to have a theory that posits that only high-precision polyhedral approximations for these exist. We will not here attempt to decide this.

Rather, we will enumerate some desirable features of the class of regions that correspond to solid objects.

1. The ontology includes the shape of any solid object that can manufactured out of plastic, stainless steel, paper, string, and so on. Of course, sufficiently complex shapes can be hard to apprehend; but it would be strange to have an ontology that ruled them out as geometrically meaningless. Geometrically, we can interpret this as the constraint that for any interior-connected geometric complex in $\mathbb{R}^3$ there is an object-region that is an arbitrarily good approximation, or at least an approximation[10] that is accurate to within the limits of perception (a fraction of a millimeter). Any such shape can be constructed by gluing together little tetrahedra.

2. Solid objects can be moved around rigidly in space. Hence, the ontology should be closed under rotation and translation. It should be closed under reflection as well; otherwise, when you look at the reflection of an object in a mirror, you could see an inconceivable shape.

---

[10]There is more than one measure of approximation for regions in Euclidean space; Hausdorff distance (see Glossary) will do fine here.

3. Solid objects deform continuously under physical processes such as thermal expansion and bending. It will certainly be convenient to view these as truly continuous changes, rather than as discrete.[11] Therefore the ontology should be rich enough to support continuous change of shape. For instance, the class of polyhedra with rational edge lengths would not satisfy this constraint.

4. Solid objects that meet in a face can be glued together. Geometrically, this means that if A and B are regions in the ontology that meet along an extended face, then the union $A \cup B$ is either in the ontology or very closely approximable within the ontology.

5. Objects can be cut or broken. It is not clear how these processes are best categorized geometrically. However you characterize these processes, they must generate shapes that are acceptable object-regions. This is a joint constraint on the theories of breaking and cutting and the definition of object-regions.

6. Invariance under scale transformations. This constraint is more doubtful; it is more a geometric convenience than the requirement of a physical theory. It is not possible to apply scale transformations to physical objects; the laws of physics are not scale invariant and the particle model of matter is not scale invariant. In a continuous model of matter, scale invariance would be natural, but could be avoided if that were desirable. For instance, one could posit that any solid object has to contain a sphere of radius 0.01 mm in order to be physically coherent.[12]

8.3   Other physically meaningful regions

There may be point sets that we wish to describe as physically meaningful that cannot be the extent of a solid object; these may well satisfy different constraints. For instance, as we will discuss in section 8.7, we may well wish to posit that the extent of a solid object cannot be a lower-dimensional region such as a geometric point, curve, or surface. However, if we wish to describe the geometry of the contact between two solid objects, then that contact can be a single point, a curve, or a two-dimensional surface, or the union of a collection of each of these. One might wish to posit that all solid objects are bounded, but, for reasoning in a terrestrial context, to idealize the atmosphere as occupying an unbounded region. One might wish to reason about the swath swept out by an object over time; what class of regions that would include depends on what kinds of motions are possible. For each of these, if they are to be included, one has to reason separately what kinds of regions are

---

[11] The amount of change that physics predicts can certainly be very small, even for reasonable sized objects undergoing perceptible changes. For instance, if a 1 cm cubed object is moving at a speed of 1 cm per second, the relativistic contraction is $5 \cdot 10^{-23}$ m.

[12] More precisely, one would want to posit that any bump or indentation contains a sphere of 0.01 mm; in the technical jargon, the region is r-regular with $r = 0.01$ mm (see Glossary).

needed and what kinds are monsters to be barred; the criteria are likely to depend to some extent on the choice of solid object regions, but may also involve further considerations.

8.4   Out and out monsters: Set of points with rational coordinates/Banach-Tarski paradox

> This is, naturally, often enough what children mean when they ask, "Is it true?" They mean, "I like this, but is it contemporary? Am I safe in my bed?" The answer, "There is certainly no dragon in England today," is all they want to hear.
>
> – J.R.R. Tolkien, "On Fairy-Stories".

We now turn to the monsters. We begin with two well-known monsters that presumably have no physical significance. Relative to some fixed coordinate system, let Q be the set of points in $[0, 1]^3$ with rational coordinates. Let B be one of the non-measurable components of the Banach-Tarski paradox. We can admit Q as a set—excluding it would inconveniently complicate the formulation of set theory. If we need the axiom of choice (which is debatable) we can admit B as a set.[13] However, it seems safe to say that neither Q nor B components have any physical significance. Therefore, for example, our robot does not have to be able to solve problems that start "Suppose that there is an object of shape Q on the table."

However, the price of excluding Q and B is, as we have said, that we have to be sure that we do not breed them inside the castle. Specifically, we have to be sure that our theory of fabricating objects does not include any means of fabricating an object of shape Q; and that our theory of slicing an orange does not allow us to slice out a piece of shape B. Those constraints certainly seem plausible; proving them, however, may not be easy.

8.5   A monster we may breed but may be able to live with: The infinitely bouncing ball

Suppose that a ball is bouncing on the ground with a partially elastic collision. Let us idealize the ball and the ground as perfectly rigid. Suppose that the coefficient of restitution is 1/2. Then each bounce is half as high as the previous one, and therefore takes time $1/\sqrt{2}$ as long as the previous bounce. The ball therefore bounces infinitely often times before coming to rest at time $\sqrt{2}/(\sqrt{2} - 1)$. Thus we have here a rather Zenonian behavior (in sense of Zeno's paradoxes) emerging from a perfectly natural physical idealization; we have bred the monster inside the castle.

---

[13] John Stillwell's chapter in this collection discusses in great depth the intricate relations between set theory and the theory of the real numbers and Euclidean space.

Of course, in reality, this analysis breaks down at the point when the time between bounces is comparable to the period of the fundamental vibrational mode of the ball. At that point you have something like a spring bouncing on the ground and the behavior is dominated by the damped harmonic oscillation of the ball itself. (It is nonetheless challenging to establish that no Zenonian behaviors are possible in this improved model.)

The question is, though, is this Zenonian behavior so monstrous that we are forced to abandon the convenient idealization of perfectly rigid objects in order to avoid it? There is a case to be made that it actually has no intolerable consequences; the velocity is well defined and finite (except at the moments of collision, but that's true of any theory of rigid objects collision) as is the momentum and the energy, and there is no actual difficulty predicting the behavior either at the instant when the bounces converge or past it. Having infinitely many events in finite time does rule out a reasoning strategy in which one always reasons from one distinguished event to the next; but that is a poor strategy whenever there are large numbers of unimportant events, even if that large number is finite. (For instance, if you shake a can half full of sand, you do not want to reason about every collision between every pair of grains of sand.)

Note finally that, if the ball has additionally some horizontal motion, then the swathe that it sweeps out will have some geometrically anomalous features. Whether this is OK depends, first on whether we are concerned about swathes at all, and second on what are the well-behavedness conditions for swathes.

8.6   Each beast is tame, but the herd is monstrous: Zeno's box of blocks

Consider the hypothetical situation shown in figure 4. An open box of depth 1 is filled with a collection of cubical blocks. The blocks in the top row have side 1/2, those in second row have side 1/4, those in the third row have side 1/8, and so on.

Is this a monster to be barred? Probably, but that is not entirely certain. Certainly a force-based formulation of Newton's theory does not work here; since there are no bricks that touch the bottom of the box, there is nothing for the bottom of the box to be exerting an upward force on. On the other hand, an energy-based formulation works fine; the total potential energy is well-defined and finite, and the current state is obviously a local minimum. Note also that, in a continuous model of solid material, each block has a perfectly reasonable shape; it is just the overall collection that is problematic.

The more serious difficulty is making sure that this kind of monster is not bred within the castle. For instance, one's theory of breaking objects or cutting objects might in some cases generate an infinite collection of objects of this kind. Someone might, for example, propose a fractal model of his beard in which there are k hairs of length 2^{-k} or some such. Then if he shaves …. Or a continuous theory of liquids might generate a fractal distribution of droplets; if these then freeze, there are infinitely many pieces of ice.

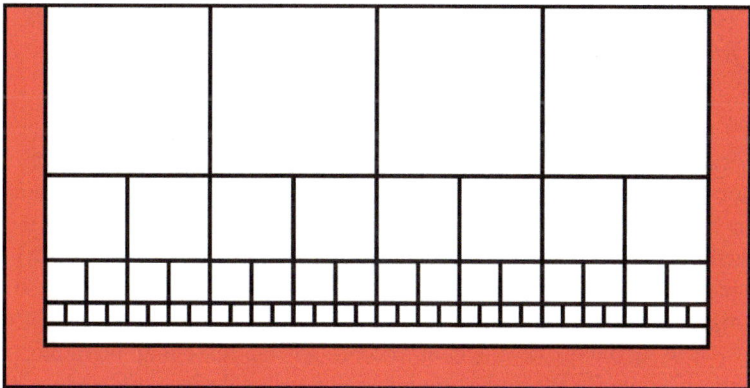

Fig. 4 Zeno's box of blocks

All these problems vanish, of course, once one switches from a continuous theory of matter to an atomic theory; but, as discussed above, the costs of avoiding a continuous theory are large.

8.7 Ghosts: Solid objects of lower-dimensional shape

Point objects, cords that are one-dimensional curves, and plates that are two-dimensional surfaces are common in the exercises for physics textbooks. Should a robot be prepared to find them in the kitchen?

I would argue that not only are these obviously impossible in an atomic theory, they are more trouble than they are worth in a continuous matter theory. For instance, suppose you drop a grain of sand on to a paper plate and it sits there for a while. You can now take it off the top of the plate, but not from the bottom. If sand and plate are modelled as objects of some thickness, this follows directly from the constraints that the interiors of object cannot overlap and that objects move continuously. However, if the sand is a point object and the plate is a two-dimensional, then the natural geometric interpretation of the sand on the plate is simply that the point is in the surface. Extra state must be added to keep track of which side the sand is on; keeping track of this state becomes more difficult if the plate's shape changes (e.g., by being folded). Characterizing the possible interactions of two cords characterized as one-dimensional curves is even more challenging. The various kinds of infinite densities associated with these objects are also a nuisance to deal with.

Therefore, it seems to be that a robot does best to consider lower-dimensional objects as purely calculational conveniences approximating a reality that is in fact solidly three-dimensional. Specifically, the qualitative behavior (the sand stays on one side of the plate) is worked out based on a three-dimensional theory. If numerical values are needed, it may be convenient to do calculations based on a lower-dimensional model.

8.8   Yoking a monster to the plow: non-regular open regions for solid objects

Perhaps the most obviously physically meaningless distinction that comes with the geometry of $\mathbb{R}^3$ is the distinction between a topologically open region, which does not contain its boundary points, and a topologically closed region, which does contain its boundary points. What could possibly be the physical difference, for any purpose, between and the closed unit sphere $\bar{B}(\mathbf{o}, 1) = \{\mathbf{x} \mid \text{dist}(\mathbf{x}, \mathbf{o}) \leq 1\}$ which includes its boundary and the open unit sphere $B(\mathbf{o}, 1) = \{\mathbf{x} \mid \text{dist}(\mathbf{x}, \mathbf{o}) < 1\}$ which does not include its boundary? By the same token it seems at first glance obvious, it is assumed in much of the AI spatial reasoning literature, that a physically meaningful region has no lower-dimensional pieces deleted or sticking out (topologically regular). What could be the physical difference between the open unit ball, and the same ball missing a point, or a line or a two-dimensional slice?

Nonetheless, there is an argument to be made that solid objects should be idealized specifically as open sets rather than closed sets, and moreover as open sets that are not necessarily topologically regular. Consider a situation where a flexible solid object is bent around back on itself; e.g., a coil of rope. One can create an idealized model of this situation by viewing the region occupied by the object to be the open set that it fills, excluding the boundaries where one part touches another. One then formulates the associated physical theory as stating that two parts $P1$ and $P2$ of object O cannot be easily separated if the boundary surface where $P1$ meets $P2$ is part of the interior of O, but may be easily separable if the boundary surface is part of the boundary of O (figure 5). For example, the rope in Figure 5 can easily be uncoiled, but not easily split into two pieces of rope.

It may seem that this is too clever by half, and that it will sooner or later get you into trouble, like having a carriage pulled by a hippopotamus. But I think it is

Fig. 5 Coil of Rope. The boundary is dashed to indicate that it is excluded from the spatial extent of the rope.

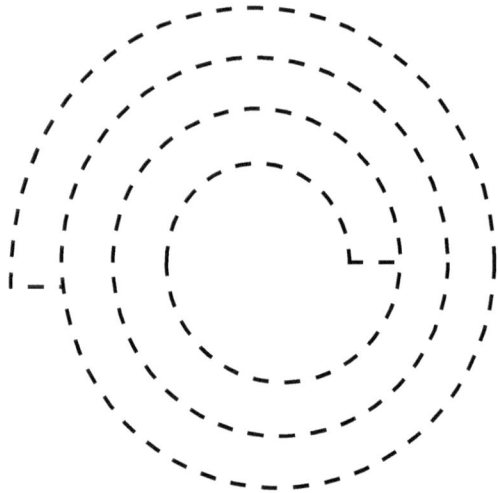

actually OK. Essentially, we're using a trick from the old days of programming, in which programmers would find a few unused bits and stuff some extra information into them. The theory of the real geometry gives us, in the distinction between including or excluding boundary points, an extra degree of freedom that we do not otherwise need for physical theories, and we're using it to encode, so to speak, the important distinction between surfaces across which an object does and does not cohere to itself.

8.9 How could one verify a choice of well-behaved regions?

Our discussion above is quite unsystematic; shapes of category A are monsters that will get you into trouble, whereas as far as I can tell it should be safe to consider shapes of category B as well behaved. How could one achieve greater confidence that a choice of well-behaved conditions will work out, ideally in terms of a theorem?

Ideally, I think, the kind of theorem that would answer this question would be along the following lines. Suppose that the programmer has worked out the following:

a. A proposed physical theory, e.g. "Newtonian physics of rigid solid objects with Coulomb friction and collisions with varying degrees of elasticity, in a uniform gravitational field." The theory does not have to be a deterministic theory; a partially specified theory is also an option. The theory may well allow exogenous influences; for instance, it may allow the possibility that an agent can move blocks around, without otherwise providing a model of the agent. If so, then the theory presumably also specifies constraints on these influences

b. A proposed collection of object-regions; e.g. "Interior-connected, bounded, topologically closed regular polyhedra." This would presumably satisfy the conditions discussed in section 8.2, or at least those that derive from physics covered in (a).

c. A collection of problems that are considered well-posed. The standard form of a well-posed problem is an initial-value problem: Given an initial situation and exogenous influences satisfying the physical constraints in (a) where objects have the shapes in (b), how does the system evolve over time? But there may also be other forms of problems to be solved, depending on the theory and its role in reasoning.

The theorem to be proved, then, would have the form of showing that all well-posed problems have a solution; in the case of initial-value problems, that there is a trajectory consistent with the problem specification and the physical theory. Such results have of course been obtained and studied in great depth for systems of time-dependent ODEs and PDEs. I do not know of any such results for the rather different kinds of physical theories that tend to arise for the household robot;

few such theories have been developed in sufficient detail to support such a theorem. In particular, the characterization of exogenous influences in a theory with a robot seems to be difficult.

This is, in any case, meta-analysis; that is, it is not reasoning that we need the robot to do, it is meta-reasoning to prove that the robot's reasoning will not arrive at a contradiction. Accomplishing it will in no ways help the robot, it will merely give us more confidence in the robot.

9 Two Objections

Some readers will find the above discussion not merely unsatisfying but well-nigh preposterous. There are, I think, two major objections.

The first objection returns to the issue we addressed in section 8.1. "These problems have got to be all pseudo-problems; it is simply not credible that building a robot that can manipulate cheese graters and string bags requires thinking about topologically closed and open sets. The problems arise from pushing on the physical idealizations too hard and in the wrong direction. The fact that when you do that, the idealization will break in a strange way is neither surprising nor important; and attempting to fix the broken theory with bubble gum and rubber bands is neither computer science nor anything else. The correct approach to the problem is to find out how an automated reasoner would use multiple idealizations, how it would choose between them and integrate them."

I would say that there is some truth in this; integrating multiple idealizations is certainly a very important problem and a very poorly understood one, and it is likely that once we have attained a better understanding of multiple models, our viewpoint on these problems will shift to some degree; major conceptual advances are apt to have that effect. But it seems to me that the seeming plausibility of the above arguments rests on a false extrapolation from people. For people, it is obvious when a physical prediction is nonsensical, and it is much easier to detect nonsense than to determine the consequences of a physical theory. Nothing is inherently obvious for a robot; a "sanity check" on a prediction is just another kind of physical reasoning. In an automated reasoner, a sanity check is just another physical theory, and in many ways a harder one to implement than a physics engine. It may have some privileged place in terms of how multiple theories are integrated, but as far as we know it is not different in kind from the theories that it is critiquing.

For example, in the bouncing ball example, the conclusion that the ball bounces infinitely many times may intuitively seem very strange, and therefore it would seem that the theory of rigid solid objects should be rejected here. Perhaps it should; but that may not be at all easy for an automated reasoner to see. After all, in many ways this case is particularly well suited to the rigid body approximation, since the collisions are all very low energy; generally, a rigid-body approximation does *better* with low energy collisions than with high energy collisions. Moreover, on standard measures, such as the mean squared distance between the prediction and behavior, the rigid body model does fine; it predicts accurately that the ball will bounce lower and lower and then come to rest. Where it does badly is in predicting the number of collisions (infinite vs. finite); but the number of collisions is not a standard measure for evaluating the quality of predictions, nor necessarily an important one. The sanity check module cannot automatically send up a red flag each time it sees an infinity in a prediction; as Zeno's paradox of Achilles and the tortoise shows, perfectly simple physical behavior can be full of infinities, depending on how it is construed.[14] In short even once we have a metatheory of multiple models, the kinds of problems that we have looked at above are likely to persist, though they may well take a different form.

I am not claiming that a programmer of a robot *has* to think through these ontological issues; only that it will be helpful to do so. As an imperfect analogy, a native speaker of English does not have to be consciously aware of the difference between regular and irregular past tense forms of verbs or the difference between a gerund and a participle; but these distinctions are useful, both in characterizing the linguistic behavior of speakers and in teaching English as a second language.

The second objection is that the above analysis is nothing more than a defense of the status quo in conventional mathematics and demonstrates only that Ernest Davis, having spent much effort over half a century becoming indoctrinated in that status quo, is now too hidebound and unimaginative to accept any alternative.

There may be some truth in that as well, but there it seems to me that the burden of proof is on the challenger. Over the course of three and a half centuries $\mathbb{R}^3$ has proved itself an extremely powerful spatial ontology for doing all kinds of physical reasoning at all scales except the extremely small and the extremely large. If an alternative can be proposed which is adequate for the purposes of a robot's physical reasoning, then the AI community, including myself, will do our best to judge it fairly. Until then, we are stuck with $\mathbb{R}^3$, which means that we have to do something about the monsters. If you want to do carpentry, then you are better off with a power saw that is powerful enough to cut wood but has inconvenient safety features than with one which avoids the potential dangers from the start but can't cut wood.

10 Beyond physical reasoning: Formal models, friends, and monsters in other aspects of automated reasoning

HORATIO: O day and night, but this is wondrous strange!
HAMLET: And therefore as a stranger give it welcome.
– *Hamlet,* Act I, scene 5

---

[14]Slight variants on standard AI reasoning techniques actually do generate a version of the Achilles paradox. Suppose that you posit there are important times, important places, and important events; when an object arrives at an important place, that is an important event; the time when an important event occurs is an important time; and the place of an object at an important time is an important place. Then, as Achilles chases the tortoise, infinitely many important events occur, by the standard argument.

Among the many subject matters that a general artificial intelligence has to deal with, reasoning about physical objects and their spatial relations is one where the relation between the real world and the mathematical model is clearest and works best. In many other areas, there either is no model at all or the gap between practice and theory is much larger.

In particular, the issue of deciding which problems an AI system needs to address and which it can reject as ill-formed or meaningless arises in many different context and forms. In natural language understanding: Does a text understanding system need to understand newly coined words, expressions, or metaphors? Does it need to deal with the Liar sentence? Does it need to be able to get out of *Jabberwocky, Finnegan's Wake*, or the first chapter of *The Sound and the Fury* what a human reader gets out of them? In computer vision: Does a computer vision system need to be able to read a political cartoon, to characterize an animal it has not seen before, to recognize the Easter Island statues depicting as people, to interpret a Magritte, a Monet, a Picasso, a Jackson Pollack? In planning: Is "Buy low, sell high" an executable plan? How about "Do something your adversary doesn't expect," or "Carry out an experiment whose result will surprise you"? In reasoning about people: Does a theory of "naïve psychology" need to subsume daydreams, nocturnal dreams, self-destructive behavior, or delusional behavior?

The ability of people to dramatically expand their horizons, to embrace what is radically new, grounding it in previous experience without reducing it to previous experience, is among the most magical characteristics of human cognition, and one of the abilities that we are furthest from attaining in AI. The ability to create mathematics and to understand mathematical entities and mathematical arguments is an extraordinary manifestation of this gift.

Glossary of Terms

Banach-Tarski paradox: It is possible to divide the sphere of radius 1 into finitely many pieces that can be rearranged to form two spheres of radius 1. The Banach-Tarski theorem can be proven given the axiom of choice. There are models of set theory that exclude the axiom of choice in which the statement is false.

Hausdorff distance: Let A and B be subsets of $\mathbb{R}^n$ (or for that matter subsets of any metric space). The Hausdorff distance between A and B is a measure of how different these two sets are. Specifically it is the maximum of two quantities: (a) the maximum (or supremum) distance from any point $a \in A$ to the closest point in B; (b) the maximum distance from any point $b \in B$ to the closest point in A. The Hausdorff distance is a metric over closed bounded subsets of $\mathbb{R}^n$.

Pythagorean triple: Three integers a, b, c such that $a^2 + b^2 = c^2$; e.g., 3,4,5 or 12,5,13.

r-regular: Let $r > 0$ be a real number. A region P in $\mathbb{R}^n$ is r-regular if, for every point **p**, there is a sphere of radius r containing **p** whose interior is entirely inside P or there is a sphere of radius r containing **p** whose interior is entirely outside P [2].

Space-filling curve: A one-dimensional curve that includes every point in an extended two-dimensional region. More precisely, a continuous function from the unit interval [0,1] onto the unit square $[0, 1] \times [0, 1]$.

Topologically regular: A topologically regular open region is equal to the interior of its closure; intuitively, it does not have any lower-dimensional pieces—points, curves, or surfaces—missing. A topologically regular closed region is equal to the closure of its interior; intuitively; it does not have any lower-dimensional pieces sticking out.

Additional Reading

R. Casati and A. Varzi, *Holes and Other Superficialities,* MIT Press, 1994.

E. Davis, "The Naive Physics Perplex", AI Magazine, vol. 19, no. 4, Winter 1998, 51–79.

A. Galton, *Qualitative Spatial Change,* Oxford U. Press, 2000.

References

1. K. Cho et al. "Learning Phrase Representations using RNN Encoder-Decoder for Statistical Machine Translation," *Conference on Empirical Methods in Natural Language Processing (EMNLP),* 2014.
2. P. Stelldinger *Image Digitization and its Influence on Shape Properties in Finite Dimensions,* IOS Press, 2008.

Mathematics and its applications

David Berlinski

Abstract If mathematics, unlike entomology, is unreasonably effective, it should be possible to say with, at least some precision, what it means for a mathematical object, structure, or theory to be applied to an object, structure or theory that is resolutely not mathematical. If it is possible to say as much, I have not found a way in which to say it. Mathematics is about mathematics; and so far as the Great Beyond is concerned, while it is obvious that mathematics is often applied, it is anything but obvious how this is done.

1 Introduction

Group theory is about groups; the theory of rings, about rings. This suggests a generalization. Mathematics is about mathematics. Mathematics is about mathematics in the sense that entomology is about bugs. Who would deny it? If mathematics is about mathematics, it is about many other things as well. No one remarks on the unreasonable effectiveness of entomology. In counting two fingers and two fingers and reaching four fingers, numbers are being applied to fingers. In what does the application consist? The idea that there is a mapping between a subset of the natural numbers and the fingers of the human hand has all of the disadvantages of an arranged marriage. It appears reasonable only to those least involved in the proceedings. To the extent that the mapping is mathematical, it cannot have fingers in its range; and to the extent that it is not, it cannot have numbers in its domain.

Children nonetheless count their fingers with ease, and so do mathematicians.

How is it done?

In finger counting, we count fingers. This might suggest that numbers are one thing, fingers, another. But the creation of numbers, as Thierry of Chartres observed, is the creation of things. One finger is one finger necessarily. It could not be two fingers. If in counting fingers, we are mapping the number one to oneness, what remains of an *application* of numbers to things? One applies to oneness in just the

---

D. Berlinski (✉)
14 rue Chanoinesse, 75004 Paris, France
e-mail: jeromed.berlinski729@gmail.com

© Springer International Publishing Switzerland 2015
E. Davis, P.J. Davis (eds.), *Mathematics, Substance and Surmise*,
DOI 10.1007/978-3-319-21473-3_6

sense in which one applies to itself. There is, after all, only one. The statement that two fingers plus two fingers make four fingers is, when divided through by fingers, simply the statement that two plus two equals four. But one cannot divide through anything by fingers; and to leave the fingers out is only to return to the observation that two plus two equals four. This is nothing to sneeze at, of course, but it is nothing about which one might wish to write home.

What is unreasonably effective in mathematics is mathematics. What is unreasonably effective beyond mathematics is, as Eugene Wigner observed, a miracle.

2 Whatever its connection to other disciplines, mathematics frequently appears to be about itself, its concepts *self*-applied. Take groups. A set **G** closed under an associative binary operation **G** X **G** → **G** is a group if **G** includes an identity and an inverse. An identity **I** returns every element to itself: a o **I** = a. An inverse returns any element to the identity: a^{-1} o a = **I**.

Groups are stable objects, important in mathematics as well as in physics. They play a role in topology, a subject devoted in large measure to the analysis of continuity. The most familiar topological space is the real line; its topology is defined by sets of open intervals. Now mathematicians, as well as philosophers, depend on the familiar upward movement of conceptual ascent, if only to rise above the smog and get a better view of things. A topological space is a particular item; this is the view from the ground. But the collection **Top** of topological spaces is a *category*; and this is the view that ascent reveals.

Categories are not simply sets. A set directs the mathematician's eye toward its elements A and B; a category, toward the morphisms **Mor**(A, B) between them. Morphisms may themselves be composed

$$\mathbf{Mor}(B, C) \text{ X } \mathbf{Mor}(A, B) \rightarrow \mathrm{Mor}(A, C),$$

subject only to the triplet of conditions that:

2.1 **Mor**(A, B) and **Mor**(A', B') are either equal ($A = A'$ and $B = B'$) or disjoint;

2.2 There is an identity morphism $id_A \in$ **Mor**(A, A) for every A in **Top**; and

2.3 Morphism composition is associative:

$$(h \, o \, g) \, o \, f = h \, o \, (g \, o \, f),$$

whenever $f \in$ **Mor**(A, B), $g \in$ **Mor**(B, C), and $h \in$ **Mor**(C, D)—this again for every $A, B, C,$ and D in **Top**.

Wherein do groups figure? They figure in algebraic topology, a subject in which algebraic objects are assigned to topological structures in such a way that topological questions may be settled by algebraic methods.

Some definitions. A *path* or *arc* (or even a curve) in a topological space **X** is a continuous mapping f of a closed interval $I = [a, b]$ into **X**. In what follows, I is always the closed interval [0, 1]. The images of a and b are the endpoints of the arc. A space is *arcwise connected* if any two points in **X** may be joined in an arc. Two paths in **X** are equivalent if one can be continuously transformed into the other.

Suppose that f and g are paths in $\mathbf{X}$ such that g starts where f ends. The product of f and g is:

$$(f \cdot g) = \begin{array}{ll} f(2t) & 0 \le t \le 1/2 \\ g(2t - 1) & 1/2 \le t \le 1. \end{array}$$

The multiplication of paths is in general not associative; but associativity is recovered when paths are grouped into equivalence classes. It is easy now to define both an identity and an inverse. Assume this done.

A path is closed (or loops) if its initial and terminal points are the same. Let x, now, be any point of $\mathbf{X}$. The set of all paths that loop from x to x satisfies group theoretic axioms; satisfying them, they form the fundamental group $\pi(\mathbf{X}, x)$ of $\mathbf{X}$ at the base point x.

A first connection between topology and algebra now emerges, like an image under darkroom developer:

2.4 If $\mathbf{X}$ is arcwise connected, then $\pi(\mathbf{X}, x)$ and $\pi(\mathbf{X}, y)$ are isomorphic for any two points x, y in $\mathbf{X}$.

The theorem's contrapositive is somewhat more revealing: No matter the pair of points in $\mathbf{X}$, if $\pi(\mathbf{X}, x)$ and $\pi(\mathbf{X}, y)$ are not isomorphic, $\mathbf{X}$ *not* arcwise connected. A topological condition has been determined by a group theoretic property.

As in all magic acts, one good trick suggests another. Let S^1, for example, be the unit circle in the real or complex plane. And let $f : I \to S^1$ be the closed path that goes around the circle just once:

$$f(t) = (\cos 2\pi t,\ \sin 2\pi t), \quad 0 \le t \le 1.$$

$\xi(f)$ is the equivalence class of f. The obvious theorem follows:

2.5 *The fundamental group* $\pi(S^1, (1, 0))$ *is an infinite cyclic group generated by* $\xi(f)$.

The proof of **2.5**, like that of **2.4**, is a matter of applying diligently the definitions; but what follows is different, altogether dramatic. Let E^n be the closed unit ball in Euclidean n-space; and let f be a continuous map of E^n into itself. Does f have a point x such that $f(x) = x$? It is a good question. The answer is provided by Brouwer's fixed-point theorem:

2.6 *Any continuous map f of E^n into itself has at least one fixed point.*

Unlike **2.5**, *this* theorem is an affirmation with a thousand faces, one of the protean declarations of mathematics.

The proof for $n \le 2$ suggests the whole. First look to $n > 0$. A subset A of a topological space $\mathbf{X}$ is a *retract* of $\mathbf{X}$ if there exists a continuous map $\gamma : \mathbf{X} \to A$ such that $\gamma(a) = a$ for every a in A. If $f : E^n \to E^n$ has no fixed points then S^{n-1} is a retract of E^n. Proceed by contraposition. S^{n-1} is *not* a retract of E^n when $n = 1$ because E^1

is connected, S^0, disconnected. Go, then, to $n = 2$. $\pi(S^1)$ is infinite cyclic; but $\pi(E^2)$ is a trivial group. It follows that S^1 is not a retract of S^2.

Categories were created with the aim of highlighting the mappings or morphisms between mathematical structures. The category **Top** of all topological spaces has already been fixed; ditto by definition the category **Grp** of all groups. A *functor* is a morphism between categories. If A and B are categories their *covariant* functor F: $A \to B$ assigns to each object a in A an object $F(a)$ in B; and assigns, moreover, to each morphism $f: A \to A$ a morphism $F(f): F(A) \to F(A)$.

The rules of the game are simple. For every a in A:

2.7 $F(id_A) = id(F_A)$,

where *id* is the identity morphism, and if $f: A \to B$ and $g: B \to C$ are two morphisms of A, then

2.8 $F(g \circ f) = F(g) \circ F(f)$.

Contravariant functors reverse the action of covariant functors; a pair, consisting of a covariant and contravariant functor, make up a *representation* functor.

Within algebraic topology, it is often useful (and sometimes necessary) to specify a topological space with respect to a distinguished point; such spaces constitute a category **Top*** in their own right. A new category does nothing to change the definition of the fundamental group, of course, but it does make for a satisfying illustration of the way in which the fundamental group may acquire a secondary identity as a functor, one acting to map a category onto an algebraic object:

$$\pi (\mathbf{X}, x) : \mathbf{Top}^* \to \mathbf{Grp}.$$

This way of looking at things indicates that the fundamental group serves not simply to mirror the properties of a given topological space, but to mirror as well the continuous maps between spaces, a circumstance not all that easy to discern from the definition itself.

These considerations were prompted by a concern with mathematics self-applied. Herewith a provisional definition, one suggested by the functorial explication of the fundamental group. If $\mathbf{X}$ and $\mathbf{Y}$ are mathematical objects, then

2.9 $\mathbf{X}$ *applies* to $\mathbf{Y}$ if and only if there are categories A and B, such that $\mathbf{Y}$ belongs to A and $\mathbf{X}$ to B, and there exists a functor F on A such that $F(\mathbf{Y}) = \mathbf{X}$.

The language of categories and functors provide a subtle and elegant descriptive apparatus; still, categories and functors are mathematical objects and the applications so far scouted are internal to mathematics.

What of the Great Beyond? The scheme that I have been pursuing suggests that mathematics may be applied to mathematical objects; it makes no provisions for applications elsewhere.

3 A mathematical theory with empirical content, Charles Parsons has written [16 p. 382], "takes the form of supposing that there is a system of *actual* objects and relations that is an instance of a structure that can be characterized mathematically" (italics mine). These are not lapidary words. They raise the right question, but by means of the wrong distinction. They misleadingly suggest, those words, that mathematical objects *without* empirical content are somehow not actual. Not actual? But surely not potential either? And if neither actual nor potential, in what sense would mathematical theories without empirical content be about anything at all? The word that Parsons wants is *physical*; and the intended contrast is the familiar one, mathematical objects or structures on the one side, and physical objects or structures on the other.[1]

The question Parsons raises about mathematics, he does not answer explicitly, his discussion trailing off irresolutely. W.V.O. Quine [17 p. 398] is more forthright. "Take groups," he writes:

> In the redundant style of current model theory, a group would be said to be an ordered pair (K, f) of a set K and a binary function f over K fulfilling the familiar group axioms. More economically, let us identify the group simply with f, since K is specifiable as the range of f. Each group, then, is a function f fulfilling three axioms. Each group is a class f of ordered triples, perhaps a set or perhaps an ultimate class. ... Furthermore, f need not be a pure class, for some of its ordered triples may contain individuals or impure sets. This happens when the group axioms are said to be applied somewhere in natural science. Such application consists in specifying some particular function f, in natural science, that fulfills the group axioms and consists of ordered triples of bodies or other objects of natural science.

Whatever else they may affirm, these elegant remarks convey the impression that mathematical concepts (or predicates) are polyvalent in applying indifferently to mathematical *and* physical objects. "[G]rouphood," Quine writes (on the same page), "is a mathematical property of various mathematical and non-mathematical functions." This is rather like saying that *cowhood is a zoological property of various zoological and non-zoological herbivores.* If there are no non-zoological cows, why assume that there are some non-mathematical groups?

Skepticism about the application of mathematics arises as the result of the suspicion that nothing short of a mathematical object will ever satisfy a mathematical predicate. It is a suspicion that admits of elaboration in terms of an old-fashioned argument.[2] Let me bring the logician's formal language L into the discussion; ditto his set K of mathematical structures. A structure's *theory* $T(K)$ is the set of sentences φ of L such that φ holds in every model $M \in K$. Let K constitute the finite groups and $T(K)$ the set of sentences true in each and every finite group. Sentences in $T(K)$, the logician might say, are *distributed* to the finite groups.

Nothing esoteric is at issue in this definitional circle. Distribution is the pattern when an ordinary predicate takes a divided reference. The logician's art is not required in order to discern that whatever is true of elephants in general is

---

[1]Parson's phrase, the "instance of a structure" is not entirely happy. Predicates have instances; properties are exemplified; structures just sit there.

[2]See [13 pp. 198–204] for interesting remarks.

necessarily true of Bruno here. It is a pattern that fractures in obvious geometric cases. Thus consider *shape*, one of the crucial concepts by which we organize sensuous experience, and the subject, at least in part, of classical Euclidean geometry.[3] Is it possible to distribute the truths of Euclidean geometry to the shapes of ordinary life—desktops, basketballs, football fields, computer consoles, mirrors, and the like?

Not obviously.

In many cases, the predicates of Euclidean geometry just miss covering the objects, surfaces, and shapes that are familiar features of experience. Euclidean rectangles are, for example, bounded by the line segments joining their vertices. Rectangles in the real world may well be finite but *un*bounded, with no recognizable sides at all because beveled at their edges.[4] The chalk marks indicating the length and width of a football field have a determinate thickness and so contain multiple boundaries if they contain any boundary at all. Euclidean rectangles are structurally unstable: small deformations destroy their geometrical character. Not so physical rectangles. Such regions of space are robust. They remains rectangular and not some other shape despite an assortment of nicks, chips, and assorted scratches. The sum of the interior angles of a Euclidean rectangle is precisely 360 degrees; the interior angles on my desk sum to more or less 360 degrees.

More or less.

These particular cases may be enveloped by a general argument. It is a theorem that up to isomorphism there exist only two continuous metric geometries. The first is Euclidean, the second, hyperbolic. The categorical model for Euclidean geometry is the field of real numbers. It follows thus that if Euclidean geometry is distributed, physical space must locally be isomorphic to $\mathbb{R}^n$.[5] It is difficult to understand how the axioms of continuity could hold for physical points;[6] difficult again to imagine a one-to-one correspondence between physical points and the real numbers.[7] How would a correspondence between the real numbers and a variety of physical points be established?

---

[3]By Euclidean geometry, I mean any axiomatic version of geometry essentially equivalent to Hilbert's original system—the one offered in chapter 6 of [5] for example.

[4]Indeed, it is not clear at all that the surface of my desk is either a two- or a three-dimensional surface. If the curved sides of the top are counted as a part of the top of the desk, the surface is a three-dimensional manifold. What then of its *rectangular* shape? If the edges are excluded, where are the desk's *boundaries*?

[5]In a well-known passage [12] Albert Einstein remarked that to the extent that the laws of mathematics are certain, they do not refer to reality; and to the extent that they refer to reality, they are not certain. I do not think Einstein right, but I wonder whether he appreciated the devastating consequences of his own argument?

[6]Quantum considerations, I would think, make it impossible to affirm any version of an Archimedian axiom for points on a *physical* line.

[7]For very interesting if inconclusive remarks, see the round-table discussion by a collection of Field medalists in [8 pp. 88–108], especially the comments of Alain Connes on p. 95.

Experimentally?

4 If distribution lapses in the case of crucial mathematical and physical shapes, it is often not by much, a circumstance that should provoke more by way of puzzlement than it does. The sum of the interior angles of a Euclidean triangle is precisely one hundred and eighty degrees—π radians, to switch to the notation of the calculus, and then simply π, to keep the discussion focused on numbers and not units.[8] This is a theorem of Euclidean geometry, a fact revealed by pure thought. Yet mensuration in the real world reveals much the same thing among shapes vaguely triangular: the sum of their interior angles appears to follow a regular distribution around the number π. The better the measurement, the closer to π the result. This would seem to suggest a way forward in the project of making sense of a mathematical application. Letting $M(\Delta)$ and $P(\Delta)$ variably denote mathematical and physical triangles, and letting niceties of notation for the moment drift, **4.1** follows as a provisional definition, one that casts a mathematical application as the inverse of an approximation.

4.1 $M(\Delta)$ *applies* to $P(\Delta)$ if and only if $P(\Delta)$ is an approximation of $M(\Delta)$.

Now approximation is a large, a general concept, and one that appears throughout the sciences.[9] It is a concept that has a precise mathematical echo. Let E be a subset of the line. A point ξ is a *limit point* of E if every neighborhood of ξ contains a point $q \neq \xi$ such that $q \in E$. The definition immediately makes accessible a connection between the existence of a limit point and the presence of convergent sequences, thus tying together a number and a process:

4.2 If $E \subseteq \mathbb{R}$ then ξ is a limit point of E if and only if there exists a set of distinct points $S = \{x_n\}$ in E itself such that $\lim_{n\to\infty} \{x_n\} = \xi$.

Approximation as an activity suggests something getting close and then closer to something else, as when a police artist by a series of adroit adjustments refines a sketch of the perpetrator, each stroke bringing the finished picture closer to some remembered standard. **4.2** reproduces within point-set topology the connection between some fixed something and the numbers that are tending toward it, S and ξ acting as approximation and approximatee. The reproduction improves on the general concept inasmuch as convergence brings within the orbit of approximation oscillating processes—those governed by familiar functions such as $f(x) = x \sin 1/x$. So long as the discussion remains entirely within the charmed circle of purely and distinctively mathematical concepts, what results is both clear and useful. The Weierstrass approximation theorem serves as an illustration:

4.3 If f is a complex valued function, one continuous on [a, b], there exists a sequence of polynomials P_n such that $\lim_{n\to\infty} P_n(x) = f(x)$ uniformly on [a, b].

---

[8] Defined as the ratio of two lengths, radians are in any case dimensionless units.

[9] See, for example, [7 pp. 56–58].

The proof is easy enough, considering the power and weight of the theorem. It is surprising that *any* complex and continuous function may be approximated by a sequence of polynomial functions on a closed and bounded interval. More to the point, **4.3** gives to approximation a precise, independent and intuitively satisfying interpretation.

Difficulties arise when this scheme is off-loaded from its purely mathematical setting. Mensuration, I have said, yields a set of numbers, but beyond any of the specific numbers, there is the larger space of possible points in which they are embedded—Ω, say. Specific measurements comprise a set of points **S*** within Ω. Relativized to the case at hand, the requisite relationship of approximation would seem now to follow:

4.4 $P(\mathbf{\Delta})$ is an approximation of $M(\mathbf{\Delta})$ if and only if π is a limit point of Ω. [15 p. 122]

4.4 is assumed even in the most elementary applications of the calculus. Talking of velocity, the authors of one text assert that "[j]ust as we approximate the slope of the tangent line by calculating the slope of the secant line, we can approximate the velocity [of a moving object] at $t = 1$ by calculating the average velocity over a small interval $[1, 1 + \Delta t]$ ..." Approximate? Approximate how? By allowing average speeds to pass to the limit, of course, the answer of analytic mechanics since the seventeenth century.

But the usefulness of **4.4** is entirely cautionary. **4.5** follows from **4.4** and **4.2**:

4.5 $P(\mathbf{\Delta})$ is an approximation of $M(\mathbf{\Delta})$ if and only if there exists a set of distinct points **S** in Ω itself such that $\lim_{n\to\infty} \{x_n\} = \pi$.

And yet **4.5** is plainly gibberish. The real world makes available only finitely many measurements and these expressed as rational or computable real numbers. There exists no set of distinct points in **S** converging to π or to anything else. Ω is a subset of the rational numbers and the definitions of point-set topology are unavailing. Taken literally, **4.5** if true implies that $P(\mathbf{\Delta})$is not—it is *never*— an approximation of $M(\mathbf{\Delta})$, however, close points in **S*** may actually come to π. **4.5** must be taken loosely, then, but if **S** and **S*** are distinct—and how else to construe the requisite looseness?—**4.1** lapses into irrelevance.

5 Symmetry is a property with many incarnations, and so a question arises as to the relationship between its mathematical and physical instances. It is group theory, Hermann Weyl [22] affirms, that provides a language adequate to its definition. Weyl's little book contains many examples and represents a significant attempt to demonstrate that certain algebraic objects have a direct, a natural, application to ordinary objects and their properties.

Let Γ be the set of all points of some figure in the plane. A permutation on Γ is a bijection $\Gamma \to \Gamma$; a given permutation f is a symmetry of Γ or an automorphism when f acts to preserve distances. The set of all symmetries on Γ under functional composition (one function mounting another) constitutes the group of symmetries $\mathbf{G}(\Gamma)$ of Γ.

Let Γ, for example, be the set of points on the perimeter of an equilateral triangle. Three sides make for three symmetries by counterclockwise rotation through 120, 240, and 360 degrees. These symmetries may be denoted as R, $R \circ R$, and $R \circ R \circ R$, which yields the identity and returns things to their starting position. There are, in addition, three symmetries D_1, D_2, and D_3 that arise by reflecting altitudes through the three vertices of the triangle. The transformations

$$\Delta_3 = \{R, R_2, R_3, D_1, D_2, D_3\}$$

describe all possible permutations of the vertices of the given triangle. These being determined, so are, too, the relevant automorphisms.

So, too, the symmetric group Δ_3.

A set **S** of symmetrically related objects is fashioned when a sequence of automorphisms is specified, as in **5.1**:

5.1

$$\begin{array}{ccc} A_1 & A_2 & A_k \\ \Gamma \to \Gamma \to \Gamma \ldots \to \ldots \Gamma. \end{array}$$

The objects thus generated form a *symmetrical sequence* **S**. This suggests the obvious definition, the one in fact favored by Weyl:

5.2 **G**(Γ) *applies* to **S** if and only if **S** is symmetrical on Γ.

So far, let me observe skeptically, there has been no escape from a circle of mathematical objects. Whatever the invigoration group theory affords, the satisfaction is entirely a matter of internal combustion. **G**(Γ) is plainly a mathematical object; but in view of **5.1**, so, too, is **S**.

Nonetheless, an extended sense of application might be contrived—by an appeal to the transitivity of application, if nothing else—if sequences such as **S** themselves apply to sequences of real objects; applying directly to **S**, Γ would then apply *in*directly to whatever **S** is itself applied. Thus

5.3 **G**(Γ) applies to **S*** if and only if **S** applies to **S*** and **G**(Γ) applies to **S**.

It is to **S*** that one must look for physical applications.

And it is at this point that any definitional crispness that **5.3** affords begins to sag. The examples to which Weyl appeals are artistic rather than physical; but his case stands or falls on the plausibility of his analysis and owes little to the choice of examples. Symmetries occur in the graphic arts most obviously when a figure, or motif, occurs again and again, either in the plane or in a more complicated space. They exist, those figures or inscriptions or palmettes—the last, Weyl's example— in space, each separate from the other, each vibrant and alive, or not, depending on the artist's skill. But in looking at a symmetrical sequence of *this* sort, **5.1** gives entirely the wrong impression. The problem is again one of distribution and confinement. **5.1** represents a symmetrical sequence generated by k operations on a *single* abstract object—Γ, as it happens. Those Persian bowmen or Greek palmettes or temple inscriptions are not generated by operations on a single figure. They are not generated at all. Each of n items is created independently and each is distinct.[10] And none is quite identical to any other.

A far more natural representation of their relationship is afforded by mappings between spaces as in:

5.4

$$\overset{f\quad g\qquad h}{X \to Y \to Z \cdots \to \ldots W.}$$

If a connection to geometry is required, X, Y, Z, and W may be imagined as point sets, similar each to Γ: f, g, and h take one space to the next. Functional composition extends the range of the mappings: $f \circ h = j$: $X \to W$. The sense in which **5.4** represents a symmetrical sequence may be expressed quite without group theory at all. Thus let the various functions be bijections; let them, too, preserve distances, so that if $x, y \in X$

$$D(x, y) = D(f(x), f(y)).$$

Each function then expresses an isomorphism. Congruence comes to be the commanding concept, one indicating that adjacent figures share a precisely defined similarity in structure.

But if **5.1** informs **5.4**, so that the sense of symmetry exhibited at **5.4** appears as group theoretical, it is necessary plainly that the following diagram must commute:

5.5

$$\overset{A_1\quad A_2\qquad A_k}{\Gamma \to \Gamma \to \Gamma \ldots \to \ldots \Gamma.}$$

$$\downarrow \qquad\qquad\qquad \downarrow$$

$$\underset{f\qquad g\qquad h}{X \to Y \to Z \ldots \to \ldots W.}$$

when $\Gamma = X$.

And obvious, just as plainly, that it never does in virtue of the fact that $X \neq Y \neq Z \ldots \neq \ldots W$.

6 If groups do not quite capture a suitable sense of symmetry, it is possible that weaker mathematical structures might. A *semigroup* is a set of objects on which an associative binary operation has been defined. No inverse exists; no identity element

---

[10]Curiously enough, this is a point that Weyl himself appreciates [23 pp. 15–17].

is required. The semigroups have considerably less structure than the groups.[11] Functional composition is itself an associative operation. Say now that **S**[X, Y, ..., Z] is any finite sequence of the point sets (or figures) X, Y, ... , Z, and let C be a collection of isomorphic mappings over **S**. **5.1**, **5.2**, and **5.3** have their ascending analogs in **6.1**, **6.2**, and **6.3**:

6.1 Isomorphisms over C form a *semigroup* **SG** under composition;

6.2 A sequence S[X, Y, ..., Z] is *symmetrical* in X, Y, ..., Z if X $\approx$ Y for every pair of elements X, Y in S[X, Y, ..., Z]; and

6.3 **SG** *applies* to **S** if and only if **S***** is symmetrical in X, Y, ..., Z.

Transitivity of application is again invoked to fashion a notion of indirect application. The application of **SG** at **6.3** makes for a weak form of algebraic animation; but it does little to dilute the overall discomfort prompting this discussion. Let me reconvey my argument. **6.1** is an abstract entity, a sequence of point sets or spaces. The symmetries seized upon by the senses obtain between palpable and concrete physical objects. Symmetries thus discerned are approximate; the discerning eye does what it does within a certain margin of error. To the extent that a refined judgment of symmetry hinges on a definition of congruence, the distances invoked by the definition are preserved only to a greater or lesser extent. Thus if x and y are points in X, and $f: X \to Y$, then

$$D(x, y) = D(f(x), f(y)) \pm \delta.$$

At **6.1**, distances are preserved precisely.[12] If we had a convincing analysis of approximation, an analysis of applicability might well follow. One rather suspects that to pin ones hopes on approximation is in this case a maneuver destined simply to displace the fog backward.

---

[11] See Eilenberg [11], from a philosophical point of view, interest in semigroups is considerable. A finite state automata constitutes the simplest model of a physical process. Associated to any finite state automata is its transition semigroup. Semigroups thus appear as the most basic algebraic objects by which change may abstractly be represented. Any process over a finite interval can, of course, be modeled by a finite state automata; but physical *laws* require differential equations. Associated to differential equations are groups, *not* semigroups. This is a fact of some importance, and one that is largely mysterious.

[12] This familiar argument has more content than might be supposed. It is, of course, a fact that quantitative measurements are approximate; physical predicates are thus *inexact*. For reasons that are anything but clear, quantitative measurements do not figure in mathematics; mathematical predicates are thus *exact*. It follows that mathematical theories typically are *unstable*. If a figure D just misses being a triangle, no truth strictly about triangles applies to D. Mathematical theories are *sensitive to their initial descriptions*. This is not typically true of physical theories. To complicate matters still further, I might observe that no mathematical theory is capable fully of expressing the conditions governing the application of its predicates. It is thus *not* a theorem of Euclidean geometry that the sum of the angles of a triangle is *precisely* 180 degrees; 'precisely' is not a geometric term. For interesting remarks, see [18].

Sections **4–6** were intended provisionally to answer the question whether a group takes physical instances. Asked provisionally, that question now gets a provisional answer:

No.

7 The arguments given suggest that nothing short of a mathematical object is apt to satisfy a mathematical predicate; it is hardly surprising, then, that within physics, at least, nothing short of a mathematical object *does* satisfy a mathematical predicate.

The systems theorist John Casti [9 pp. 22–25] has argued that mathematical modeling is essentially a relationship between a natural system $\mathbf{N}$ and a formal system $\mathbf{F}$. Passage between the two is by means of an *encoding* map $\zeta: \mathbf{N} \to \mathbf{F}$, which serves to associate observables in $\mathbf{N}$ with items in $\mathbf{F}$. The idea of an encoding map is not without its uses. The encoding map, if it exists at all, conveys a natural into a mathematical world: $\zeta: \mathbf{N} \to \mathbf{M}$. For my purposes, the map's interest lies less with its ability to convey natural into mathematical objects, but in the reverse. If an encoding map exists, its *inverse* $\zeta^*: \mathbf{M} \to \mathbf{N}$ should serve to demarcate at least one sense in which mathematical objects receive an application.

The argument now turns on choices for ζ^*. Within particle physics (but in other areas as well), $\mathbf{M}$ is taken as a group, $\mathbf{N}$ as its representative, and ζ^* understood as the action of a group homomorphism. Such is the broad outline of group representation theory. Does this scheme provide a satisfactory sense of application? Doubts arise for the simplest of reasons. If ζ^* does represent the action of a group homomorphism, surely $\mathbf{N}$ is for that reason a mathematical object? If this is so, the scheme under consideration has in a certain sense overshot the mark, the application map, if it is given content, establishing that every target in its range is a mathematical object.

Consider a single particle—an electron, say—on a one-dimensional lattice; the lattice spacing is b [20]. The dynamics of this system are governed by the Hamiltonian

7.1 $\mathbf{H} = p^2/m + V(x),$

where m measures the mass of the electron and p its momentum. V is the potential function and satisfies the condition that:

7.2 $V(x + nb) = V(x)$

for every integer n. The system that results is *symmetrical* in the sense that translations $x \to x' = x + nb$ leave $\mathbf{H}$ unchanged. Insofar as they are governed by a Hamiltonian, any two systems thus related behave in the same way.

A few reminders. Within quantum theory, information is carried by state vectors. These are objects that provide an instantaneous perspective on a system, a kind of snapshot. Let Q be the set of such vectors, and let /y> and /y'> be state vectors related by a translation. The correspondence /|y> → /|y'> may itself be expressed by a linear operator T in Q:

7.3 $/y> \to /|y'> = T/y>,$

— this for every state vector /y>.

Not *any* linear operator suffices, of course; physical observables in quantum theory are expressed as scalar products $<F/y>$ of the various state vectors. It is here that old-fashioned numbers make an appearance and so preserve a connection between quantum theory and a world of fact and measurement. Unitary linear operators preserve scalar products; they preserve, as well, the length and angle between vectors. To the extent that **7.3** takes physically significant vectors to physically significant vectors, those operators must be unitary; so too the target of T.

Let us step backward for a moment. Here is the Big Picture. On the left there is a symmetrical something; on the right, another something. Symmetrical somethings of any sort suggest group theory, and, in fact, the set of symmetry operations on a lattice may be described by the discrete translation group $G(T^D)$. Those other somethings comprise the set $\{T\}$ of unitary linear operators. $\{T\}$ constitutes a representation of the symmetry operations on **H**; it resembles an inner voice in harmony.

Now for a Close Up. First, there is the induction of group theoretic structure on the alien territory of a set L of linear operators in a vector space Q. L becomes a group $G(L)$ under the definition of the product of two operators A and B in L:

7.4 $Cx = A(B(x))$.

The identity is the unit operator. And every operator is presumed to have an inverse.

Next an arbitrary group **G** makes an appearance. The homomorphism

7.5 $h: G \rightarrow G(L)$

acts to establish a representation of **G** in $G(L)$, with $G(L)$ its representative. In general, group theory in physics proceeds by means of the group representation.[13] In the example given, $G(T^D)$ corresponds to **G**; $\{T\}$ to L; and given an h such that

7.6 $h: G(T^D) \rightarrow G\{T\}$,

$G\{T\}$ corresponds to $G(L)$.

7.5 has an ancillary purpose: It serves to specify the *applications* of group theory to physics in a large and general way. **7.6** makes the specification yet more specific. The results are philosophically discouraging (although not surprising). **G** and $G(L)$ are mathematical objects; but then, so are $G(T^D)$ *and* $G\{T\}$. The real (or natural) world intrudes into this system only via the scalars. If there is any point at which mathematics is applied directly to anything at all, it is here. But these applications involve only counting and measurement. This is by no means trivial, but it does suggest that the encoding map carries information of a severely limited kind. An application of group theory to physics, on the evidence of **7.5** and **7.6**, is not yet an application of group theory to anything physical: so far as group representation goes, the target of every mathematical relation is itself mathematical; and as far

---

[13]For a more general account, see [6]. For a (somewhat confusing) discussion of the role of groups in physics, see [14].

as quantum theory goes, those objects marked as physical by the theory—the *range* of Casti's encoding map—do not appear as the targets of any sophisticated mathematical application.

This conclusion admits of ratification in the most familiar of physical applications. Consider the continuous rotation group **S0(2)**. The generator J of this group, it is easy to demonstrate, is $R(\phi) = e^{-i\phi J}$, where $R(\phi)$ is, of course, a continuous measure of rotation through a given angle. **S0(2)** is a Lie group; its structure is determined by group operations on J near the identity. So too its representations. Thus consider a representation of **S0(2)** defined in a finite dimensional vector space V. $R(\phi)$ and J both have associated operators $R(\phi)^*$ and J^* in V. Under certain circumstances J^* may be understood as an angular momentum operator in the sense of quantum mechanics. This lends to J a certain physical palpability. Nonetheless, J^* is and remains an operator in V, purely a mathematical object in purely a mathematical space. The same point may be made about the actions of **SU(2)** and **SU(3)**, groups that play a crucial role in particle physics. **SU(2)**, for example, is represented in a two-dimensional abstract isospin space. The neutron and the proton are regarded as the isospin up and down components of single nucleon. **SU(2)** defines the invariance of the strong interaction to rotations in this space. But **SU(2)** is a mathematical object; so, too, *its* representative. Wherever the escape from a circle of mathematical concepts is made, it is surely not here.

8 It is within mathematical physics that mathematics is most obviously applied and applied moreover with spectacular success, the very terms of description—*mathematical* physics—suggesting one discipline piggy-backed upon another. Still, to say that within quantum field theory or cosmology, mathematics has been a great success is hardly to pass beyond the obvious. A success in virtue of *what*? The temptation is strong to affirm that successes in mathematical physics arise owing to the application of mathematical to physical objects or structures, but plainly this is to begin an unattractive circle. It was this circumstance, no doubt, that prompted Eugene Wigner to remark that the successes of mathematical physics were an example of the 'unreasonable effectiveness of mathematics.'

The canonical instruments of description within mathematical physics are ordinary or partial differential equations. In a well-known passage, Hilbert and Courant asked under what conditions "[a] mathematical problem ... corresponds to physical reality." By a problem they meant an equation or system of equations. Their answer was that a system of differential equations corresponds to the physical world if unique solutions to the equations exist, and that, moreover, those solutions depend continuously on variations in their initial conditions, [10 p. 227].[14] Existence and uniqueness are self-evident requirements; the demand that solutions vary

---

[14]The notion of a solution to a differential equation is by no means free of difficulties. Consider a function $f(x) = Ax$, and consider, too, a tap of the sort that sends f to $g(x) = Ax + \mu x$, where μ is small. Do f and g represent two functions or only one?

continuously with variations in their initial conditions is a concession to the vagaries of measurement:

> The third requirement ... is necessary if the mathematical formulation is to describe observable natural phenomena. Data in nature cannot possibly be conceived as rigidly fixed; the mere process of measuring them involves small errors. ... Therefore, a mathematical problem cannot be considered as realistically corresponding to a physical phenomena unless a variation of the given data in a sufficiently small range leads to an arbitrary small change in the solution.

Following Hadamard, Hilbert and Courant, call a system of equations satisfying these three constraints *well-posed*.

Well-posed problems in analysis answer to a precise set of mathematical conditions. Consider a system of ordinary first-order differential equations expressed in vector matrix form:

8.1 $d\mathbf{x}/dt = f(\mathbf{x}, t), \mathbf{x}(a) = b.$

Existence, uniqueness and continuity depend on constraints imposed on $f(\mathbf{x}, t)$. Assume thus that R is a region in $<\mathbf{x}, t>$. $f(\mathbf{x}, t)$ is *Lipschitz continuous* in R just in case there exists a constant $k > 0$ such that:

$$f(x_1, t_1) - f(x_2, t_2) \le k |x_1 - x_2|.$$

Here (x_1, t) and (x_2, t) are points in R.

Assume, further, that f is continuous *and* Lipschitz continuous in R; and let δ be a number such that $0 < \delta < 1/k$. And assume, finally, that u and v are solutions of **8.1**. Uniqueness now follows, but only for a sufficiently small interval (the interval, in fact, determined by δ):

8.2 If u and v are defined on the interval $|t - t_0| \le \delta$, and if $u(t_0) = v(t_0)$, then $u = v$.

What of existence? Let $u_1, u_2, \ldots, u_n$ be successive approximations to **8.1** in the sense that

$$u_0(t) = x_0$$

$$u_{k+1}(t) = x_0 + \int_{t_0}^{t} f(x, u_k(t)) dt, \quad k = 0, 1, 2, \ldots$$

Suppose now that f is continuous in R: $|x - x_0| \le a$, $|t - t_0| \le b$, $(a, b) > 0$; and suppose, too, that M is a constant such that $f(t, x) < M$ for all (t, x) in R. Let I be the interval $|x - x_0| \le h$, where h is the minimum of $\{a, b/M\}$. The Cauchy–Peano theorem affirms that:

8.3 The approximations $u_1, \ldots, u_n$ converge on I to a solution $\mathbf{u}$ of **8.1**.

8.3 is purely a *local* theorem: it says that solutions exist near a given point; it says nothing whatsoever about wherever else they may exist. The theorem is carefully hedged. And for good reason. There are simple differential equations for which an embarrassing number of solutions exist, or none at all. The equation $y^2 + x^2 y' = 0$ is an example. Infinitely many solutions satisfy the initial condition $y(0) = 0$. No solution satisfies the initial condition $y(0) = b$, if $b \neq 0$. At <0, 1>, this equation fails of continuity.

The Cauchy–Peano theorem does not apply.

8.3 may be supplanted by a global existence theorem, but only if f is Lipschitz continuous for *every* t in an interval I. It follows then that successive solutions are defined over I itself.

There remains the matter of continuity. Let u be a solution of **8.1** passing through the point (t_0, x_0); and let u^* be a solution passing through (t_0', x_0'). Both u and u^* pass through those points in R. The requisite conclusion follows, preceded by a double condition:

8.4 If for every $\epsilon > 0$, there is a $\delta > 0$ such that u and u^* exist on a common interval $a < t < b$; and if $a < t' < b$, $|t - t'| < \delta$, $|t_0 - t_0'| < \delta$, $|x_0 - x_0'| < \delta$, then $|u(t) - u(t)^*| < \epsilon$.

To the extent that **8.1** satisfies the hypotheses of **8.2, 8.3,** and **8.4,** to *that* extent **8.1** is well-posed.

The concept of a well-posed problem in analysis is interesting insofar as it specifies conditions that are necessary for applicability; but however necessary, they are, those conditions, hardly sufficient. How could they be? Like any other equation, a differential equation expresses an affirmation: some unknown function answers to certain conditions. The Cauchy–Peano theorem establishes that for a certain class Φ of differential equations, a suitable function exists. The elements of the theory $\mathbf{T}(\Phi)$ satisfied in models of $\mathbf{T}(\Phi)$ are true simply in virtue of being elements of $\mathbf{T}(\Phi)$.

But true of *what*? Surely not physical objects? This would provide an access to the real world too easy to be of any use.

Like so many other mathematical objects, a differential equation is dedicated to the double life. If the first is a matter of the solutions specified by the equation, the second involves the induction of form over space. The simple differential equation:

8.5 $df/dt = Af(t)$

provides a familiar example. The association established between t and $f(t)$ creates a coordinate system. The set of points evoked—t, on the one hand, $f(t)$, on the other—is the *phase* (or state) *space* of the equation. Ax evidently plays A against each of its phase points. This implies that Ax represents the rate of change of x at t. Rates of change evoke tangent lines and slopes. To each point in the $<t, f(t)>$ plane, a differential equation—*the* differential equation—assigns a short line segment of fixed slope. A phase space upon which such lines have been impressed is a *direction* or *lineal* field.

Imagine now that the plane has been filled with curves tangent at each point to the lines determining a lineal field. The set of such curves fills out the plane. And each curve defines a differential solution $u(t) = x$ to a differential equation, for plainly $du(t)/dt = Au(t)$ in virtue of the way in which those curves have been defined.

It is thus that a differential equation elegantly enters the geometrical scene. Nothing much changes in the interpretation of **8.1** itself. The lineal field passes to a vector field, but the induction of geometrical structure on an underlying space proceeds apace. The system of equations

8.6
$$dX/dt = y$$
$$dY/dt = -x$$

assigns to every point in the $<X, Y>$ plane a vector $<y, -x>$. The trajectory or *flow* of a differential equation corresponds to the graph of those points in the plane whose velocity at $<x, y>$ is simply $<y, -x>$. Trajectories have a strange and dreamy mathematical life of their own. The Poincaré–Bendixson theorem establishes, for example, that a bounded semi-trajectory tends either to a singular point or spirals onto a simple closed curve.

An autonomous system of n differential equations is one in which time has dwindled and disappeared. Points in space are n-dimensional, with R^n itself the collection of such points. On R^n things are seen everywhere as Euclid saw them: R^n is an n-dimensional Euclidean vector space. On a differential manifold, things are seen locally as Euclid saw them; globally, functions and mappings may be overtaken by weirdness. Thus the modern definition of a dynamical system as a pair $<\mathbf{M}, \mathbf{V}>$ consisting of a differential manifold $\mathbf{M}$ and its associated vector field $\mathbf{V}$. To every differential equation there corresponds a dynamical system.

On the other hand, suppose that one starts at a point x of $\mathbf{M}$. Let $g^t(x)$ denote the state of the system at t. For every t, g defines a mapping $g^t: \mathbf{M} \to \mathbf{M}$ of $\mathbf{M}$ onto itself. Suppose that $g^{t+s} = g^t g^s$. Suppose, too, that g^0 is an identity element. That family of mappings emerges as a one parameter group of transformations. A set $\mathbf{M}$ together with an appropriate group of transformations now acquires a substantial identity as a flow or dynamical system. In the obvious case, $\mathbf{M}$ is a differential manifold, g a differential mapping. Every dynamical system, so defined, corresponds to a differential equation. Since g^t is differentiable, there is—there must be—a function v such that $dg^t/dt = v(g^t)$. A differential equation may thus be understood as the action of a one parameter group of diffeomorphisms on a differential manifold.

The attempt to assess the applicability of differential equations by looking toward their geometric interpretation runs into a familiar difficulty. Neither direction fields nor manifolds are anything other than mathematical structures. And the association between groups and differential equations suggests that insofar as the applicability of differential equations must be defended in terms of the applicability of groups, the result is apt to be nugatory. There is yet no clear and compelling sense in which groups are applicable to anything at all.

This may well suggest that the applicability of differential equations turns not on the equations, but on their solutions instead. An icy but invigorating jet now follows. The majority of differential equations cannot be solved by analytic methods. Nor by any other methods. This is a conclusion that physicists will find reassuring. Most differential equations cannot be solved; and they cannot solve most differential equations.

9 The exception to this depressing diagnosis arises in the case of *linear* differential equations; such systems, V. I. Arnold remarks [2 p. 97], constitute "the only large class of differential equations for which there exists a definitive theory."

These are encouraging words.

Linear algebra is the province of linear mappings, relations, combinations, and spaces. Instead of numbers, linear algebra trades often in vectors, curiously rootless objects, occurring in physics as arrows taking off from the origins of a coordinate system, and in the calculus as directed line segments. In fact, vectors are polyvalent, described now in one formulation, now in another, the various definitions all equivalent in the end.

If $x_1, x_2, \ldots, x_n \in R^n$, with $\{c_i\}$ a set of scalars, dying to be attached, the vector $c_1 x_1 + \ldots + c_k x_k$ is a *linear combination* of the vectors $x_1, x_2, \ldots, x_k$—linear because only the operations of scalar multiplication and vector addition are involved, a combination because something is being heaped together, as in a Caesar salad. A set of vectors $\{x_1, x_2, \ldots, x_k\}$ is independent if $c_1 x_1 + \ldots + c_k x_k = 0$ implies that $c_1 = c_2 = \ldots = c_k = 0$; otherwise, dependent, the language itself suggesting that among the dependent vectors, some are superfluous and may be expressed as a linear combination of those that remain.

This makes for an important theorem, one serving to endow vectors with an exoskeleton. Suppose that $S \subset R^n$ and let E be the set of all linear combinations of vectors in S. Then S is said to be spanned by E. This is a definition. It follows that E is the intersection of all subspaces in R^n containing S. Theorem and definition may, of course, be reversed. A basis B of a vector space X is an independent subset of X spanning X. This, too, is a definition.

Herewith that important theorem:

9.1 If B is a basis of X, then every x in X may be uniquely expressed as a linear combination of base vectors.

The representation, note, exists in virtue of the fact that B spans X. It is unique in virtue of the fact that B is independent. Thus in R^2, every vector $x = <x_1, x_2>$ exists first in its own right, and second as a linear combination of basis vectors $x = x_1 e_1 + x_2 e_2$. The unit vectors $e_1 = <1, 0>$ and $e_2 = <0, 1>$ constitute the standard basis for R^2.

The theory in which Arnold has expressed his confidence constitutes a meditation on two differential equations. The first is an inhomogeneous second-order linear differential equation

9.2 $d^2 x/dx + a_1(t) \, dx/dt + a_2(t) x = b(t)$;

the second, a reduced version of the first

9.3 $d^2 x/dx + a_1(t) dx/dt + a_2(t) x = 0$.

It is the play between these equations that induces order throughout their solution space. Let $L(x)$ denote $x'' + a_1(t) x' + a_2(t) x$ so that **9.2** is $L(x) = b(t)$. That order is in evidence in the following theorems; and these comprise the theory to which Arnold refers.

For any twice differential functions u_k and constants c_k

9.4 $L(c_1u_1(t)+c_2u_2(t)+ \ldots +c_mu_m(t)) = c_1L(u_1(t) + c_2L(u_2(t) + \ldots + c_mL(u_m(t)).$

9.4 follows directly from the fact that **9.2** is linear.

9.5 If u and w are solutions to **9.2**, then $u - w$ is a solution to **9.3**.

$L(u(t) = b(t) = L(w(t))$. But then $L[u(t) - w(t)] = L(u(t)) - L(w(t)) = b(t) - b(t) = 0$.

9.6 If u is a solution of **9.2**, and w a solution of **9.3**, then $u + w$ is a solution of **9.3**.

Say that $L(u(t)) = b(t)$; say too that $L(w(t)) = 0$. $L\{u(t) + w(t)\} = L(u(t)) + L(w(t)) = b(t) + 0 = b(t)$.

Now let u be *any* solution of **9.3**:

9.7 Every solution of **9.3** is of the form $u + w$, where w is a solution of **9.4**.

Let v be a solution to **9.3**. By **9.5** it follows that $v - u = w$, where w is a solution of **9.4**.

The *general* solution of **9.3** may thus be determined by first fixing a single solution v to **9.3**, and then allowing w to range over all solutions to **9.4**, an interesting example in mathematics of a tail wagging a dog.

Suppose now that u is a solution of **9.4**:

9.8 If $u(t_0) = 0$ for some t_0 and $u'(t_0)$ as well, then $u(t) = 0$.

But $u(t) = 0$ is already a solution of **9.4**.

9.9 If $u_1, \ldots, u_m$, are solutions of **9.4**, then so is any linear combination of $u_1, \ldots, u_m$.

This follows at once from **9.7**.

9.10 If u_1 and u_2 are linearly independent solutions of **9.4** then every solution is a linear combination of u_1 and u_2.

Note that **9.9** affirms only that linear combination of solutions to **9.4** are solutions to **9.4**; but **9.10**, that *all* solutions of **9.4** are linear combinations of two linearly independent solutions to **9.4**.

Let u be any solution of **9.4**. The system **E** of simultaneous equations

$$u_1(0)x_1 + u_2(0)x_2 = u(0)$$
$$u_1'(0)x_1 + u_2'(0)x_2 = u'(0)$$

has a non-vanishing determinant. **E** thus has a unique solution $x_1 = c_1$ and $x_2 = c_2$. If $v(t)$ is $c_1u_1(t) + c_2u_2(t)$ it follows that $v(t)$ is a solution of **9.4** because it is a linear combination of solutions to **9.4**. But $v(0) = u(0)$ and $v'(0) = u'(0)$; thus $v = u$ by the existence and uniqueness theorem for **9.2**.

It is **9.10** that provides the algebraic shell for the theory of linear differential equations, *all* solutions of **9.4** emerging as linear combinations of u_1 and u_2. u_1 and u_2, and thus forming a *basis* for **S**, which is now revealed to have the structure of a finite dimensional vector space. This in turn implies the correlative conclusion that solutions of **9.3** are all of the form $w + c_1u_1(t) + c_2u_2(t)$, where u_1 and u_2 are linearly independent solutions of **9.4**.

A retrospective is now in order. The foregoing was prompted by the desire to see or sense the spot at which a differential equation or system of equations applies to anything beyond a mathematical structure. Well-posed differential equations are useful in the sense that if their solutions did not exist or were not unique, they would not be useful at all. Continuity is less obvious a condition, whatever the justification offered by Hilbert and Courant. Still, there is nothing in the idea of a well-posed problem in analysis, or a well-posed system of equations, that does more than indicate what physically relevant systems must have. What they do have, and how they have it, this remains unstated, unexamined, and unexplored. The qualitative theory of ordinary differential equations has the welcome effect of turning the mathematician's attention from their solutions taken one at a time to all of them at once. The imagination is thus enlarged by a new prospect, but the rich and intriguing geometrical structures so revealed does little, and, in fact, it does nothing, to explain the coordination between equations and the facts to which they are so often applied. So far as the linear differential equations go, V.I. Arnold is correct. There is a theory, and a theory, moreover, that has a stabilizing effect across the complete range of linear differential equations. This is no little thing. But while the theory draw a connection between linear equations and linear algebra, so far as their applications go, the connection is internal to mathematics, falling well within the categorical definitions of section **2**.

The spot at which a differential equation or system of equations applies to anything beyond a mathematical structure?

I'm just asking.

10 Consider a physical system **P**. A continuously varying physical parameter ξ is given, one subject intermittently to measurement, so that $g(t) = \xi$ is a record of how much, or how little, there is of ξ at t. Or in the case of bacteria, how many of them there are. If the pair $<P, \xi>$ makes for a physical system, there is by analogy the pair $<D, f>$, where **D** is a differential equation, and f its solution.

Let us suppose that for some finite spectrum of values, $f(t_k) \cong g(t_k)$.

The example that follows is the very stuff of textbooks. Having made an appearance at **8.7**, the equation

10.1 $df/dt = Af(t)$

is simple enough to suggest that its applications must be transparent if any applications are transparent. Solutions are exponential: $x = f(t) = Ke^{At}$. The pair $<D, f>$, where **D** just is **10.1** and f its solution, makes for a differential model.

Can we not say over an obvious range of cases, such as birth rates or the growth of compound interest, that there is a very accessible sense of applicability to be had in the play between differential equations and the physical processes to which they apply? Let me just resolve both $g(t) = \xi$ and $f(t) = Ke^{At}$ to $t = 0$, so that for the mathematician, **10.1** appears as an initial value problem, and for the biologist or the bacteria, as the beginning of the experiment. The voice of common sense now chips in to claim that for $t = 0$, and for some finite spectrum of values thereafter,

10.2 $<\mathbf{D}, f>$ *applies to* $<\mathbf{P}, \xi>$ if and only if $f(t_k) \cong g(t_k)$.

I do not see how **10.2** could be faulted if only because the relationship in which $<\mathbf{D}, f>$ *applies to* $<\mathbf{P}, \xi>$ is loose enough to encompass indefinitely many variants. One might as well say that the two structures are coordinated, connected, or otherwise companionably joined. If one might as well say any of that, one might as well say that differential equations are very often useful and leave matters without saying why.

But **10.2** raises the same reservations about the applicability of mathematics as considerably more complicated cases. It is a one man multitude. If the inner structure of $<\mathbf{D}, f>$ and $<\mathbf{P}, \xi>$ were better aligned, one could replace an unclear sense of applicability by a mathematical mapping or morphism between them. Far from being well-aligned, these objects are not aligned at all. The function $g(t) = \xi$ is neither differentiable nor continuous; barely literate, in what respect does it have anything to do with $<\mathbf{D}, f>$? The function f, on the other hand, is differentiable and thus *continuous*. Continuous functions take intermediate values; in what sense does it have anything to do with $<\mathbf{P}, \xi>$? There is no warm point of connectivity between them. Differential and physical structures are radically unalike.

In that case, why should $f(t_k) \cong g(t_k)$?

We are by now traveling in all the old familiar circles.

11 "To specify a *physical* theory," Michael Atiyah writes in the course of a discussion of quantum field theory, "the usual procedure is to define a Lagrangian or action $\mathbf{L}$." A Lagrangian $\mathbf{L}(\varphi)$ having been given, where φ is a scalar field, the partition function P of the theory is described by a Feynmann functional integral. "These Feynmann integrals," Atiyah writes with some understatement, "are not very well defined mathematically … ." [3 p. 3].

The parts of the theory that *are* mathematically well defined are described by the axioms for topological quantum field theory. A topological **QFT** in dimension d is identified with a functor $\mathbf{Z}$ such that $\mathbf{Z}$ assigns i) a finite dimensional complex vector space $\mathbf{Z}(\Sigma)$ to every compact oriented smooth d-dimensional manifold Σ; and ii) a vector $\mathbf{Z}(Y) \in \mathbf{Z}(\Sigma)$ to each compact oriented $(d + 1)$ dimensional manifold Y whose boundary is Σ. The action of $\mathbf{Z}$ satisfies involutory, multiplicative, and associative axioms. In addition, $\mathbf{Z}(\varnothing) = C$ for the empty d-manifold.

"The *physical* interpretation of [these] axioms," Atiyah goes on to write, is this: … "[F]or a closed $(d + 1)$ manifold Y, the invariant $\mathbf{Z}(Y)$ is the partition function given by some Feynmann integral … ." It is clear that $\mathbf{Z}(Y)$ is an invariant assigning a complex number to any closed $(d + 1)$ dimensional manifold Y (in virtue of the fact that the boundary is empty) and clear thus that $\mathbf{Z}$ and P coincide.

This definition has two virtues. It draws a relatively clear distinction between parts of a complex theory; and it provides for an interpretation of the mathematical applications along the lines suggested by Section **2**. It is less clear, however, in what sense P is a *physical* interpretation of $\mathbf{Z}$, the distinction between $\mathbf{Z}$ and P

appearing to an outsider (this one, at any rate) to have nothing whatsoever to do with any relevant sense of the physical, however loose. The distinction between the mathematical and the physical would seem no longer to reflect any intrinsic features either of mathematical or physical objects, things, or processes, with *physical* a name given simply to the portions of a theory that are confused, poorly developed, largely intuitive, or simply a conceptual mess.

In quite another sense, the distinction between the mathematical and the physical is sometimes taken as a reflection of the fact that mathematical objects are quite typically general, and physical objects, specific, or particular. The theory of differential equations provides an example. The study of specific systems of equations may conveniently be denoted a part of theoretical physics; the study of generic differential equations, a part of mathematics. But plainly there is no ontological difference between these subjects, only a difference in the character of certain mathematical structures. And this, too, is a distinction internal to mathematics.

The project of determining a clear senses in which mathematics has an application beyond itself remains where it was, which is to say, unsatisfied.

12 "To present a theory is to specify a family of structures," Bas van Fraassen has written, [21 p. 64][15]

> its models; and secondly, to specify certain parts of those models (the empirical substructures) as candidates for the direct representation of observable phenomena. The structures which can be described in experimental and measurement reports we can call appearances: the theory is empirically adequate if it has some model such that all appearances are isomorphic to empirical substructures of that model.

Some definitions. A language L is a structure whose syntax has been suitably regimented and articulated—variables, constants, predicate symbols, quantifiers, marks of punctuation. A language in *standard formulation* has the right kind of regimentation. A model $\mathbf{M} = <D, f>$ is an ordered pair consisting of a non-empty domain D and a function f. It is f that makes for an *interpretation* of the symbols of L in $\mathbf{M}$. Predicate symbols are mapped to subsets of D; relation symbols to relations of corresponding rank on D. The general relationship of language and the world model theory expresses by means of the concept of satisfaction. This is a relationship that is purely abstract, perfectly austere. Formulas in L are satisfied in $\mathbf{M}$, or not; sentences of L are true or false in $\mathbf{M}$. Languages neither represent nor resemble their models. The scheme is simple.

There are no surprises.

Let $\mathbf{T}$ be a theory as logicians conceive things, a consistent set of sentences; and a theory furthermore that expresses some standard (purely) mathematical theory—the theory of linear differential operators, say. If $\mathbf{T}$ has empirical content, it must have empirical consequences—φ, for example:

---

[15]See also [4] for a very detailed treatment of similar themes.

12.1 $T \mathrel{|\!\!-} \varphi$.

But equally, if **T** has empirical content, some set of sentences $\mathbf{T}(E) \subseteq \mathbf{T}$, must express its empirical *assumptions*. Otherwise, **12.1** would be inscrutable. Subtract $\mathbf{T}(E)$ from **T**. The sentences $\mathbf{T}(M) \subseteq \mathbf{T}$ that remain are purely mathematical.

Plainly

12.2 $\mathbf{T} = \mathbf{T}(M) \cup \mathbf{T}(E)$.

And plainly again

12.3 $\mathbf{T}(M) \cup \mathbf{T}(E) \mathrel{|\!\!-} \varphi$,

whence by the deduction theorem,

12.4 $\mathbf{T}(M) \mathrel{|\!\!-} \mathbf{T}(E) \rightarrow \varphi$.

The set of sentences $\Theta = \mathbf{T}(E) \rightarrow \varphi$ constitutes the *empirical hull* of **T**.

If model theory is the framework, the concept of a mathematical application resolves itself into a relationship between theory and model and so involve a special case of satisfaction. $\mathbf{T}(M)$ as a whole is satisfied in a set theoretic structure **M**; Θ, presumably, in another structure **N**, its domain consisting of physical objects or bodies. But given **12.4**, Θ is also satisfied in *any* model **N** satisfying $\mathbf{T}(M)$. **N** is a model with a kind of hidden hum of real life arising from the elements in its domain. An application of mathematics, if it takes place at all, must take place in the connection between **M** and **N**. An explanation of this connection must involve two separate clauses, as in **12.5**, which serves to give creditable sense to the notion of a mathematical application in the context of model theory:

12.5 $\mathbf{T}(M)$ applies to **N** if i) $\mathbf{T}(M)$ is satisfied in **N**; and, ii) **N** is a sub-model of **M**.

This leaves one relationship undefined. Sub-models, like sub-groups, are not simply substructures of a given structure. If **N** is a sub-model of **M**, the domain D' of **N** must be included in the domain D of **M**; but in addition:

i) Every relation R' on D' must be the restriction to D' of the corresponding R on D; and
ii) ditto for functions; and moreover,
iii) every constant in D' must be the corresponding constant in D.

This definition has an undeniable appeal. The mathematical applications find their place within the antecedently understood relationship between theories and their models. This does not put mathematics directly in touch with the world, but with its proxies instead. The parts of the definition cohere, one with the other. It is obviously necessary that $\mathbf{T}(M)$ has empirical consequences. Otherwise there would be no reason to talk of applications whatsoever. It is necessary, too, that **N** be a sub-model of **M**; otherwise the connection between what a theory implies and the structures in which it holds would be broken. Finally, it is necessary that $\mathbf{T}(M)$ be satisfied in **N** as well as **M**; otherwise what sense might be given to the notion that $\mathbf{T}(M)$ *applies* to any empirical substructure of **M** at all? Those conditions having been met, a clear sense has been given to the concept of a mathematical application.

This is somewhat too optimistic. **M**, recall, is a mathematical model, and **N** a model that is not mathematical: the elements in *its* domain are physical objects. The assumption throughout is that in knowing what it is for a mathematical theory to be satisfied in **M**, the logician knows what it is for that same theory to be satisfied in **N**. In a purely formal sense, it must, that assumption, be true; the *definition* of satisfaction remains constant in both cases. What remains troubling is the question whether the conditions of the definition are ever met. The definition of satisfaction, recall, proceeds by accretion. A sentence **S** is satisfied in **N** under a given interpretation of its predicate symbols **S[F,G,...,H]**. The interpretation comes first. In the case of a pure first-order language, it is the predicate symbols that carry all of the mathematical content.

Were it antecedently clear that **S[F,G, ... ,H]** admits of physical interpretations, why did we ever argue?

13 Under one circumstance, the question whether a mathematical theory is satisfied in a physical model may be settled in one full sweep. A theory **T** satisfied in *every* one of the sub-models of a model **M** of **T** is satisfied in particular in the empirical sub-models of **M** as well. It must be. Demonstrate that **T** *is* satisfied in every one of its sub-models and what remains, if **12.5** is to be justified, is the correlative demonstration that **T** is satisfied in a model containing empirical structures as sub-models. What gives pause is preservation itself.

The relevant definition:

13.1 A theory *T* is preserved under sub-models if and only if *T* is satisfied in any sub-model of a model of *T*.

In *any* sub-model, note. Preservation under sub-models is by no means a trivial property.

Given **13.1**, it is obvious that preservation hinges on a sentence's quantifiers. A sentence is in prenex form if its quantifiers are in front of the matrix of the sentence; and universal if it is in prenex form and those quantifiers are universal. It is evident that every sentence may be put into prenex form.

Herewith a first theorem on preservation:

13.2 If φ is universal and *N* is a sub-model of *M*, then if φ is satisfied in *M* it is satisfied as well in *N*.

The proof is trivial.

13.1 takes a more general form in **13.3**:

13.3 A sentence φ is preserved under sub-models if and only if φ is logically equivalent to a universal sentence.

Again, the proof is trivial.

13.3 is, in fact, a corollary to a still stronger preservation theorem, the only one of consequence. A theory **T** has a set of axioms A just in case A and **T** have the same consequences; those axioms are *universal* if each axiom is in prenex normal form, with only universal quantifiers figuring. A theory derivable from universal axioms is itself universal. Let **T**, as before, be a theory. What follows is the Los-Tarski theorem:

13.4 **T** *is preserved under sub-models if and only if* **T** *is universal.*

13.4, together with **12.1** mark a set of natural boundaries of sorts. **12.1** indicates *what* is involved in the concept of an application; **13.4** specifies *which* theories are apt to have any applications at all. The yield is discouraging. Group theory is *not* preserved under sub-models; *neither* is the theory of commutative rings, *nor* Peano arithmetic, *nor* Zermelo Fraenkel set theory, *nor* the theory of algebraically closed fields, *nor* almost anything else of much interest.

14 When left to his own devices, the mathematician, no less than anyone else, is apt to describe things in terms of a natural primitive vocabulary. Things are here or there, light or dark, good or bad. The application of mathematics to the world beyond involves a professional assumption. And one that is often frustrated. Sheep, it is worthwhile to recall, may be collected and then counted; not so plasma. Set theory, I suppose, marks the point at which a superstitious belief in the palpability of things gives way. Thereafter, the dominoes fall rapidly.

If some areas of experience seem at first to be resistant to mathematics, there is yet a doubled sense in which mathematics is inexpungable, a feature of every intellectual task. The idea that there is some arena in which things and their properties may be directly apprehended is incoherent. Any specification of the relevant arena must be by means of some theory or other; there is no describing without descriptions. But to specify a theory is to specify its models. And so mathematics buoyantly enters into areas from which it might have been excluded, if only for purposes of *organization*.

Mathematics makes its appearance in another more straightforward sense. Every intellectual activity involves a certain set of basic and ineliminable operations of which counting, sorting, and classification are the most obvious. These operations may have little by way of rich mathematical content, but at first cut they appear to be amenable to formal description. It is here that the empirical substructures that van Fraassen evokes come into play. Mention thus of empirical substructures is a mouthful; let me call them *primitive models*, instead, with *primitive* serving to emphasize their relative position on the bottom of the scheme of things, and *models* reestablishing a connection to model theory itself. The primitive models are thus a mathematical presence in virtually every discipline, both in virtue of their content—they are models, after all; *and* in virtue of their form—they deal with basic mathematical operations.

Patrick Suppes envisages the primitive models as doubly finite: Their domain is finite; so are all model-definable relations. Those relations are, moreover, qualitative in the sense that they answer to a series of *yes* or *no* questions asked of each object in the domain of definition [19]. This definition reflects the fact that in the end every chain of assertion, judgment, and justification ends in a qualitative declaration. There it is: The blotting paper is red, or it is not; the balance beam is to the right, or it is not; the rabbit is alive, or it is dead. But now a second step. A physical object is any object of experience; and objects of experience are those describable by primitive *theories*. Primitive theories are satisfied in primitive models.

The salient feature of a primitive model is a twofold renunciation: Only finitely many objects are considered; and each object is considered only in the light of a predicate that answers to a simple *yes* or *no*. Such are the primitive properties F_1, ..., Fn. A primitive model may also be described as any collection C of primitive properties, together with their union, intersection, and complement.

There is yet another way of characterizing primitive theories and their models. A boolean-valued function f is one whose domain is the collection of all n-tuples of 0 and 1, and whose range is $\{0, 1\}$. Such functions are of use in switching and automata theory. Their structure makes them valuable as instruments by which qualitative judgments are made and then recorded. The range of a boolean-valued function corresponds to the simple *yes* or *no* (true or false) that are features of the primitive models; but equally, the domain corresponds either to qualitative properties or to collocations of such properties. A primitive theory, on this view, is identified with a series of Boolean equations; a primitive structure, with a Boolean algebra.

Set theory provides still another characterization of the primitive models, this time via the concept of a generic set. The generic sets are those that have only the members they are forced to have and no others. Forcing is atomistic and finite, the thing being done piecemeal. Thus suppose that L is a first-order language, with finitely many predicates but indefinitely many constants. By the extension $L^*(S_1, ..., S_n)$ of L, I mean the language obtained by adjoining the predicate symbols S_n to L. A *basic sentence* of L^* has the form $k \in S_n$ or $k \notin S_n$. A finite and consistent set of basic sentences ξ constitutes a *condition*. A sequence of conditions ξ_1, ..., ξ_n is *complete* if and only if its union is consistent, and, moreover, for any k and n, there exists an n such that $k \in S_n$ or $k \notin S_n$ belongs to ξ_n. A complete sequence of conditions determines an associated sequence of sets $S_1, ... S_n$:

$$k \in S_n \iff (\exists m)\,(k \in S_n \text{ belongs to } \xi_n)\,.$$

The path from conditions to sets runs backwards as well as forwards.

Sets have been specified, and sequences of sets; conditions, and sequences of conditions. The model structure of L^* is just the model $M^* = <D, f>$, where f maps $S_1, ..., S_n$ onto $S_1, ..., S_n$. There is a straightforward interpretation of the symbolic apparatus. The conditions thus correspond to those *yes* or *no* decisions that Suppes cites; that they are specified entirely in terms of some individual or other belonging to a set is evidence of the primacy of set formation in the scheme of things.

The specification of sets by means of their associated conditions is a matter akin to enumeration. A given set S is *generic* if in addition to the objects it has as a result of enumeration, it has only those objects as members it is forced to have. Forcing is thus a relationship between finite conditions and arbitrary sentences of L, the sentences in turn determining what is truly in various sets. The definition proceeds by induction on the length of symbols. What it says is less important than what it implies. Every sentence about a generic set is decidable by finitely many sentences of the form $k \in S_n$ or $k \notin S_n$. *Finitely* many, note, and of an *atomic* form.

Whatever the definition of primitivity, theories satisfied in primitive models admit of *essential application*. This is a definition:

14.1 *T applies essentially* to M if and only if M is primitive and *T* is satisfied in M.

Such theories apply directly to the world in the sense that no other theories apply more directly. Counting prevails, but only up to a finite point; the same for measurement. The operation of assigning things to sets is suitably represented; and in this sense one has an explanation of sorts for the universal feeling that numbers may be directly applied to things in a way that is not possible for groups or Witten functors. It is possible that the operation of assigning things to sets is the quintessential application of mathematics, the point that dizzyingly spins off all other points. But even if assigning things to sets is somehow primitive, the models that result are themselves abstract and mathematical.

The concept of a primitive model does not itself belong to model theory. The primitive models have been specified with purposes other than mathematics in mind. Nor is it, that concept, precisely defined, if only because so many slightly different structures present themselves as primitive. Nonetheless, the primitive models share at least one precisely defined model-theoretic property. A number of definitions must now be introduced:

Let L be a countable language and consider a theory T:

14.2 A formula $\varphi(x_1, x_2, \ldots, x_n)$ is *complete* in T if and only if for every other formula $\chi(x_1, x_2, \ldots, x_n)$, either $\varphi \supset \chi$ or $\varphi \supset \sim \chi$ holds in T; and

14.3 A formula $\theta(x_1, x_2, \ldots, x_n)$ is completable in T if and only if there is a complete formula $\varphi(x_1, x_2, \ldots, x_n)$ such that $\varphi \supset \theta$ holds in T.

The definitions of complete and completeable formulas give rise, in turn, to the definitions of atomic theories and their models:

14.4 A theory T is *atomic* if and only if every formula of L consistent with T is completeable in T; and

14.5 A model M is *atomic* if and only if every relation in its domain satisfies a complete formula in the theory T of M.

It follows from these definitions that every finite model is atomic; it follows also that every model whose individuals are constant is again atomic. Thus a first connection between empirical substructures and model theory emerges as a trivial affirmation:

14.6 Every primitive model is atomic.

The proof is a matter of checking the various definitions of the primitive models, wherever they are clear enough to be checked.

The real interest of atomic models, however, lies elsewhere. The relationship of a model to its sub-models is fairly loose; not so the relationship of a model to its *elementary* sub-models. Consider two models, N and M, with domains D' and D. A first-order language L is presumed throughout:

14.7 The mapping $f: D' \to D$ is an *elementary embedding* of N into M if and only if for any formula $\varphi(x_1, x_2, \ldots, x_n)$ of L and n-tuples $a_1, \ldots, a_n$ in D', $\varphi[a_1, \ldots, a_n]$ holds in N if and only if $\varphi[fa_1, \ldots, fa_n]$ holds in M.

Given **14.7**, it follows that the target of f in M is a sub-model of M; the elementary sub-models are simply those that arise as the result of elementary embeddings. From the perspective of first-order logic, elementary sub-models and the models in which they are embedded are indiscernible: No first-order property distinguishes between them. The models are *elementarily equivalent*.

Another definition, the last. Let N be a model and $T(\mathrm{N})$ its theory:

14.8 N is a *prime model* if and only if N is elementarily embedded in every model of $T(\mathrm{N})$.

14.6 establishes trivially that every primitive model is atomic. It is also trivially true that primitive models are countable. But what now follows is a theorem of model theory:

14.9 If N is a countable atomic model then N is a prime model.

Assume that N is a countable atomic model; $T(\mathrm{N})$ is its theory. Say that $A = \{a_0, a_1, \ldots, \}$ constitutes a well-ordering of the elements in the domain of N. Assume that M is any model of $T(\mathrm{N})$. Suppose that F is a complete formula satisfied by a_0. It follows that $(\exists x)F$ follows from $T(\mathrm{N})$; it follows again that there is a b_0 among the well-ordered elements B of M that satisfies F. Continue in this manner, exhausting the elements in A. By **14.7**, going from A to B defines an elementary embedding of N into M. The conclusion follows from **14.8.**

14.6 establishes that the primitive models are among the atomic models; but given the very notion of a primitive model, it is obvious that any primitive model must be countable. It thus follows from **14.9** that

14.10 Every primitive model is prime.

In specifying a relationship between the primitive and the prime models, **14.10** draws a connection between concepts arising in the philosophy of science and concepts that are model-theoretic. There is a doubled sense in which **14.10** is especially welcome. It establishes the fact that the primitive models are somehow at the bottom of things, in virtue of **14.8** the *smallest* models available. And it provides a *necessary* condition for a theory to have empirical content. Recall van Fraassen's definition: "[A] theory is empirically adequate if it has some model such that all appearances are isomorphic to empirical substructures of that model." A mathematical theory T has empirical content just in case T has a prime model.

Such is the good news. The bad news follows. **14.10** does little—it does *nothing*—to explain the relationship, if any, between those mathematical theories that are not primitive and those that are. As their name suggests, the primitive models are pretty primitive. The renunciations that go into their definition are considerable. There is thus no expectation that any mathematical theory beyond the most meager will be definable in terms of a primitive theory. Let us go, then, to the next best thing. Assume that most mathematical theories are satisfied in models with primitive sub-models. **14.6** might then suggest that such theories apply to elements in a primitive sub-model if they are satisfied in a primitive sub-model. But those mathematical theories that are not preserved under sub-models generally will not be preserved generally under primitive sub-models either. The primitive theories and their models are simply too primitive.

There is no next best thing.

15 Conclusion

The argument that mathematics has *no* application beyond itself satisfies an esthetic need: It reveals mathematics to be like the other sciences and so preserves a sense of the unity of intellectual inquiry. Like any argument propelled by the desire to keep the loose ends out of sight, this one is vulnerable to what analysts grimly call the return of the repressed. Mathematics may well be akin to zoology: Yet the laws of physics, it is necessary to acknowledge, mention groups, and not elephants. And mathematical theories in physics are strikingly successful. Alone among the sciences, they permit an uncanny epistemological coordination of the past, the present, and the future. If this is not evidence that in some large, some irrefragable sense, mathematical theories apply to the real world, it is difficult to know what better evidence there could be.

That mathematical objects exist is hardly in doubt. What else could be meant by saying that there exists a natural number between three and five? Where they exist is another matter. The mathematical Platonist is often said to assert that mathematical objects exist in a realm beyond space and time, but since this assertion involves a relationship that is itself both spatial and temporal, it is very hard to see how it could be made coherently. The idea that mathematical objects are the free creations of the human mind, as Einstein put it, is hardly an improvement. If the numbers are creations of the human mind, then it follows that without human minds, there are no numbers. In that case, what of the assertion that there is a natural number between three and five? It is true now; but at some time before the appearance of human beings on the earth, it must have been false. The proposition that there exists a natural number between three and five cannot be both true and false, and so it must be essentially indexical, its truth value changing over time. That Napoleon is alive is accordingly true during his life and false before and afterwards. But if the proposition that there exists a natural number between three and five is false at some time in the past, the laws of physics must have been false as well, since the laws of physics appeal directly to the properties of the natural numbers. If the laws of physics were once false, of what use is any physical retrodiction—any claim at all about the distant past? Perhaps then mathematical assertions are such that once true, they are always true? This is a strong claim. On the usual interpretation of modal logics, it means that if P is true, then it is true in every possible world. Possible worlds would seem no less Platonic than the least Platonic of mathematical objects, so the improvement that they confer is not very obvious.

Various accounts of mathematical truth and mathematical knowledge are in conflict. The truths of mathematics make reference to a domain of abstract objects; they are not within space and they are timeless. Contemporary theories of knowledge affirm that human agents can come to know what they know only as the result of a causal flick from the real world. It is empirical knowledge that is causally evoked. Objects that are beyond space and time can have no causal powers.

To the extent that mathematical physics is mathematical, it represents a form of knowledge that is not causally evoked. To the extent that mathematical physics is not causally evoked, it represents a form of knowledge that is not empirical. To the extent that mathematical physics represents a form of knowledge that is not empirical, it follows that the ultimate objects of experience are not physical either.

What, then, are they? As a physical subject matures, its ontology becomes progressively more mathematical, with the real world fading to an insubstantial point, a colored speck, and ultimately disappearing altogether. The objects that provoke a theory are replaced by the enduring objects that sustain the theory. Pedagogy recapitulates ontology. The objects treated *in* classical mechanics, to take a well-known example, are created *by* classical mechanics. Unlike the objects studied in biology, *they* have no antecedent conceptual existence. In V.I. Arnold's elegant tract, for example, a mechanical system of *n* points moving in three-dimensional Euclidean space is defined as a collection of *n* world lines. The world lines constitute a collection of differentiable mappings. Newton's law of motion is expressed as the differential equation $x'' = F(x, x', t)$ [1].

Nothing more.

Mathematics is not applied to the physical world because it is not applied to anything beyond itself. This must mean that as it is studied, the physical world becomes mathematical.

References

1. V.I. Arnold, *Mathematical Methods of Classical Mechanics* (New York: Springer Verlag, 1980), ch. 1.
2. V.I. Arnold, *Ordinary Differential Equations* Cambridge, Massachusetts: The MIT Press, 1978
3. M. Atiyah, *The Geometry and Physics of Knots*. New York: Cambridge University Press, 1990.
4. W. Balzer, C. Moulines and J. Sneed, *An Architectonic for Science* .Dordrecht: D. Reidel, 1987.
5. H. Behnke, F. Bachman, and H. Kunle, editors, *Fundamentals of Mathematics*. Cambridge, Massachusetts: The MIT Press, 1983.
6. H. Boerner, *Representation of Groups*. New York: American Elsevier Publishing Company, 1970.
7. T. Brody, *The Philosophy behind Physics*. New York: Springer-Verlag, 1993.
8. C. Casacuberta and M. Castellet, eds., *Mathematical Research Today and Tomorrow*. New York: Springer-Verlag, 1991.
9. J. Casti, *Alternate Realities*, New York: John Wiley, 1988.

10. R. Courant and D. Hilbert, *Methods of Mathematical Physics*, Volume II. New York: John Wiley, 1972.

11. S. Eilenberg, *Automata, Languages, and Machines*, Volume A (New York: Academic Press, 1974).

12. A. Einstein, *Geometrie und Erfahrugn*, Berlin: Springer 1921.

13. Marceau Feldman, *Le Modèle Géometrique de la Physique* Paris· Masson, 1992.

14. V. Guillemin and S. Sternberg, *Variations on a Theme by Kepler*, Providence, R.I. American Mathematical Society, 1990.

15. R. Larson, R. Hostetler, and B. Edwards, *Calculus*. Lexington, Massachusetts: D.C. Heath & Company, 1990.

16. Charles Parsons, 1986. 'Quine on the Philosophy of Mathematics,' in Edwin Hahn and Paul Arthur Schilpp, *The Philosophy of W.V. Quine*. LaSalle, Illinois: Open Court Press.

17. W.V.O Quine, "Reply to C. Parsons" in E. Hahn and P.A. Schilpp (eds.), *The Philosophy of W.V. Quine*. LaSalle Illinois: Open Court Press, 1986.

18. J. T. Schwartz, 'The Pernicious Influence of Mathematics on Science,' in M. Kac, G.C. Rota, J. T. Schwartz (eds.) , *Discrete Thoughts*. Boston: Birkhäuser Boston, 1992, pp. 19–25.

19. D. Scott and P. Suppes. "Foundational aspects of theories of measurement." *Journal of Symbolic logic* (1958): 113–128.

20. W. Tung, *Group Theory in Physics*. Philadelphia: World Scientific, 1985.

21. B. C. Van Fraassen, *The Scientific Image*. Oxford: The Clarendon Press, 1980.

22. H. Weyl, *The Classical Groups*. Princeton: The Princeton University Press, 1946.

23. H. Weyl, *Symmetry*. Princeton: The Princeton University Press, 1952.

.

Nominalism, the nonexistence of mathematical objects

Jody Azzouni

Abstract Nominalism is the view that, despite appearances, there are no mathematical entities. The ways that nominalism is both compatible with there being mathematical truths and falsehoods, and compatible with mathematical truths being valuable in scientific applications are explored in this paper. Some of the purely psychological reasons for why nominalism is so hard to believe in will also be discussed.

1 The advantages of Platonism

There is an amazingly large number of different positions in the philosophy of mathematics—almost as many as there are professionals who have written on the subject area. But as far as the ontology of mathematics is concerned—as far as the question of whether or not mathematical objects exist—every position boils down to one of two.

(i) Mathematical objects exist.
(ii) Mathematical objects don't exist.

For purely historical reasons, and to adhere to common nomenclature, call Platonism the position that mathematical objects exist. Call nominalism the position that mathematical objects don't exist. Between these two choices, Platonism is the far more natural position to adopt for two reasons. I'll discuss the first reason now, and take up the second one in a couple of paragraphs. The first reason is that the language of mathematics is pretty much the same as that of ordinary language. That is, it shares its grammar and much of its vocabulary with ordinary language. The language of Euclidean geometry, for example, is just ordinary language plus some additional specialized vocabulary that's been refined from ordinary life, for example, *point*, *line*, *triangle*, and so on.

J. Azzouni (✉)
Department of Philosophy, Tufts University, Miner Hall, Medford, MA 02155, USA
e-mail: jodyazzouni@mindspring.com

© Springer International Publishing Switzerland 2015
E. Davis, P.J. Davis (eds.), *Mathematics, Substance and Surmise*,
DOI 10.1007/978-3-319-21473-3_7

The reason that this is relevant to the issue of whether mathematical objects exist or not is that the noun phrases that occur in natural language and that occur in mathematics play exactly the same grammatical and semantic roles. Here is an example:

(iii) There are finitely many grains of sand.
(iv) There are infinitely many prime numbers.

It's very natural to think of (iii) as indicating not only that there are a finite number of grains of sand, but also that these grains of sand *exist*. It seems weird, after all, to think that grains of sand don't exist but there are nevertheless only finitely many of them. In the same way, it's very natural to think of (iv) as indicating not only that there are infinitely many primes, but also that these primes exist as well. Surely (iii) and (iv) are very similar in meaning—so similar in meaning that if one indicates that the objects it is about exist then the other one surely indicates the same thing.

I mentioned that there are two reasons for Platonism being a much more natural position than nominalism. The second is this. (iii) and (iv) are both true. This is something that the ancient Greeks, Plato in particular, were very sensitive to. If something doesn't exist, here's a way to put the point, then how is it possible to say something true about it? For that matter, how is it possible to say something *false* about it? What, after all, would it be that we were saying something true (or false) *about*?

So that's one big issue. It seems that mathematical statements, generally, are true (or false, for example, $2 + 2 = 5$) but this can only be because they are about particular objects. They are true if they describe the mathematical objects they are about correctly ($2 + 2 = 4$) but otherwise they are false ($2 + 2 = 5$). If mathematical statements aren't about mathematical objects then it's hard to see on what grounds we could sort them into the ones that are true and the ones that aren't.

There is a second point to make about this that connects truth to the empirical application of mathematics. Ancient Euclidean geometry already exemplifies this important aspect of mathematics, one that's more dramatically illustrated in contemporary physics, and in many other sciences as well. This is that mathematics is used in scientific applications. The way that mathematical statements being *true* bears on this fact about the scientific application of mathematics is this. Our empirical theories—ones in physics, for example—are usually deeply intertwined with mathematics. These empirical theories, however, are ones that we draw implications from. A major scientific event (for example, the discovery of Bell's theorem) is often the derivation of an important scientific result from a scientific theory, a derivation that involves a great deal of mathematics. But the steps of any derivation have to be truth-preserving. That is, what makes valuable the derivation of a result from a scientific theory is that if the scientific theory is true, then the result derived must be true as well. This truth-preserving nature of scientific inference, however, will be short-circuited if mathematical statements aren't true. For anything can follow from what's false: something false or something true. But this means that we wouldn't know that a result derived from a scientific theory were true if that derivation involved mathematical steps.

Some philosophers have entertained the idea that mathematical statements are part of one or another mathematical game, that there can be something like "truth in a game," and that this is the kind of truth that we have, say, when we use a system of mathematical axioms, either in pure mathematics or in empirical applications. So, instead of saying that $1 + 1 = 2$ is true, we can instead say that $1 + 1 = 2$ is true-in-arithmetic, although we can go on to say that it isn't *actually* true. The suggestion is that this is just like what happens with fictional truth. We can say that Hamlet is the Prince of Denmark is true-in-the-play-Hamlet, although of course it isn't actually true.

The problem with this suggestion is that it forces all the empirical results that follow from our scientific theories using applications of mathematics to yield result that are only true-in-mathematics, and not simply true, as we need all our empirical results to be. Imagine, for example, that D is some scientific theory, one that we take to be true. Let M be some arithmetic that we don't take to be true, but only true-in-arithmetic. Suppose we can derive from M+D some important empirical result C. We aren't licensed to assert that C is true, which is what we need, but only that C is true-in-arithmetic, a much weaker result. Imagine, for example, that we calculate the area of a rug based on its boundaries. It's a peculiar conclusion to draw that results about the rug's area are only true-in-arithmetic and not true outright. The size of a rug being true-in-arithmetic and not true outright sounds just like the tricky sort of thing a dishonest salesperson might say who was trying to sell us a rug that's different in size from what it's been advertised to be.

The upshot. Here are some of the reasons to think that mathematical objects exist and that mathematical statements are about those existing objects. First, mathematical statements are true and false the way that other statements normally are. Second, in general, statements seem to be true and false precisely because they describe accurately (or inaccurately) the properties of the objects they are about. Since mathematical statements are true and false, they must be about the objects they describe accurately (or inaccurately). Finally, mathematical statements are ones that we use in empirical science. This last point is what prevents us from replacing the apparent truth and falsity of mathematical statements with something weaker like truth-in-arithmetic and falsity-in-arithmetic, or the like.

2 Given the above, why would anyone think nominalism was a sensible position?

If mathematical language is just like ordinary language, and presuming the importance of mathematical statements being true (or false) as well as the importance of the empirical applications of mathematics, why would anyone want to be a nominalist? Why would anyone want to deny that mathematical objects exist? The primary reason is that there don't seem to be any good candidates for mathematical objects among the things that we already take to exist. Let's consider Euclidean geometry again, specifically Book 2 of Euclid's *Elements*, which is where Euclid describes the construction of various diagrammatic proofs of results like: the sum of the interior angles of a triangle is one hundred and eighty degrees. A possibility for the ontology of Euclidean geometry is that the mathematical objects, triangles, rectangles, points, lines, and so on, that are proved to have the properties shown in that book are the actual diagrammatic items—or classes of them—that are drawn during the construction of the proofs of the properties of these things.

One problem with this suggestion is that the mathematical objects, points, lines, and so on, are traditionally described in ways that seem to rule out this possibility: points are described as having no dimension, lines are described as having one dimension and as being perfectly straight, and so on. This is obviously false of anything we draw: drawn lines—regardless of how sharp the pencil is—have various thicknesses, and are genuinely fuzzy, for example. Points have (at least) two dimensions, and arguably three, as well as having fairly irregular contours.

The real problem with this suggestion, however, isn't merely that points, lines, triangles, rectangles, and so on, are described in ways that rule out their being drawings of one sort or another. It's that drawings of triangles, lines, and so on, don't have the properties that these items are *proven* in Euclidean geometry to have. A drawing of a triangle, for example, doesn't have angles that sum to exactly one hundred and eighty degrees simply because the lines composing it aren't straight. There are other similar failures. For example, since diagrammatic points have dimensions, various uniqueness results (like any two lines intersect in one and only one unique point) are simply false, unless we attempt to define, for example, a point as the maximal area that two lines overlap on. (Doing this, of course, is going to cause lots of problems elsewhere in Euclidean geometry because, on this kind of view, points will have different sizes relative to the lines they occur in.)

This is one sort of issue we have with finding suitable objects for mathematical statements to be about. The candidate objects usually fail to have the properties that are attributed to mathematical objects by a mathematical discipline. There is a second problem that's more important, however. This is that any candidates for mathematical objects have to be ones that the mathematical statements in question are clearly *about*. To make the issue clear, consider the case of an empirical theory about a certain kind of micro-organism. What's needed to force the sentences of that theory to be about a *specific* kind of micro-organism (and not something else entirely) is more than a lot of theoretical descriptions of that kind of organism. The descriptions provided by a theory—especially an empirical theory—are always incomplete and can apply to more than one kind of thing. This is why we can always empirically discover new facts about the kinds of things that an empirical theory is about, results that go, often, far beyond what the theory itself says about the things in question.

What makes an empirical theory about a certain kind of micro-organism a theory about *that kind* of micro-organism isn't theory at all. It's the facts about how scientists—as it were—make contact with the micro-organisms in question. In other words, it's the instrumental and perceptual facts about how scientists empirically study, say, *specific* micro-organisms. The micro-organisms in question, for example, may be ones that live in the human stomach, and ones that have certain effects on humans that we can measure and study. Or it can be that we take samples (of various types) and we can actually see, using certain instruments, the specific micro-organisms that we're theorizing about. In general, it's this *making-contact* with *particular* micro-organisms and not other micro-organisms (or something else entirely) that forces our sentences to be about *those* micro-organisms. Furthermore, it's this making-contact with these specific micro-organisms that allows our theories about these specific micro-organisms to be wrong. It could be, for example, that the theory in question places them in one microbial family but we discover by instrumentally interacting with them that in fact they belong to a different microbial family. The terms naming the microbes in the theory about them succeed in referring to those microbes despite the theory being false only because of the instrumental and perceptual connections we have to *those* microbes (and not to other microbes).

The problem of applying to *mathematics* this making-contact explanation, for what kinds of entities an empirical theory is about, however, is that there isn't anything like instrumental or perceptual connections to objects that cements what mathematical terms refer to. Mathematics is, as it were, pure theory all the way down. Here's a somewhat idealized picture of how we practice mathematics: we set down a set of axioms and then derive some consequences of those axioms. We could imagine an empirical subject matter being axiomatized in a similar way, an axiomatization of a theory of a kind of cellular structure, for example. In that case, we could write down a set of characterizations of these structures, and then derive results about the structures from those characterizations. But the characterizations have to be localized to particular cellular structures to begin with and not to some other ones precisely because we instrumentally or perceptually fix on those particular structures to initially test our characterizations on. It's this first step that's simply missing from mathematical practice.

So the deep point supporting nominalism is this. There doesn't seem to be a role in mathematical practice for mathematical objects. This absence of a role is in the sense that nothing in how we do mathematics makes contact with mathematical objects in a way that we can understand as forcing mathematical terms to refer to those (specific) objects, and not to some other set of objects.

3   How does the nominalist handle the problem of the truth of mathematical statements?

My job in this paper is to defend nominalism. So I won't consider how the Platonist might try to respond to the challenge to the position that I raised in section 2. I'm going to instead take that challenge as motivating nominalism, and take up the first set of challenges to nominalism that I described in section 1. I'll defend nominalism by describing some of the responses available to nominalists.

Let's start with the first problem I raised for the nominalist. This is the problem that mathematical statements seem to be true, and false, and not merely "true-in-mathematics" and "false-in-mathematics," or something like that. $2 + 2 = 4$, for example, is *simply* true, and $2 + 2 = 5$ is *simply* false. How is this possible if numbers don't exist?

Thinking about other subject areas quickly dispels this concern because a little thought shows that there are *lots* of statements that are true (or false) but are nevertheless about things that we take not to exist. Here is a handful of examples (it will be easy for the reader to think of lots more).

> Hamlet is depicted as the Prince of Denmark in Shakespeare's play *Hamlet*.
> Hamlet is depicted as the Prince of Latvia in Shakespeare's play *Hamlet*.
> James Bond is depicted as a spy for the United Kingdom in Ian Fleming's novels.
> James Bond is depicted as a spy for the Soviet Union in Ian Fleming's novels.

Notice that despite the nonexistence of Hamlet, the first statement is true of Hamlet, although the second isn't. Similarly, despite the nonexistence of James Bond, the first statement about James Bond is true, and the second isn't. Indeed, it's clear that we often speak about nonexistent beings, and say all sorts of true (and false) things about them. Here are a few more examples.

> There are as many prominent Greek gods as there are Greek goddesses.
> Most of the people I dream of at night are far more interesting than real people.
> Scientific history would have been quite different if phlogiston had existed.

In section 1, I mentioned the possibility of truth-in-fiction, that a statement can be true in a fiction although it isn't genuinely true. Notice that none of these examples are ones of truth-in-fiction, or anything similar. To say that Hamlet is the Prince of Denmark is to say something that—strictly speaking—is false, although it can be taken to be true-in-the-play-Hamlet. But this isn't the case with my first example above, "Hamlet is depicted as the Prince of Denmark in Shakespeare's play *Hamlet*." This sentence is simply true: it's making a claim about a fictional character, Hamlet, and the role of that character in one of Shakespeare's plays. It's not instead a fictional statement that's part of the play itself, as "Hamlet is the Prince of Denmark," must be taken. It's easy to see that all of my other examples are the same, although they are also about objects that don't exist. For these sentences, too, are simply true (or false), and not true-in-a-fiction (or false-in-a-fiction), or true-of-a-mythology, or true-of-a-dream, or anything like that.

That so many statements, that are about things that don't exist, are true (and false) raises interesting and important questions in philosophy of language—specifically in the science of semantics. One simple result that follows is this: our ubiquitous practice of talking about the nonexistent shows that the little story I gave in section 1 about how truth and falsity work is wrong. That a statement is true because it describes accurately what it's about, and otherwise a statement is false, are facts that only hold of those statements that are about things that exist. Truth and falsity work in some other way entirely when it comes to things that don't exist. And mathematical statements are true and false in the same way that other statements about things that don't exist are true and false. This is what the nominalist will claim.

I mentioned in section 1 that mathematical statements being true is important because so much scientific work involves the inference of results from scientific theories, and inference has to be truth-preserving. If the mathematical statements, that are used in empirical science, are true—as the nominalist claims—then this important fact about scientific inference doesn't put the Platonist in a better position than the nominalist, as it appeared to in section 1. The Platonist may nevertheless claim that the nominalist still has a problem with the application of mathematics. This is because it seems odd that statements that are about things that don't exist have any useful empirical applications *at all*. Here is a way of posing the problem. If a statement is about something then its usefulness in applications makes sense to begin with. It's about something, and it describes that something accurately. That's what makes its application to that something valuable. But if a statement is only about what doesn't exist, it's hard to see how it could be valuably applied to *anything*. I take up this challenge to nominalism in the next two sections.

4 An example of a Platonist explanation of the value of applied mathematics

What's being asked for is a kind of explanation. We have a subject area, a particular branch of mathematics, say, and the question is why that branch of mathematics is valuable for certain empirical applications. By using the phrase "for certain empirical applications," I'm signaling that I don't intend to give a global explanation for the value of *all* applied mathematics, one that explains *all at once* why mathematics is empirically valuable. I don't intend to do this because I don't think a global explanation for all of successful applied mathematics is possible—or even reasonable to expect. Explanations for why particular branches of mathematics are valuable for specific applications turn on details about the mathematics in question and on details about the specific applications. But the way the challenge to nominalism was posed at the end of the last section makes it hard to see how *any* branch of mathematics can be valuable for any specific application if there are no mathematical objects. So presenting how this works for a specific case will be valuable. Having said this, I should add that I do intend to offer a kind of explanation that is generalizable to more cases than the one I apply it to.

My strategy is to look again at the case of applied Euclidean geometry, but I'll first consider how a Platonist explanation for its value goes. Then, in the following section, I'll investigate whether there is anything in that explanation that requires mathematical objects to actually exist, as the Platonist presumes. To anticipate, I'm going to show that in fact there is nothing about Euclidean objects existing that the Platonist explanation for the value of applied Euclidean geometry turns on. So the nominalist can comfortably borrow the Platonist's explanation for the value of applied Euclidean geometry.

So let's consider Euclidean geometry. The Platonist, recall, will claim that points, lines, and triangles exist, and that this is true of all the items referred to by terms in the language of Euclidean geometry. These things aren't, however, the diagrammatic items that we draw to prove various properties of points, lines, triangles, and so on. Instead, the real points, lines, triangles, and so on, exist somewhere else. Outside of space and time, say. That's one view, originating in Plato perhaps, but still popular today. There is a second option, however, one that's common among philosophers and certain philosophizing physicists. This is to say that Euclidean objects (or, rather, their relativistic/quantum-mechanical/string-theoretical/etc. generalizations) are located in one or another spacetime-like structure. Spacetimes, for example, are taken by many philosophers and physicists to be composed of points. And everything else that can be composed out of points—triangles, of one or another sort, for example—therefore exists in the various spacetimes as well.

Even if a spacetime itself is taken to be composed of some sort of physical item—quantum foam for example—it still turns out that mathematical objects end up appearing as the physically basic units on views like this. For the strings of string theory, for example, live in a (many-dimensional) space, and are themselves composed of coordinate items that are points, at least as far as mathematical purposes are concerned.

Let's return to the simpler case of Euclidean geometry. On any Platonist view of the location of the mathematical objects of Euclidean geometry, it's still true that the *actual objects* that Euclidean geometry is applied to, rugs for example, don't have the properties that mathematical proof shows Euclidean objects to have. How then is Euclidean geometry nevertheless empirically valuable?

The answer, of course, is that although the actual shapes that Euclidean geometry is applied to aren't themselves *Euclidean shapes*—they aren't (for example) bounded by curves of one dimension—they nevertheless approximate Euclidean objects in a way that enables the application of Euclidean geometry to them to be successful. A square rug, for example, isn't *exactly* square. At best it's squarish because its borders are ragged and fuzzy in various ways. Despite this, the deviations from the exactly square can be (roughly) measured, and correspondingly, the deviations of the actual area of the rug from the idealized result Euclidean geometry predicts can be measured as well.

It's worth noting that this approximation story has wide—although probably not universal—application. The reason that Newtonian physics is so valuable, despite its strict falsity, turns on the fact that the slower objects move the closer their physical properties approximate to the predictions of Newtonian physics. So this approximation story goes well beyond providing explanations for applied *mathematics*.

5   How the nominalist can borrow the Platonist's story about the usefulness of applied mathematics

Here is what the nominalist is required to do to borrow the Platonist's story about the usefulness of Euclidean geometry. The value of approximation has to be separated from the notion that what's going on is that the properties of empirical objects (for example, the properties of a rug) approximate the idealized properties of mathematical objects (for example, a square on a plane).

How is this to be managed? Consider a triangle that you've drawn on a piece of paper. What it has to mean (according to the nominalist) that the triangle that you've drawn approximates a Euclidean triangle *can't be* that it resembles a Euclidean triangle. This is because something that's real can't resemble something that doesn't exist. What could that even mean? Things that don't exist don't resemble anything because they *aren't* anything.

What it must mean can be illustrated instead this way: The narrower and straighter you draw actual lines, and the more carefully you measure the interior angles of the actual lines you draw, *the closer the result will be to one hundred and eighty degrees.* This is true about everything you can prove in Euclidean geometry about Euclidean figures: there are ways of drawing *real* figures so that the results of doing so will more and more closely approximate results from Euclidean geometry. That is, we explicitly shift the explanation of the value of Euclidean geometry, that real objects approximate in their properties Euclidean objects, to one in terms of the *theorems* of Euclidean geometry. That is, what we can empirically demonstrate as true of empirical objects can be made to approximate the theorems of Euclidean geometry. We've shifted the grounds of the explanation from ontology to theoremhood. Notice that crucial to this explanation is that the theorems of Euclidean geometry are true. But it's already been demonstrated that the nominalist has access to the true theorems of Euclidean geometry without being forced to take the terms in those theorems to refer to actual objects.

So, as just noted, this explanation for why Euclidean geometry is empirically valuable avoids treating Euclidean objects as ones that exist (in one or another weird way)—in particular, this explanation avoids talking about nonexistent entities resembling real things. Plato, by contrast, hypothesized that circles on Earth resembled "perfect" circles—the ones studied in Euclidean geometry. This would also explain why Euclidean geometry is empirically valuable, as I indicated in the last section. If things in domain B resemble things in domain A, then studying the things in domain A will shed light about the things in domain B. But an explanation like that requires the objects in both domains to *exist*. I've given a different explanation: it doesn't focus on the objects of Euclidean geometry at all—which is a good thing, the nominalist thinks, since those objects don't exist. It focuses instead on the truths of Euclidean geometry. It focuses on the sentences of Euclidean geometry and not on what the sentences are supposedly about. And it explains why the *truths* about the objects we draw on (relatively flat) pieces of paper or in sand, or whatever, are going to approximate the *truths* of Euclidean geometry under certain circumstances.

Notice what has been done here. We're starting, say, with a drawing of a triangle on a piece of paper that it's empirically profitable to analyze the properties of in terms of Euclidean geometry. This triangle, however, *isn't* a Euclidean triangle precisely because it's a drawing on a piece of paper, and so it doesn't have the standard properties that Euclidean triangles are taken to have. The question is: why is it valuable to treat the drawing, when empirically applying Euclidean geometry to the triangle, *as if* it is a Euclidean triangle? The Platonic answer isn't that the drawing of a triangle *is* a Euclidean triangle because everyone can agree that it *isn't*. The Platonic answer is that Euclidean triangles exist, and the drawing of a triangle sufficiently resembles Euclidean triangles so that treating the drawing *as if* it's a Euclidean triangle is valuable. My alternative answer is this: Consider a truth of Euclidean geometry: The sum of the interior angles of a triangle sums to one hundred and eighty degrees. We can apply this theorem to a drawing, by treating its angles as if they are Euclidean angles (bounded by straight lines), and the result will more closely approximate the answer: one hundred and eighty degrees, the sharper and straighter we draw the boundaries of the triangle.

So far, in this paper, I've focused on what might be called the real obstacles to adopting nominalism. These are issues about the nature of applied mathematics, and the nature of mathematical truth. If the nominalist can't make sense of mathematical truth or of the success of applied mathematics, on a view of mathematical objects not existing, then that's a genuine deficit for nominalism—perhaps a fatality for the position. I'm going to turn now to what might be described as a softer consideration against nominalism. This is that the way that we think about mathematics seems to require that we think about mathematical objects of some sort—numbers, functions, spaces, and so on. The question, therefore, is whether the psychological need to think about mathematics in terms of kinds of objects is any reason to think that nominalism must be wrong. The question is whether the fact, if it is a fact, that we have to think through our mathematics in terms of mathematical objects requires that mathematical objects exist.

6   How we think about nonexistent objects

As before, it pays off to initially discuss a nonmathematical area of discourse where we think about objects that we antecedently recognize not to exist. When we are fantasizing, making up stories, dreaming, or otherwise thinking up or thinking about imaginary beings, for example, our psychological methods of thinking about real objects is the only mental tool we have to manage this.

The mental tool we use is an interesting capacity for "object-directed thinking," one that we can detach the operations of from the real things that we normally think about, and that we can continue to use *even though we recognize that there are no objects involved any longer*. And, along with this, we can also borrow our ordinary ways of talking "about" objects when fantasizing (for example). That's why we can successfully speak of "thinking about" hobbits or elves or dragons, even though we know—or most of us do, anyway—that there are no hobbits or elves or dragons.

Strictly speaking, to think about a hobbit isn't any different from thinking about a dragon. That is to say, as far as the hobbit and the dragon are concerned, because there are no such things, thinking about one isn't any different from thinking about the other. In both cases—strictly speaking—*we are thinking about nothing at all.* And nothing at all isn't different from nothing at all. In the same way, to think about a hobbit isn't any different—strictly speaking—from thinking about two hobbits. For one nothing isn't different from two nothings. Nevertheless, because we import the ways we think about real objects to how we think about what doesn't exist, we experience thinking about a hobbit *as thinking about a different thing from thinking about a dragon.* Notice how strange this is. After all, if there are no hobbits and no dragons, then to think about a hobbit isn't to think about a different thing than to think about a dragon *because no things are involved at all.*

Nevertheless, we *experience* our talking about (or thinking about) *hobbits* as different from our thinking about *dragons*—even though there is nothing we are thinking about (or talking about) in both cases. This is reasonably described as a cognitive illusion, in particular, as an aboutness illusion. When thinking about something we recognize not to exist, such as a fantasy character, despite our recognition that it doesn't exist, we nevertheless experience that thinking as thinking *about* a particular thing.

What's really going on when we are thinking about dragons and hobbits? Well, there are different words, and different concepts that we use. Strictly speaking (since there are no hobbits or dragons), there are only thought vehicles and language vehicles involved, and nothing else. I just spoke of aboutness illusions. Aboutness illusions misfocus our minds. Consider the two sentences.

(v) Pegasus was believed to be a flying horse.
(vi) Hercules was believed to be a flying horse.

Instead of our feeling the difference between talking about Pegasus, as we do when we utter (v), and talking about Hercules, as we do when we utter (vi), as a matter of differences in the truth values of *sentences* in which the words "Pegasus" and "Hercules" appear, we instead feel the difference as due to there being different *objects* we are talking about that have different properties. This feeling occurs simultaneously with the awareness that there are no objects such as Pegasus and Hercules—that they are complete *fictions.*

What makes (v) and (vi) differ in truth value, of course, isn't that they are about different objects with different properties. What makes them differ in truth value is that they allude to a mythological tradition in which certain beliefs (about Hercules and Pegasus, in particular) circulated. Objects don't come into it at all; thinking they do—even for a moment—is to fall prey to aboutness illusions.

Unfortunately, telling ourselves (or others) all this stuff about *aboutness illusions* won't eliminate our experience of these illusions that arises whenever we "transact" with the nonexistent. These illusions are like optical illusions: No matter how much we stare and stare at an optical illusion, we can't make it go away just by saying to ourselves (for example): "I know those lines are the same length even though they appear not to be." Here too, we'll always have the overwhelming cognitive impulse to experience our thinking "about" Pegasus and our thinking "about" Hercules as kinds of thinking about objects—and objects that are different. Even when we know it isn't true. There is no way to escape these aboutness illusions. Not for us. Not for us humans.

It seems that the above yields a promissory note. This is that I've described us as thinking about what doesn't exist by psychologically borrowing the object-directed mental tool that we use when we think about real objects. But is there any other way to manage this? I've already suggested a way that it could be done, although not by us. Consider the following two statements.

> Hobbits are depicted in the *Lord of the Rings* as being very short.
> Wizards are depicted in the *Lord of the Rings* as being very short.

These statements have different truth values, the first is true and the second is false. We humans naturally think about these two sorts of sentences as respectively about two kinds of objects, hobbits and wizards. The truth of the first sentence is due to the fact that hobbits are indeed depicted as quite short in the *Lord of the Rings*. The falsity of the second sentence is due to the fact that wizards, instead, are depicted as quite tall in the *Lord of the Rings*. So we naturally think of one sort of object—hobbits—being depicted as short, whereas wizards—another kind of object—are depicted as being tall. But we could have been a different kind of creature, one that never thinks about objects when it recognizes that terms, like "hobbit" and "wizard," don't refer to anything. Instead, this creature would instead automatically think about the sentences of a novel, and that certain sentences appear in that novel and not other sentences. If such a creature understood "depiction" in a novel, for example, simply as the appearance of certain sentences in that novel, then that creature might think that "Hobbits are depicted in the *Lord of the Rings* as being very short," means only that certain sentences appear in the novel, such as "The average hobbit is about four feet." That creature might find it peculiar to think of hobbits as any sort of object at all, as opposed to there just being a term "hobbit" that appears in sentences with specific truth values.

I should make this point as well. Novelists, when writing novels, and the rest of us, when reading novels, almost always *read through* the sentences of the novel to what the sentences depict. That is, we always read novels by employing the object-directed thinking that arises because the sentences of the novel make us think of characters and their properties—and often in great detail. But this should not suggest to anyone that fictional characters therefore exist. Rather, it should be seen the way I've presented it in this section. We engage in a complex psychological process of thinking of objects that we simultaneously recognize don't exist. It isn't that because we must think of the characters of a novel as having specific properties (and not other properties) that we should therefore think that these characters must have some sort of existence. This would exactly be the sort of peculiar mistake that, usually, only philosophers make.

I conclude this paper, in the next section, by drawing on behalf of the nominalist the same conclusion about the mathematical objects that mathematicians (and others) think of when doing mathematics.

7 Object-directed thinking in mathematics

Broadly speaking (and with a bit of inaccuracy), mathematical thinking comes in two forms, what we might describe as geometric thinking and as algebraic thinking. By algebraic thinking, I mean a kind of "object-blind" manipulation that we often engage in when maneuvering formulas into a suitable form to extract certain results from them. Although we do think of numerals as names for numbers, it's often the case that complex numerical manipulations are similarly "object-blind," especially when "short-cut" algorithms, for example, for multiplication and division, are employed. Manipulating code according to rules is yet a third similar sort of experience. In all these cases, we aren't thinking of mathematical objects that we conceive of in a certain way but instead, and more directly, we are thinking of the various kinds of formulas we are writing down on a piece of paper (or manipulating on a computer screen).

The second kind of mathematical thinking—broadly speaking—is geometric. In this case, we *are* thinking of the mathematical language we are manipulating as diagrammatically depicting the mathematical objects we are attempting to prove results about. This is the case, of course, with Euclidean geometry; but this kind of thinking shows up with conceptual proofs about, say, manifolds, or the diagram-chasing that takes place in topology or knot theory.

There are mixed cases of course. And, it might even be argued that attention to the "form" of a discourse, as we do with the mechanical manipulation of mathematical formulas (according to rules) is reflected somewhat in fiction writing, especially with those authors who are concerned with the formal properties of their sentences—such as rhythm, for example.

Informal-rigorous mathematical proof almost always has conceptual elements that induce object-directed thinking. I mean by this that there is always an aspect to it where it is most natural for us to be thinking of a kind of object that we are trying to prove has certain properties or other. The most familiar case of this, of course, is number theory. But it's important that informal-rigorous mathematical proofs can in principle be replaced by formal derivations, where each step follows mechanically from previous steps, and all the assumptions—implicit and explicit—employed in such a proof can be axiomatized. This shows, as I indicated in section 2, that objects play no essential role in mathematical practice. The fact that we think about mathematical objects the way we do (and when we do) is an interesting psychological fact that's very important. It's just not very important for *ontology*.

An Aristotelian approach to mathematical ontology

Donald Gillies

Abstract The paper begins with an exposition of Aristotle's own philosophy of mathematics. It is claimed that this is based on two postulates. The first is the embodiment postulate, which states that mathematical objects exist not in a separate world, but embodied in the material world. The second is that infinity is always potential and never actual. It is argued that Aristotle's philosophy gave an adequate account of ancient Greek mathematics; but that his second postulate does not apply to modern mathematics, which assumes the existence of the actual infinite. However, it is claimed that the embodiment postulate does still hold in contemporary mathematics, and this is argued in detail by considering the natural numbers and the sets of ZFC.

1 Introduction. The Problem of Mathematical Ontology

The problem of mathematical ontology is the problem of whether mathematical entities exist, and, if so, how they exist. It can be illustrated by considering the example of the natural numbers, i.e. 1, 2, 3, ... , n, ... Do such numbers exist? Simple considerations suggest that we should recognise their existence. For example, anyone, if asked whether there are numbers between 4 and 7, would reply: "Yes, there are, namely 5 and 6". But, if there are numbers between 4 and 7, it follows logically that there are numbers, i.e. that numbers exist. On the other hand, admitting that numbers exist often gives rise to feelings of unease, since numbers appear to be curious shadowy entities very different from familiar everyday objects such as trees and boulders, tables and chairs, cats and dogs, other people, etc.

These feelings of unease have led some philosophers to argue that numbers do not after all exist, since talk of numbers is always reducible to talk of more

D. Gillies (✉)
Department of Science and Technology Studies, University College London,
Gower Street, London WC1 6BT, UK
e-mail: donald.gillies@ucl.ac.uk

© Springer International Publishing Switzerland 2015   147
E. Davis, P.J. Davis (eds.), *Mathematics, Substance and Surmise*,
DOI 10.1007/978-3-319-21473-3_8

familiar things. This view is often called *reductionism*,[1] and is perhaps most easily explained by considering statements about the average Englishman. Let us suppose it is true that p, where p = the average Englishman has 1.87 children. It is reasonable to suppose that every true statement, such as p, apparently about the average Englishman, is reducible to a logically equivalent statement in which no mention is made of the average Englishman. For example, we can construct a statement (call it p*) logically equivalent to p as follows. Let n = the number of Englishmen, and m = the number of children of Englishmen, then define p* as m/n = 1.87. The reductionist argues that accepting p as a true statement does not commit us to accepting the existence of that shadowy pseudo-entity, the average Englishman with his curious family, for p is reducible to a logically equivalent statement p* which does not mention the average Englishman. Similarly the reductionist would claim that statements apparently about numbers are reducible to logically equivalent statements, which do not mention numbers. Thus we do not have to accept that numbers really exist, and these curious shadowy entities fortunately disappear.

Of course it is not so easy to establish reductionism regarding the natural numbers, as it is to establish reductionism regarding the average Englishman. Can we really, for every statement about numbers, construct a logically equivalent statement, which does not mention numbers? And, if we can, do these logically equivalent statements mention other entities, which are just as suspect as numbers? Despite these difficulties, there have been several ingenious attempts to carry out the reductionist programme regarding numbers. However, the degree of their success remains controversial.

The ancient Greeks are noted for their studies of mathematics and philosophy. So we would expect them to take an interest in the problem of mathematical ontology, and this is indeed the case. Plato's best-known theory is his postulation of a world of forms, and he introduced this theory partly in order to deal with the problem of mathematical ontology. His most famous account of the world of forms is in *The Republic* [27]. According to Plato this world consists of objective ideas which have an eternal existence outside space and time, and which cannot be perceived by any of the usual five bodily senses. Our minds, however, after a long and rigorous intellectual training in philosophy and other subjects, can gain knowledge of the world of forms, which Plato therefore calls the intelligible world. Plato mentions as part of his intelligible world, the objects of arithmetic and geometry, namely (*Republic*, VI, 510): "odd and even numbers, or the various figures and the three kinds of angle". Mathematical statements are true on Plato's account, if they correspond to what is really the case in the world of forms.

---

[1] Actually reductionism is only one form of *nominalism*, which is the general view that mathematical entities do not exist. Nominalism is defended in Azzouni [3]. It is often connected to the so-called *fictionalist* view according to which mathematics is similar to literary fictions such as Shakespeare's *Hamlet*. Azzouni discusses this view in his 2015, and it is also discussed in section 7 of the present paper.

Platonism is not confined to the ancient world, but has had its adherents in more recent times. Two very distinguished modern philosophers of mathematics who have adopted a Platonic position are Frege and Gödel. Yet the majority of modern philosophers have felt uneasy about Platonism. The postulation of a mysterious transcendental world of objective ideas outside space and time has seemed to them too strange and metaphysical to be acceptable. The doubts of such philosophers must have arisen in ancient Greek times as well, because Plato's own most brilliant student, Aristotle, made an attempt to bring the world of forms back to the material world.

Aristotle did not adopt the nominalist position, which has been developed in more recent times. He agreed with Plato that both the forms and mathematical objects had an objective existence. However, Aristotle did not agree with Plato that the forms and mathematical objects existed in a transcendental world outside of space and time. He argued that they did exist, but in the familiar material world.

I will give a more detailed account of Aristotle's views on mathematical ontology in the next section, but they can be summarised, in my words rather than his, as follows. Mathematical objects and other abstract entities do indeed exist objectively, but not, as Plato claimed, in a transcendental world which is separate from the material world. They exist rather embodied in the material world. For example (an example of mine rather than Aristotle's), the number five exists embodied in quintuples such as the toes of my left foot. Wittgenstein gives a very elegant illustration of this position, writing [32, p. 29]:

> "A man may sing a song with expression and without expression. Then why not leave out the song – could you have the expression then?"

Wittgenstein's point is that the expression cannot exist without the song. It has to be embodied in the way the song is actually sung. This is a particular instance of the general Aristotelian thesis concerning abstract entities.

Aristotelianism in modern philosophy of mathematics has not been such a popular approach to mathematical ontology as either Platonism or reductionism. However, there have been some adherents of this position. Franklin's 2014 [16] book is a striking recent example. My own interest in Aristotelianism in mathematics began in the mid-1980s. I was then studying the views of Maddy [22] and Chihara's criticisms of them [12]. At the same time, I was reading a good deal of Aristotle in order to prepare a lecture course, and it occurred to me that Aristotle's approach might help to develop the positions of Maddy and Chihara. I have written various papers on this theme over the years, and I hope in the present paper to put all this work together in a more complete and finished form.

There is, however, a difficulty concerning the application of Aristotelianism to contemporary mathematics. Aristotle's analysis of the problem of mathematical ontology was undertaken for ancient Greek mathematics, and this is very different from contemporary mathematics. Many points in Aristotle's account, most notably his discussion of the nature of infinity, were quite appropriate for ancient Greek mathematics, but are contradicted by the mainstream assumptions of current mathematics. Still, it seems to me that the general Aristotelian approach, as just characterised, can be applied to contemporary mathematics.

To deal with this issue, the plan of the paper is as follows. In the next section (2), I will give a detailed account of Aristotle's own views on mathematical ontology, and will argue that they were quite appropriate for ancient Greek mathematics. Then in section 3, I will examine the important respects in which contemporary mathematics differs from ancient Greek mathematics, and the changes in the Aristotelian position, which these differences make necessary. The rest of the paper (sections 4–7) will then develop a detailed account of an Aristotelianism suitably modified for current mathematics.

2 Aristotle on Mathematical Ontology

Aristotle's views on this question were developed from a criticism of Plato. He states Plato's position on mathematical objects as follows (*Metaphysics*, I.6, 987^b 14–18):

> "Further, besides sensible things and Forms he (i.e. Plato – D.G.) says there are the objects of mathematics, which occupy an intermediate position, differing from sensible things in being eternal and unchangeable, from Forms in that there are many alike, while the Form itself is in each case unique."

At first this passage is puzzling because, as we argued earlier, Plato in *The Republic* [27] seems to locate mathematical objects in the world of forms, and not as intermediates. However, Aristotle supplies an argument, which explains what is going on here. Let us consider the equation $2 + 2 = 4$. Here we find two 2 s which are added together. However, there is only one form of two. This is what Annas calls *the Uniqueness Problem*, and she illustrates it with a geometrical example. Annas [1, p. 25]:

> "The real argument for geometrical intermediates is … the Uniqueness Problem. A theorem mentions two circles intersecting."

However there can be only one form of *the* Circle. Indeed in *The Republic*, Plato speaks (VI, 510) [27] of "*the* Square and *the* Diagonal". It thus seems likely that Aristotle gives a correct account of Plato's later views. The Uniqueness Problem could have come to light in discussions in the Academy, and this may have led Plato to change his mind regarding mathematical objects. Later in the paper, I will discuss the Uniqueness Problem briefly from a modern point of view.

In fact if Plato really changed to the view that mathematical objects are intermediates, this would have been a development of ideas already expressed in *The Republic* [27]. In the section on the divided line, Plato locates mathematical objects in the world of forms, but nonetheless stresses that they are in that part of the world of forms, which is closest to the visible (or sensible) world. This is because mathematicians make use of objects in the visible world such as groups of pebbles or diagrams when thinking about mathematical objects. As Plato says (VI, 510):

"The diagrams they draw and the models they make are actual things … but now they serve as images, while the student is seeking to behold those realities which only thought can apprehend."

Aristotle criticises Plato's theory of forms and his theory of mathematical objects in the same way. This common criticism is that abstract entities do not exist separately from, but rather in, the material world. I will first give two passages in which this approach is applied to forms. Let us start with *Metaphysics*, M(XIII).2, 1077^b 4–8 [5]:

"For if there are no attributes over and above real objects (e.g. a moving or a white) then white is prior in definition to white man, but not in reality, since it cannot exist separately but only together with the compound (by compound I mean the white man)."

I take Aristotle here as accepting the premise that "there are no attributes over and above real objects"; so there is no entity 'movement' or 'a moving' apart from moving bodies. Similarly whiteness cannot exist separately, but only in compounds such as a white man, or a white snowball.

This point of view is further expounded in the following passage (*Metaphysics* M(XIII).3, 1077^b 23–27) [5]:

"… there are many statements about things merely as moving, apart from the nature of each such thing and their incidental properties (and this does not mean that there has to be either some moving object separate from the perceptible objects, or some such entity marked off in them) …"

The point here is that we can consider a body *qua* moving without having to postulate some abstract entity 'a moving' separate from the body.

Let us now turn from forms to mathematical objects. First of all it will be useful to examine what Aristotle meant by 'mathematical object'. He is very clear about this saying (*Metaphysics*, M(XIII).1, 1076^a 18–19) [5] that by "mathematical object … I mean numbers, lines, and things of that kind". Aristotle did not recognise the existence of negative, rational, real or complex numbers. So by 'number' he means more or less what we have called 'natural numbers'. However, there is a slight difference between his numbers and our natural numbers, i.e. 1, 2, 3, …, n, …. The difference is that Aristotle, in common with many ancient Greek thinkers, did not regard '1' as a number. As he says (*Physics*, IV.12, 220^a 27) [6]: "The smallest number, in the strict sense of the word 'number', is two". The reason for this is explained at *Metaphysics*, X.1, 1053^a 30, where he says: "number is a plurality of units". The unit itself, i.e. 1, is not a number, or to put it another way (*Metaphysics*, N(XIV).1, 1088^a 4–6) [5]. "one means a measure of some plurality, and number means a measured plurality … Thus there is good reason for one not to be a number". Of course, the ancient Greeks did not have 0 as a number either, although today, it is just as acceptable to begin the natural numbers with 0 as with 1.

This clarifies what Aristotle meant by numbers. The next point is that he definitely regarded numbers as existing. Indeed he says (*Metaphysics*, M(XIII).3, 1077^b 31–33) [5]:

"So since it is true to say without qualification not only that separable things exist but also that non-separable things exist (e.g. that moving things exist), it is also true to say without qualification that mathematical objects exist ..."

Indeed, speaking of mathematical objects, he says (*Metaphysics* M(XIII).1, 1076^a 36) [5]:

"So our debate will be not whether they exist, but in what way they exist."

So in what way do mathematical objects exist? Aristotle sometimes formulates his view by saying that such objects are not separate from things, which exist in the perceptible world. He explains this with an analogy to male and female in the following passage (*Metaphysics*, M(XIII).3, 1078^a 5–9) [5]:

"Many properties hold true of things in their own right as being, each of them, of a certain type – for instance there are attributes peculiar to animals as being male or as being female (yet there is no female or male separate from animals). So there are properties holding true of things merely as lengths or as planes."

In some places Aristotle appears to contradict himself on this point, since he speaks of mathematicians separating mathematical objects, which were supposed to be not separate. Thus he writes (*Physics*, II.2, 193^b 24–194^a 10) [6]:

"The next point to consider is how the mathematician differs from the physicist. Obviously physical bodies contain surfaces and volumes, lines and points, and these are the subject-matter of mathematics. ... Now the mathematician, though he too treats of these things, nevertheless does not treat of them as the limits of a physical body; nor does he consider the attributes indicated as the attributes of such bodies. That is why he separates them; for in thought they are separable from motion, and it makes no difference, nor does any falsity result, if they are separated. ... geometry investigates physical lines but not *qua* physical ..."

However, it is obvious that there is no contradiction here, since Aristotle is clear that mathematical objects are separable in thought but not in reality. Still he speaks rather confusingly of separating what is not separate in the following passage which is a good illustration of his general position (*Metaphysics*, M(XIII).3, 1078^a 21–30) [5]:

"The best way of studying each object would be this: to separate and posit what is not separate, as the arithmetician does, and the geometer. A man is one and indivisible as a man, and the arithmetician posits him as one indivisible, then studies what is incidental to a man as indivisible; the geometer, on the other hand, studies him neither as man nor as indivisible, but as a solid object. ... That is why geometers speak correctly: they talk about existing things and they really do exist ..."

Suppose an ancient Greek arithmetician is studying men. He might want, for example, to count the number of men taking part in a running race. Each man is regarded as one and indivisible, or, to put it another way, the unit used to measure the plurality is a whole man. With this unit, the number taking part in the running race might be 6. A geometer might also study a particular runner *qua* solid object. He might measure the proportions of the runner's body to see whether they corresponded to the proportions of ideal beauty, whether the runner *qua* solid object could be fitted into a circle and square, etc.

Aristotle summarises his position by claiming that mathematical objects really exist, but they do not exist in a separate realm, as Plato claimed. They cannot be separated from familiar material objects, or, rather, they can only be separated from such objects in thought but not in reality. Using a slightly different terminology from Aristotle himself, I will refer to this as the *embodiment postulate*, which is the postulate that mathematical objects exist not in a separate world, but embodied in the material world.

Now there is a difficulty with the embodiment postulate, which Aristotle himself recognised. Numbers do appear to be exactly embodied in the material world. For example, the number of toes on my left foot is exactly 5. Provided the ancient Greek arithmetician had counted correctly, the number of runners in the race would have been exactly 6, and so on. If we turn to geometrical entities, however, it seems that they are embodied in the material world approximately but not exactly. Thus, for example, a stretched string is approximately, but not exactly, a geometrical straight line. Aristotle gives a number of examples of this type in *Metaphysics*, III.2, 997^b 35–998^a 6:

> "neither are perceptible lines such lines as the geometer speaks of (for no perceptible thing is straight or round in the way in which he defines 'straight' or 'round'; for a hoop touches a straight edge not at a point, but as Protagoras used to say it did, in his refutation of the geometers), nor are the movements and spiral orbits in the heavens like those of which astronomy treats, nor have geometrical points the same nature as the actual stars."

All this is very convincing. Hoops or wheels are approximately circles, and a ruler is approximately a straight edge. If this were exact, then the ruler would touch the hoop or the wheel in one and only one point. Obviously this is not the case. As regards the orbits of the planets, we can consider the modern position. The orbit of a planet, such as Mars, is a very good approximation to an ellipse, but it is not exactly an ellipse. There are small perturbations in the orbit, owing to the gravitational attractions of the other planets of the solar system. In many astronomical calculations, stars can be treated as geometrical points, but they are obviously not precisely points. I will call this problem the *approximation problem*. It obviously raises a problem for Aristotelianism. If geometrical entities are only approximately embodied in material things, this suggests that they may, after all, really exist in a separate Platonic realm.[2]

That concludes my account of Aristotle's embodiment postulate. There is, however, another important part of his philosophy of mathematics, which is his treatment of infinity. I will now give an account of this. Aristotle's general position regarding the infinite is that it has a potential existence but not an actual existence. He states this very clearly in the following passage (*Physics*, III.6, 206^a 17–21) [6]:

> "the infinite has a potential existence.

---

[2] Andrew Gregory stressed the importance of the approximation problem to me in conversation. It is one of the factors, which have led him to prefer Platonism to Aristotelianism regarding abstract entities. The approximation problem is also mentioned by Azzouni in section 2 of his 2015. Azzouni regards it as a strong argument in favour of nominalism.

But the phrase 'potential existence' is ambiguous. When we speak of the potential existence of a statue we mean that there will be an actual statue. It is not so with the infinite. There will not be an actual infinite."

Aristotle clarifies this further by giving a definition of the infinite (*Physics*, III.6, 207^a 6–8) [6].

"Our definition is then as follows:
   A quantity is infinite if it is such that we can always take a part outside what has been already taken."

Aristotle illustrates this definition by applying it to the infinity of numbers and to the infinity of magnitude, i.e. the continuum. He distinguishes the infinite by addition from the infinite by division. Numbers are infinite by addition, but not by division (*Physics*, III.7, 207^b 2–14) [6]:

"in number there is a limit in the direction of the minimum, and ... in the other direction every assigned number is surpassed. ... The reason is that what is one is indivisible ... Hence number must stop at the indivisible ... But in the direction of largeness it is always possible to think of a larger number ... Hence this infinite is potential, never actual ... its infinity is not a permanent actuality but consists in a process of coming to be ... "

The numbers are infinite by addition, because, given any number however large, it is always possible to produce a larger number by adding one. They are not, however, infinite by division, since one, the unit, is taken in arithmetic to be indivisible.

We might expect magnitude or the continuum to be infinite both by addition and division. For consider a finite line segment AB. We can always extend AB to C, where AC is longer than AB. This would seem to give the infinite by addition. Similarly if we divide AB by a point C between them so that AC and CB are both shorter than AB, we do have to stop at that point, but can divide AC by a point C' between them so that AC' and C'C are both shorter than AC. This makes AB infinite by division. Surprisingly, however, Aristotle claims that magnitude is infinite by division, but not by addition and thus is the opposite of numbers. He writes (*Physics*, III.7, 207^b 15–21) [6]:

"With magnitudes the contrary holds. What is continuous is divided *ad infinitum*, but there is no infinite in the direction of increase. ... it is impossible to exceed every assigned magnitude; for if it were possible there would be something bigger than the heavens."

This shows where the problem lies. For Aristotle the Universe is finite and bounded by the sphere of the fixed stars. Suppose a geometrical proof required extending a finite line segment AB to a point C such that AB = BC. If AB where so large that 2AB were bigger than the diameter of the Universe, this extension would not be possible, because C would have to lie outside the Universe. At first sight this seems to create a problem for mathematicians, who, in the course of their proofs do want to extend a line as much as they like. However, Aristotle replies that this apparent difficulty can always be overcome by shrinking the original diagram to a size where the required extension of the line becomes compatible with the dimensions of the Universe. As he says (*Physics*, III.7, 207^b 27–34) [6]:

"Our account does not rob the mathematicians of their science ... They postulate only that the finite straight line may be produced as far as they wish. It is possible to have divided in the same ratio as the largest quantity another magnitude of any size you like. Hence, for the purposes of proof, it will make no difference to them to have such an infinite instead, while its existence will be in the sphere of real magnitudes."

Aristotle's notion of the potential infinite does indeed seem to be adequate for ancient Greek mathematics, even for the developments of Euclidean geometry and Diophantine equations which occurred after his lifetime. Arithmeticians required numbers as high they pleased, but did not need to consider the sequence of numbers as a completed whole. Geometers in their proofs needed to extend lines as much as they wanted, and to divide them as much as they wanted, but they never needed to postulate infinitely long lines. Aristotle's potential infinity would have fitted ancient Greek geometry better if he had been prepared to allow that the Universe was potentially infinite. However, his device of shrinking the diagram made his potential infinities adequate for the geometers. It might be thought that the actual infinite could appear in ancient Greek geometry if it was allowed that a line is composed of points. However, Aristotle denied explicitly that this was the case (*Physics*, VI.1, 231ᵃ 25–26) [6]:

"a line cannot be composed of points, the line being continuous and the point indivisible."

There is, however, one seeming exception to the claim that the actual infinity does not occur in ancient Greek mathematics.[3] This occurs in Archimedes' treatise on *The Method*. In his proof of Proposition 1, Archimedes considers a triangle CFA. He takes an arbitrary point O on AC, and draws a line MO parallel to AF and meeting CF in M. He then says [4, p. 572]: "the triangle CFA is made up of all the parallel lines like MO". Aristotle had denied that a line can be composed of points, but here Archimedes gives a two-dimensional version of the same principle by claiming that a triangle is composed of parallel lines. All the parallel lines like MO which make up the triangle CFA constitute an actual infinity which here appears in a work of ancient Greek mathematics.

However, Archimedes says at the end of his proof of Proposition 1 (p. 572):

"Now the fact here stated is not actually demonstrated by the argument used; but that argument has given a sort of indication that the conclusion is true. Seeing then that the theorem is not demonstrated, but at the same time suspecting that the conclusion is true, we shall have recourse to the geometrical demonstration which I myself discovered and have already published."

Archimedes' treatise on *The Method* was considered by him as a work of heuristics rather than genuine mathematical demonstration. Its results indicated that some theorems might be true, but did not give genuine proofs of them. So, although Archimedes did indeed consider the actual infinite, it was only as a heuristic device, and not as part of legitimate mathematics. Thus our conclusion that ancient Greek mathematics did not use the actual infinite remains true, despite this example from

---

[3]This was pointed out to me by Silvio Maracchia and Anne Newstead. See Maracchia [25].

Archimedes. Archimedes use of the actual infinite in this particular context is the exception, which proves the rule.

If we turn to contemporary mathematics, however, the picture changes. As I will show in the next section (3), contemporary mathematicians work within a framework of assumptions, which imply that the actual infinite exists.[4] So, for contemporary mathematics, one of the postulates of Aristotle's philosophy of mathematics, namely the postulate that there exists only the potential, but not the actual, infinite, must be dropped. We can, however, still hold on to his other postulate—the embodiment postulate. This, however, has to be applied in a different, and, in some respects, more complicated way. It has to be shown, for example, that some actually infinite sets are embodied in the material world. This is no easy matter for sets of very high cardinality, such as we find in Cantor's theory of the transfinite. As a preliminary to examining these problems in detail, we need to explain how contemporary mathematics differs from ancient Greek mathematics, and this will be done in the next section.

3 How Contemporary Mathematics differs from Ancient Greek Mathematics

As we have seen, when Plato and Aristotle examined the question of mathematical ontology, they considered, as mathematical objects, numbers and geometrical entities, such as points, lines, figures, etc. If we want to examine the ontology of contemporary mathematics, what should we consider as the relevant mathematical objects? This is a difficult question, since contemporary mathematics is much more complicated than ancient Greek mathematics and makes reference to many curious objects. However, we can cover most, though perhaps not all, objects of contemporary mathematics by limiting ourselves to just two kinds of object. These are: (i) *the natural numbers*, which are more or less the same as the numbers of Plato and Aristotle, except that we have 0 and 1 at the beginning of the sequence; and (ii) *sets* in the sense characterised by one of the accepted systems of axiomatic set theory. There are in fact two leading systems of axiomatic set theory, namely ZSF, named after its developers Zermelo, Skolem and Fraenkel, and NBG, named

---

[4]Exactly when the actual infinity entered mathematics in a significant way is an interesting but difficult historical question. The years between 1500 and 1800 saw the rise of algebra, analytic geometry and calculus. Did these developments introduce the actual infinite? Ladislav Kvasz has suggested plausibly (personal communication) that the actual infinite is already in Descartes. Jeremy Gray's contribution to the present volume has some points relevant to this question. Gray says [20]: "At no stage did Apollonius, or any other Greek geometer, generalise a construction by speaking of points at infinity". However, he goes on to show that points at infinity were introduced by Desargues in 1639. Perhaps this could be considered as an example of the actual infinity, though this is not clear. In what follows, however, I will not discuss this question in detail, but limit myself to the developments from the 1860s, which gave rise to contemporary mathematics.

after its developers von Neumann, Bernays and Gödel. ZSF and NBG are more or less equivalent, but ZSF is perhaps slightly simpler, and, maybe for this reason, seems to be the one, which is more generally used nowadays. If we add to ZSF the axiom of choice, which again is usually done nowadays, we get the axiom system ZSFC, which is standardly abbreviated to ZFC. ZFC is the most commonly adopted version of axiomatic set theory at present, and there is even an almost canonical version of ZFC, which is the one formulated by Cohen in his 1966 [13]. I will therefore consider the ontology of sets as they appear in ZFC. It should be added that ZFC takes as its underlying logic classical 1st order predicate calculus.

The importance of the concept of set in modern mathematics is undeniable, but what is curious is that this concept was very little used by mathematicians until the second half of the 19th century. Mathematicians dealt with magnitudes, and it was thought that these could be represented geometrically, usually as lengths along a straight line. However, a movement grew up in the 1860s to arithmetise analysis—that is to purge analysis of any dependence on geometrical concepts, and to base it on purely arithmetical concepts. The result of this movement was what was called the *arithmetisation of analysis*. It consisted essentially of replacing the old theory of 'quantities' or 'magnitudes', which could be represented geometrically, by a new theory of real numbers, which were defined in terms of rational numbers. Since rational numbers could be defined as ordered pairs of integers, and integers as ordered pairs of natural numbers, this based the theory of real numbers on that of natural numbers, that is on traditional arithmetic. Several mathematicians produced definitions of real numbers in terms of rational numbers around 1870, but the ones, which became best known, were those of Cantor and Dedekind, which were published in 1872.

The problem with the arithmetisation of analysis was that, while it did indeed eliminate geometric considerations, it did not found analysis purely on arithmetic, but rather upon arithmetic and set theory. Every one of the various definitions of real number involved a consideration of sets containing an infinite number of rationals. Now such infinite sets could well be regarded as just as problematic as, if not more problematic than, the geometrical evidences which had been eliminated. It was a question of out of the geometric frying pan into the set theoretic fire. Moreover, an infinite set of rationals is an example of an actual infinity, and so, if we acknowledge the existence of such sets, we have to abandon Aristotle's restriction of infinity to the potential infinity.[5]

Despite, or perhaps because of, these problems, the introduction of actually infinite sets proved very stimulating for the development of mathematics. Cantor worked out his new theory of the infinite in the two decades following 1872, and published a complete account of it in his papers of 1895 and 1897 [10]. His friend Dedekind, with whom Cantor corresponded, also worked on the theory of infinite

---

[5]The change from the potential to the actual infinite is also discussed in Stillwell [29], particularly section 3. Stillwell's chapter also gives more details about the set theoretic matters which are discussed briefly in the remainder of this section.

sets in this period. The title of Dedekind's book of 1888 [15] (*Was sind und was sollen die Zahlen?*) suggests that it is a book about natural numbers, but in fact Dedekind has a lot to say in the book about sets (or systems, as he called them). Some of the principles regarding systems, which he formulated, later on became axioms for set theory. Moreover Dedekind stated and at least partly proved some important theorems regarding infinite sets.

A mathematical theory of infinite sets was thus developed between 1872 and 1897, but then disaster struck when contradictions were discovered in this new theory. The first contradiction to be published was Burali-Forti's paradox of the greatest ordinal, which appeared in 1897. Then came Cantor's paradox of the greatest cardinal. Developing the reasoning behind these contradictions, Russell discovered that it led to a contradiction within logic itself—Russell's paradox.

Despite the shock generated by the discovery of these paradoxes, it did not take mathematicians long to find a way round them. In 1908 Zermelo published an axiomatic version of set theory, which was designed to avoid the contradictions while still allowing the theory of infinite sets of Dedekind and Cantor to be developed. Actually Zermelo's original axiomatisation did not allow the full theory of Dedekind and Cantor to be developed, and further work was needed. Skolem and Fraenkel replaced Zermelo's axiom of separation by the axiom of replacement in 1922, and in 1925 von Neumann showed a way in which Cantor's transfinite ordinal and cardinal numbers could be introduced. However, by 1930 axiomatic set theory had been created in much the same form as it exists today.

ZFC is no longer compatible with Aristotle's theory of the infinite as always potential, because it explicitly postulates an axiom of infinity, stating that an infinite set exists. This is Axiom VII in Zermelo's 1908 paper. Twenty years earlier, Dedekind had tried to prove the existence of an infinite set, but his 'proof' was much criticised by, among others, Russell; and, as it seemed to involve reasoning very similar to that which had led to the contradictions, it was abandoned. As infinite sets were needed for the definition of real numbers, there was nothing for it but to assume their existence as one of the axioms. Moreover, this axiom is independent of the other axioms of set theory, since these are all satisfied in the universe of finite sets.

From the axiom of infinity it follows that the set of all natural numbers (N say) = {1, 2, 3, ... , n, ...} exists. Thus the natural numbers are no longer a potentially infinite sequence 1, 2, 3, ..., n, ... There is an actually infinite set of them (N). In this respect therefore contemporary mathematics goes beyond ancient Greek mathematics. Aristotle's theory of the infinite as potential is no longer adequate for mathematics, and, if contemporary mathematics is accepted, we have to accept the existence of the actual infinity.

I will next argue that nearly all the objects of contemporary mathematics can be defined in terms of natural numbers and the sets of ZFC. This justifies limiting ourselves to these two types of object, when studying the ontology of contemporary mathematics. While this may not be entirely adequate, at least it makes a good start with the problem. I will begin my survey with a classic text written by two leading mathematicians, namely Birkhoff and MacLane, *A Survey of Modern Algebra*. This was first published in 1941 [8] shortly after set theory had become established. The book then went through many editions and its approach is still that which is generally used today. The authors begin with the integers i.e. $0, \pm 1, \pm 2, \ldots$, $\pm n, \ldots$, though later in the book they remark that they could just as easily have started with the positive integers, i.e. the natural numbers. Rational numbers are introduced as ordered pairs of integers, and real numbers by Dedekind cuts (Ch. IV, section 5, pp. 97–99). Complex numbers are next introduced as ordered pairs of reals. At a more abstract level, groups are characterised as sets of elements on which is defined an operation satisfying various laws. An operation is of course a function, and functions can be defined in terms of sets. A similar approach is adopted for other abstract concepts such as rings, fields, Boolean algebras, etc.

What about geometry? The most natural approach to geometry is the analytic approach in which a point is defined as an n-tuple of real numbers; and lines and other figures are defined by equations. For $n = 1$, 2 or 3, diagrams still remain a useful guide, but such "geometric evidences" now have a subordinate role. Let us consider, for example, a line segment of unit length. The points of such a segment would be real numbers x such that $0 \leq x \leq 1$. Such a line segment, a continuum, would consist, in the contemporary approach, of the set of its points—a set which is of course well-defined within ZFC. Thus contemporary mathematics contradicts Aristotle's claim (already quoted) that (Physics VI.1 231ᵃ 25) [6] "a line cannot be composed of points". Lines in contemporary geometry are sets of points, and are, in this sense, composed of points.[6]

Analysis is developed in the same way as algebra, with abstract concepts such as topologies being defined in terms of sets. Another striking example is probability theory. In 1933 [21] Kolmogorov published his axioms of probability, which he formulated in a set theoretic framework. Essentially the same approach is still adopted today, over 80 years later, by those who work on mathematical probability.

These are my arguments for accepting the natural numbers and ZFC as fundamental to the ontology of contemporary mathematics. Let us now consider what arguments might be raised against this position.

The first counter-argument makes use of Gödel's incompleteness theorems. These show, it could be said, that ZFC is not adequate for contemporary mathematics. Let us consider the implications for ZFC of the two theorems, starting with the second incompleteness theorem. This theorem shows that the consistency of ZFC cannot be proved without using assumptions, which are stronger than those of ZFC. Now this is undoubtedly the case, but is it a serious problem? Mathematicians have been working on ZFC for over 80 years now, but no contradiction has yet come to light. This surely suggests that ZFC is indeed consistent. Suppose moreover that a contradiction in ZFC was discovered in the next few years. When the contradictions in the infinite set theory of Dedekind and Cantor came to light in 1897, it took only a few decades for the damage to be repaired, and their theory to be put on

---

[6]For a detailed and informative account of the similarities and differences between Aristotle's theory of the continuum, and those of modern mathematics, see Newstead [26].

a footing, which has held good until today. Surely if a new contradiction came to light, the damage it caused would again be repaired quite quickly. Of course, in the contemporary world, we have to give up the old idea that mathematics is completely certain. There are many reasons for thinking that mathematical knowledge, like all other kinds of knowledge, is doubtful to some degree.

Let us now turn to Gödel's first incompleteness theorem. This states that, if ZFC is indeed consistent, then there is a theorem of arithmetic which cannot be proved within ZFC, but which can be proved by an informal argument outside ZFC. This result does indeed show that ZFC is not adequate for the whole of contemporary mathematics. Even leaving aside Gödel's first incompleteness theorem, there are parts of contemporary mathematics, which do seem to go beyond ZFC. Some set theorists have suggested strengthening ZFC by adding new axioms. Usually these are stronger forms of the axiom of infinity, which postulate the existence of very large transfinite cardinals. Some category theorists have made the more radical suggestion that the whole framework of set theory is unsatisfactory and should be replaced by a category theoretic framework. More recently it has been suggested that homotopy type theory might provide a more satisfactory framework for contemporary mathematics than ZFC. Moreover some very complicated proofs of modern mathematics seem to make implicit use of assumptions, which are stronger than those of ZFC. Such a claim has, for example, been made about Andrew Wiles' proof of Fermat's Last Theorem. What can be said about all these objections to regarding the sets of ZFC as fundamental for the ontology of contemporary mathematics?

Well, first of all, it should be observed that ZFC is unlikely to remain an adequate framework for mathematics for all time. Euclidean geometry provided an adequate framework for mathematics for many centuries, but eventually developments in mathematics meant that Euclidean geometry was no longer adequate as an overall framework, though it still continues to be used in some contexts. A similar fate is likely to overtake ZFC at some time in the future. However, it should be added that such revolutionary changes in mathematics occur relatively rarely. During most of its history "normal" mathematics has been conducted within a framework of assumptions, which could be called, using Kuhn's term, a paradigm. Nowadays despite all the dissatisfaction of category theorists and others, that paradigm is largely defined by ZFC. If a mathematical proof can be carried out within ZFC, nearly all mathematicians would accept it as valid. If the proof required assumptions, which went beyond ZFC, it would appear more doubtful. The success and resilience of ZFC over a period of more than eighty years is somewhat surprising if we reflect on the rather inauspicious circumstances in which it was first developed.

So far I have considered objections to ZFC, which consider ZFC to be not strong enough to provide a framework for contemporary mathematics. However, there is another set of objections, which regard ZFC as too strong. ZFC, as I pointed out above, was designed as a system within which Cantor's theory of transfinite numbers could be developed. However, many mathematicians of the Cantor's time, and also some today, regard Cantor's theory as very dubious, not genuine mathematics, and perhaps even as meaningless. An example of a famous mathematician who opposed Cantor's ideas is Poincaré, who wrote in 1908 [28] (p. 195): "*There is no actual infinity. The Cantorians forgot this, and so fell into contradiction*". Several mathematicians, notably Brouwer, who were sceptical about Cantorian set theory, tried to develop forms of mathematics which would prove to be a viable alternative to the mainstream approach based on ZFC and standard 1st-order logic. Brouwer developed intuitionistic mathematics, Poincaré's neo-Aristotelian approach led to predicative mathematics, and there are other forms of constructive mathematics besides. These new systems of mathematics are not useless by any means, because they introduced new ideas, which have been made use of by the mainstream. However, Brouwer's revolutionary objective of replacing standard mainstream mathematics by another kind of mathematics proved to be a failure. This is another illustration of how difficult it is to carry out revolutions. Though revolutions do occur from time to time, many would-be revolutionaries are doomed to disappointment.

Another objection to the approach I am adopting might be raised, this time by strong advocates of ZFC. They might argue that there is little point in considering the natural numbers separately from ZFC, since the natural numbers can be defined within ZFC as the finite ordinals. This is the approach taken by Cohen who writes [13, p. 50]:

"a very reasonable position would be to accept the integers as primitive entities and then use sets to form higher entities. However, it can be shown that even the notion of an integer can be derived from the abstract notion of a set, and this is the approach we shall take. Thus in our system, all objects are sets. We do not postulate the existence of any more primitive objects."

However, Gödel's first incompleteness theorem does seem to me to provide an argument against the view that all mathematical objects are sets. Stating the implications of this theorem of Gödel rather informally, we can say that the natural number concept has the ability to go beyond any characterisation of it using a formal system. The ultimate reason is that all the parts of any formal system (F say) can be labelled using the natural numbers, and this labelling shows a way in which the natural numbers can be used outside F, and so go beyond any formal characterisation of natural number within F. So the natural number concept contains within itself a principle by which natural extensions of the concept can always be found. The set concept lacks this capacity for natural self-extension. This is a reason for considering among fundamental mathematical objects, natural numbers as well as sets.

Our discussion of contemporary mathematics has shown that one part of Aristotle's philosophy of mathematics must be given up in the modern context, namely his claim that infinity is always potential. Our project therefore is to take the other part of his philosophy of mathematics, namely the embodiment postulate, which states that mathematical entities do exist, but they exist as embodied in the material world and not in a separate realm. In the rest of the paper I will argue for this claim for natural numbers and sets, as characterised by ZFC. Naturally the hardest part of this claim to establish will be that infinite sets and Cantor's transfinite numbers are embodied in the material world. Before we come to this, however, it will be as well to consider first the simpler case of finite sets, and this will be dealt with in the next section.

4   Finite Sets

Our project then is to show that numbers and sets are embodied in the material world. To carry this out, it may be as well to begin with a brief explanation of what will be meant by 'material world'. Some authors use 'physical world' as a synonym for 'material world', but I intend to use 'material world' in a broader sense. The physical world can be characterised as the world studied by physics. However, the material world should include not just the physical world, but the natural world or world of nature. The natural world is the world studied by any of the natural sciences, that is by chemistry, biochemistry, biology, as well as by physics.

The next important point about the material world is that it consists not just of objects, such as electrons, stones, molecules, cells, plants or animals, but of objects standing in various relationships to each other.[7] These relationships are to be regarded as just as real as the objects themselves. However, relations are abstract, and so abstract entities have been introduced into the material world.

An example from biology illustrates this point of view. A plant growing in the ground, and an insect flying through the air, may, at first sight, seem two distinct and separate material things, but yet in reality they are strongly interconnected. Many flowers cannot survive without being pollinated by insects while some insects cannot survive without obtaining nectar from flowers. Note also that these relationships have nothing to do with human consciousness since they existed long before the appearance of men and women.

Let us now pass from relationships to sets. As a result of objective relations between them, some things are bound together to form sets, which really exist in the material world. This can again be usefully illustrated by a biological example, namely a colony of bees. This consists of one queen, 40,000 to 50,000 workers, and a few hundred drones. Well-known relationships exist between these various types of bee. The queen lays the eggs. The workers collect nectar, turn it into honey, and tend the young. The drones have the sole function of fertilising the queen and are, with the approach of winter, expelled (see [11], Ch. 1). Once again these relationships have nothing to do with human consciousness, since, as Chauvin writes (1961, Introduction, p. 11):

> "Ants and bees were already in existence 40 million years ago at least, and scarcely differ from those we know today. … And *Homo sapiens* has had hardly 150,000 years of existence …"

---

[7]Cf. Wittgenstein [30]: "The world is the totality of facts, not of things".

Other examples of such naturally occurring sets are: the planets of the solar system, the electrons in an oxygen molecule, the oxygen molecules in the Earth's atmosphere, the leaves of a tree, the trees of a wood, the cells of a living organism, and so on.

So far we have considered the natural world, but some aspects at least of the social world of human beings can be considered as part of the material world. Here, of course, caution is needed, since humans have sensations, feelings, thoughts and consciousness, which, according to many philosophers, are not material. Indeed some higher animals have sensations and feelings, and perhaps thoughts and consciousness as well. So we must also be careful how much of the biological world we consider as material. Still there are many social events, much of which can be considered as definitely material in character. Consider again our earlier example of a running race between 6 men in ancient Greece. This involved 6 male human bodies in rapid motion during a particular time period. It therefore had a distinctly material aspect. In general many aspects of the social world, and most aspects of the biological world, can justly be considered as part of the material world.

In the world of nature, we gave examples of situations in which the relations between various things bind them together into naturally occurring sets. The same phenomenon occurs in the social world, though here it is perhaps better to speak of socially constructed sets rather than naturally occurring sets. A good example of a socially constructed set in the social world is a nation. The inhabitants of a modern nation state are connected together by a complicated series of relationships. These relationships, though not physical in any direct sense, are nonetheless very real, since, for example, they enable one nation state to wage war against another, as Germany did against France in 1870.

Another example of a socially constructed set is the set of Russian billionaires. This set was constructed socially in the last 20 to 25 years by somewhat curious social processes. However, this set is now definitely in existence and exerts an influence on the rest of humanity in many striking ways.

The naturally occurring sets and social constructed sets, which we have so far considered, are all embodied in the material world as our Aristotelianism requires. Benacerraf in his 1973 [7] (see particularly p. 415) draws attention to a problem in Platonism. Suppose mathematical entities exist in a transcendental world outside space and time, it is not at all clear how we can interact causally with them. But if causal interaction between them and us is impossible, how can we get to know anything about them? This could be called: 'Benacerraf's causal interaction problem'. It is clear that Aristotelianism offers a solution to this problem, for, if mathematical entities are embodied in the material world then we can interact causally with them. Indeed all the naturally occurring and socially constructed sets so far mentioned can interact causally with human beings.

One particular kind of causal interaction is perception. So we can raise the question of whether it is possible to perceive sets. Now clearly some sets are not perceptible. No one can perceive, at a glance so to speak, the whole of the German nation or the set of oxygen molecules in the atmosphere. However, there are some sets, which do seem to perceptible, such as the leaves on a tree or the runners in a race. Maddy in her 1980 argues for the view that it is possible to see a set. She considers [22, p. 178] the case of a cook looking at three eggs in an egg carton and argues that the cook sees a set of three eggs. This seems to me correct, but Chihara in his 1982 makes a striking objection to it. Chihara considers the example of his desk, which has been cleared of all objects except an apple. According to Maddy, when he looks at this apple, he is seeing not just the physical object, but also an abstract object namely the unit set of the apple. However, the unit set of the apple appears to generate exactly the same sensations as the physical apple. Why therefore should we ever claim that we are seeing such an abstract object in addition to the physical object? I will now try to answer this objection of Chihara's by first giving a more detailed analysis of the process of perception.[8]

Let us take the example of a man looking at a saltcellar on a table. At first sight this seems to be a very simple case of a physical object interacting causally with a human. In fact, however, what interacts causally with the man is a whole complex of physical objects standing in various relations to one another. This becomes clear if we suppose that there is also a vase of flowers on the table. If the saltcellar is to the left of the vase, the man will see both objects. If the vase is in front of the saltcellar, the man will only be able to see the vase. In the first case, the man can as validly say: "I see that the saltcellar is to the left of the vase" as "I see the saltcellar". In other words it is just as correct to say that the man is seeing the relationship between the saltcellar and the vase (an abstract entity) as that he is seeing the saltcellar (a physical object). What is really affecting the man's senses is an interrelated complex of physical objects. If he focuses on just one of these and says: "I see a saltcellar", he is, in effect, picking out one element from the complex. To regard the saltcellar as independent of the other objects to which it is related is a distortion. This distortion does not lead us too far wrong in macro-physics, but may be at the root of some of the problems about the micro-world.

This analysis is connected with the familiar point that all observation is theory laden. Thus when our observer attributes his sensations of sight, and perhaps touch and taste as well, to an enduring physical object (a saltcellar), he is interpreting his sensations in terms of a theoretical scheme involving physical objects located in space and time. This theoretical scheme is by no means unsophisticated, and, if it is legitimate to use it to interpret sensory experience, why should it not be legitimate, on some occasions, to use an enriched theoretical scheme, which involves relations between physical objects, or sets of physical objects, as well as just physical objects. The use of such an enriched interpretative theory might be justified if it could be claimed that, for example, a set of physical objects (as opposed to the physical objects themselves) had a causal efficacy in the process under investigation.

---

[8]What follows is my own attempt to answer Chihara's objection. Maddy's rather different answer to the objection is to be found on pp 150–154 of her 1990 book which gives an overall account of her views on the philosophy of mathematics at that time. It is interesting that she describes her position as [24, p. 158]: "more Aristotelian than Platonistic".

With this in mind, let us return to the controversy between Maddy and Chihara. Suppose Maddy is in the middle of baking a cake, which requires two eggs. She opens the egg carton, and observes to her horror that there is only one egg. Now she will have to go to the shop to buy some more. Maddy really sees the egg as a unit set since she is comparing it with the desired couple. The unit set of the egg has, moreover, a causal efficacy since it has led Maddy to form the plan of leaving her house to buy some more eggs.

At this moment the doorbell rings. It is Chihara, who has come to visit his former pupil. As he enters the kitchen he notices an egg lying on the table, but, knowing nothing of the cake-making plans, he sees it simply as a physical object. For Chihara, it is indeed no more than a physical object. This example shows that two different observers viewing the same scene can nonetheless see different objects. What is really affecting their senses is a complex of interrelated physical objects. One observer can quite legitimately pick one element out of this complex, while the other can, with equal legitimacy, pick a different element out of the complex.

In this section, I have argued that some finite sets are indeed embodied in the material world. Such sets can interact causally with human beings, and indeed it may be possible to perceive sets in some cases. In the next section, I will argue that the same holds for natural numbers.

5   Natural Numbers

I take numbers to be properties of sets. If a set is embodied in the material world, then the number of its members, being a property of the set can also be considered as embodied in the material world.

Maddy also holds that numbers are properties of sets (see her [23], p. 502), but she argues that there is an important difference between numbers and sets, and this must now be discussed. I have argued, in agreement with Maddy, that some sets at least can be assigned a definite spatio-temporal location. Maddy argues, however [23, pp. 501–2], that numbers cannot be located in space and time in the same way as such sets, because numbers are universals. This point can be explained by adapting an example of Maddy's to British conditions. Suppose on a Saturday afternoon that games of football are taking place (one hopes peacefully) up and down the land. Each team, or set of eleven men, has a definite spatial location at a given time, but the number eleven does not seem to have such a spatial location. It is manifested as much by one team in one ground as by another team in another ground. It would therefore be a mistake to say that a particular number has a definite spatio-temporal location. There does, however, seem to be another possibility; we could say that a number is instantiated at a particular spatio-temporal location if there is a set there to which it applies, and that therefore the number 11 is scattered about space-time (like water, or hydrogen) rather than having no space-time position at all, i.e. being outside space-time.

This approach gives an answer to the uniqueness problem which, as we saw earlier, seems to have pre-occupied Plato and Aristotle. The uniqueness problem is that there should only be one number 2, but that in the equation $2 + 2 = 4$, there seem to be two numbers 2. The answer suggested by the analysis of the previous paragraph is that there is indeed only one number 2, but that it can be instantiated by different sets. The equation $2 + 2 = 4$ makes an implicit reference to sets. It says that if 2 is instantiated by set S_1, and also by set S_2, where S_1 and S_2 are disjoint, then 4 is instantiated by $S_1 \cup S_2$. The apparent plurality of the number 2 is really the plurality of the underlying sets, which are implicitly assumed.

The next question is whether we can see numbers. I have already argued that we can see some relations and some sets. Can we also see some numbers? Let us take a concrete example. Suppose I am looking at a bunch of three bananas. I can see various relationships between the bananas. They are attached to the same stalk, for example. These relationships bind the bananas together into a set, and I have argued that this can correctly be considered as visible. But is the number three also visible in this situation? This question is a difficult one, but I am inclined to answer: "Yes". Is it not legitimate to say that I see the redness of that rose, or that I see water in that pond? But then I ought also to be allowed to say that I see the number three.

Even if this is correct, however, it is neither necessary nor plausible to maintain that all numbers can be directly perceived by the senses. This is clearly not true of, for example, the number of Germans, or the number of oxygen molecules in the Earth's atmosphere. Moreover we often have excellent reasons for believing in the existence of objects, which cannot be directly apprehended by the senses. Such objects may, for example, be postulated by scientific theories, which have been very rigorously tested and confirmed in a whole variety of circumstances. Indeed this is really the situation as regards many numbers.

6   Infinity: (i) Denumerable Sets and the Continuum

I have so far considered finite sets and natural numbers. A critic might object: "Your Aristotelian account may look plausible in such elementary cases, but what about the more sophisticated entities of advanced mathematics? How, in particular, are you going to deal with infinite sets of very high cardinality?" Indeed Gödel says explicitly [19, p. 483]: " ... the objects of transfinite set theory ... clearly do not belong to the physical world". "Surely", it might be argued by our hypothetical critic, "Gödel is correct here. It might be claimed that finite numbers such as 1, 2, ... are embodied in the material world, but we cannot, without absurdity, maintain that $\aleph_1, \aleph_2, \ldots, \aleph_n, \ldots, \aleph_\omega, \ldots$ exist in the material world". "On this rock", our critic might conclude, "your Aristotelianism is bound to founder. After all, Aristotle himself only dealt with the potential infinity, and never with the actual infinity".

My reply to this objection is that transfinite set theory does not constitute a difficulty for Aristotelianism. On the contrary, the approach developed here affords a very simple and plausible account of transfinite set theory.[9]

In order to introduce infinite sets into the material world, I will argue that continuous intervals of points are physically real. The points could be points of space, or of time, or of space-time. Now when should we regard an entity as physically real, or as a constituent of the physical world? My answer to this question is as follows. Suppose some of the best theories of physics, which we have, i.e. theories which are very well confirmed by observation and experiment, contain symbols referring to the entity in question, then we should regard the entity as part of the physical world. To put the point another way, we should regard as physically real those entities, which are postulated by our best confirmed theories of physics.

Now almost all of the best theories of physics of the present day introduce a four vector (x, y, z, t) where x, y, z are said to be spatial co-ordinates and t a temporal co-ordinate. Moreover it is always assumed that x, y, z, t are continuous variables which take real numbers as values. From this it seems reasonable to conclude that spatio-temporal continua are physically real, just as we conclude that electrons or atoms are physically real.

It is important to note here that there is no *a priori* reason why space and time should be continua. After all it was once thought that energy was continuous, but then the quantum theory was introduced and proved more successful at explaining physical phenomena. Thus there is no reason why a quantum theory of space and time should not, one day, be introduced and prove more successful than the present continuum theory at explaining physical phenomena. If, and when, that happened we should have to give up our present assumptions regarding physical reality—which are anyway tentative and conjectural. In fact, however, no quantum theory of space and time has, so far, proved at all successful. On the contrary the assumption that space and time are continuous appears in all the well-confirmed theories of physics, and so is itself a very well-confirmed assumption. Thus we have excellent grounds for supposing that spatio-temporal continua are physically real.

Against this it might be argued that in fluid mechanics it is customary to assume that water and other liquids are continuous, although we know perfectly well that they are composed of molecules. Similarly, it could be claimed, we should regard the assumption that space and time are continuous merely as an approximation, and not as telling us anything about physical reality.

My reply is that the two cases are not analogous. In the case of fluid mechanics, we have a deeper theory, which tells us that the fluid is composed of molecules, and that the assumptions of continuity are only approximations, which, moreover, break down in situations which can be specified. In the same way, we should not, in the light of modern physics, regard Newtonian gravitational forces acting at a distance, as physically real, because we have a deeper theory (Einstein's General Relativity)

---

[9]In his [17], Franklin discusses how to deal with "large infinite numbers" from an Aristotelian point of view. His approach is somewhat different from the one developed here.

which shows that Newtonian theory is only an approximation, and which, moreover, specifies situations in which considerable divergences from Newtonian theory will occur. Now, if there was a deeper quantum theory of space and time, then the continuous theory of space and time would be in the same situation as fluid mechanics or Newtonian theory; but this is not the case. The continuum theory of space and time is everywhere successful, and, as things stand at present, there is no deeper theory, which corrects it. We are thus justified, so I claim, in taking the continuum theory of space and time as a guide to what is actually the case in the real physical world.

These then are my reasons for supposing that spatio-temporal continua are physically real, and are, moreover, of the character assumed by physical theories, that is to say consist of sets of points which are representable by real numbers. This, however, gives us in the material world, infinite sets whose cardinality is that of the continuum, i.e. c, where

$$c = 2^{\aleph_0}$$

It is now an easy matter to define denumerable sets, which are embodied in the material world. For example, the real number π occurs in many contexts in physics, and so can be regarded as embodied in the physical world. Therefore the set of all digits of π, a denumerable set, is also embodied in the physical world. So we can say that both $\aleph_0$ and c occur in the material world.

This line of argument also answers an objection, which might be raised to our earlier treatment of finite sets and natural numbers. It could be said that the examples given only establish the material embodiment of relatively small finite sets and natural numbers. What about natural numbers, which are much greater than the number of particles in the Universe as so far observed? Such numbers and the corresponding finite sets are certainly assumed by mathematicians, but they may lack any material embodiment of the kind we considered in sections 4 and 5. This is true enough, but of course there will obviously be a set of points whose members have any specified natural number, however large. Hence this difficulty is overcome.

So far we have given an Aristotelian account of the infinite numbers $\aleph_0$ and c, but what of the other Cantorian alephs $\aleph_1, \aleph_2, \ldots, \aleph_n, \ldots, \aleph_\omega, \ldots$? They need to be dealt with in a rather different manner, which will be explained in the next section.

7   Infinity: (ii) Other Transfinite Cardinals

In order to deal with the other transfinite cardinals, it will be helpful to take into account another aspect of mathematical entities. So far I have emphasised that natural numbers are embodied in the world of nature. However, such numbers can also be considered as human social constructions. They were constructed by devising symbols such as '1', '2', '3', $\ldots$, and giving these symbols meaning. Regarding the nature of meaning, I will assume the view, which Wittgenstein presents in his later philosophy. Wittgenstein's theory is that [31, p. 20]: "the meaning of a word is its use in the language", and he further analyses language as consisting of a number of interrelated language games. By a 'language game' he means some kind of rule-guided social activity in which the use of language plays an essential part. He himself introduces the concept as follows [31, p. 5]:

> "I shall also call the whole, consisting of language and the actions into which it is woven, the 'language game'."

And again [31, p. 11]:

> "Here the term 'language-*game*' is meant to bring into prominence the fact that the *speaking* of language is part of an activity, or of a form of life."

Wittgenstein illustrates his concept of language game by his famous example involving a boss and a worker on a building site [31, p. 3]. The boss shouts, e.g. 'slab', and the worker goes off and fetches a slab. Wittgenstein's point is that the meaning of the word 'slab' is given by its use in the activity carried out by boss and worker. Actually Wittgenstein's term 'language game' seems hardly appropriate, since his first example of a language 'game' is not a game at all, but work. It would seem preferable to speak of 'language activities'—thus leaving it open whether the activity in question is work or play.

If we accept this theory, it is easy to give an account of how one sort of abstract entity—meaning—is constructed by human activity. For some sign S to acquire meaning, it suffices that S comes to have a generally accepted use in a social activity or activities. Thus by setting up social activities in which signs play an essential part, human beings create a world of abstract entities (meanings). These meanings are the product of human social activity. So, for example, by introducing the numerals '1', '2', '3', '4', '5', ..., and giving them a use in various social activities, numbers are created. If numbers have been created, and the boss on the building site shouts 'five slabs', the worker will (probably) bring five slabs. In this way the construction of numbers assists in the construction of houses. Numbers are a sort of abstract tool.[10]

However, there are some limitations to the view of mathematical entities as human constructions. This issue can be clarified by introducing another important idea from the philosophy of language. This is Frege's distinction between sense and reference [18].

Frege illustrates this distinction by his well-known example of the morning star and the evening star. The phrases 'morning star' and 'evening star' have different senses. The first phrase has the same sense as 'the star which sometimes shines very brightly in the early morning' while the sense of the second phrase is the same as 'the star which sometimes shines very brightly in the early evening'. However, our astronomical knowledge tells us that both phrases, despite their difference in sense, refer to the same celestial body, namely the planet Venus.

---

[10]The point of view of this paragraph is rather similar to that of Avigad in his 2015. Avigad writes [2]: "What I am advocating is a view of mathematics as a linguistic artefact, something we have designed, and continue to design, to help us get by in the world".

If we accept Frege's distinction, then it is clear that Wittgenstein's later theory of meaning deals with sense, but not at all with reference. Consider the phase 'The Sun'. By being given a use in a host of human social activities, this phrase acquires a meaning or sense. In virtue of this sense, the phrase refers to a large incandescent body. However, this large incandescent body has nothing to do with human social activities. It existed long before there were human beings, and its existence has been little affected by human life.

Let us now try to relate this to the Aristotelian view of mathematical ontology, which has been developed so far. It is one of the simpler claims of chemistry that in the air there are a large number of oxygen molecules of the form O_2. The set of electrons in an oxygen molecule is thus an example of a naturally occurring set. Moreover, the set of electrons of a particular oxygen molecule in the air is an embodiment of the number 16. Yet oxygen molecules existed before there were human beings, and human social activity has not altered the number of electrons in an oxygen molecule. This shows that number is not a purely human construction, but is also an aspect of the non-human world of nature. If therefore we are going to give an adequate account of the existence of numbers, we must combine, in some way, the view of numbers as human constructions with the view of them as aspects of the non-human world of nature.

To do so, let us consider the question: "Did human beings construct the number 16?" One could say in reply that human beings did invent the system of signs of which '16' is part and by using this sign system, they gave a meaning or sense to '16'. Thus humans did construct the sense of '16', or the concept of 16. However, in virtue of the way the natural world is, this sense picks out a reference in the natural world. It refers, for example, to the number of electrons in an oxygen molecule. The reference of '16' existed and exists in the natural world quite independently of human beings, and to this extent the number '16' was not a human construction. Note, however, that the sense of '16', or the concept of 16, are also embodied in the material world. They are embodied in the human social activities, which give the term '16' meaning, and so are embodied in a part of the social world. Thus the Aristotelian account applies to the sense of '16' just as much as to its reference, though in a different way.

We can now use these ideas about sense and reference to give our account of the Cantorian alephs $\aleph_1, \aleph_2, \ldots, \aleph_n, \ldots, \aleph_\omega, \ldots$. It is useful to compare these alephs to the natural numbers $1, 2, 3, \ldots, n, \ldots$. In both cases we have symbols, such as '16' in the first case and '$\aleph_5$' in the second. These symbols acquire meanings or senses by being used in language activities. The symbols of the finite numbers all have references. For example the number 16 is embodied in the naturally occurring set of electrons in an oxygen molecule. By contrast the transfinite alephs have as yet no application in physics, which would give them a reference in the material world. So the transfinite alephs have no reference. Thus for example '$\aleph_5$' has a sense because it has a use within the language activity of Cantorian set theory. This activity may have few participants, but it is nonetheless a perfectly definite social activity carried out in accordance with clear and explicit rules. On the other hand '$\aleph_5$' has no reference.

Let us next examine the consequences of this analysis for the question of mathematical truth. An Aristotelian view of mathematical entities allows the correspondence theory of truth to be extended from physics to mathematics. The electrons of an oxygen molecule exist in the material world, and statements about them are true if they correspond to what is the case in the material world. Similarly, since in the Aristotelian view, numbers and sets are embodied in the material world, a statement about numbers or sets is true if it corresponds to what is the case in the material world. Note that this account of mathematical truth allows us to defend the law of excluded middle without appealing to Platonism.

It is a consequence of this view of mathematical truth that there are no truths about the transfinite cardinals $\aleph_1, \aleph_2, \ldots, \aleph_n, \ldots, \aleph_\omega, \ldots$.[11] However, this is not a defect of this approach, but an advantage rather. All we need do, as far as these transfinite cardinals are concerned, is to speak not of a proposition being true but of its being a theorem in a particular version of set theory. Thus the correct claim that

$$\aleph_4 + \aleph_5 = \aleph_5$$

would no longer be regarded as a truth about transfinite cardinals in the way that

$$4 + 5 = 9$$

is a truth about finite cardinals. It would be regarded as a theorem of the development of the theory of transfinite cardinals within say ZFC. The advantage of this approach is that many theorems of the theory of transfinite cardinals and ordinals do in fact depend on what version of set theory is being used. A theorem may hold within one version of set theory but not another. Thus it is better to make the appeal to some underlying set theory explicit rather than to use an objective notion of truth, which does not apply in this case.

This view of mathematical truth sheds light on the question of whether the continuum hypothesis will ever be shown to be true or false. Cohen showed in his 1966 that the continuum hypothesis is both consistent with and independent of the other axioms of ZFC. The continuum hypothesis states that $\aleph_1 = c$, and so it contains a term ('$\aleph_1$') which lacks a reference. It does not therefore have a truth value at the moment, and will only acquire one if '$\aleph_1$' does acquire a reference by, for example, being given an application in physics. There is no guarantee, however, that the transfinite cardinals $\aleph_1, \aleph_2, \ldots, \aleph_n, \ldots, \aleph_\omega, \ldots$ will ever be used in

---

[11] I am here assuming Frege's view that if a referring expression in a sentence lacks reference, then the proposition expressed by that sentence lacks a truth value.

physics or some other successful theory. So the continuum hypothesis could remain undecidable for the indefinite future.[12]

There are several general consequences of this approach, which need to be made explicit. To begin with, on the present Aristotelian view, the axioms of ZFC cannot be regarded as true, but only as acceptable. If the axioms of ZFC were true, then any consequence of those axioms would have to be true as well. However, we have argued that $\aleph_4 + \aleph_5 = \aleph_5$ should be regarded as neither true nor false rather than as true, even though it is a consequence of the axioms of ZFC. The axioms of ZFC are acceptable in roughly the following sense. If a statement (S say) follows from the axioms of ZFC, and if S only refers to entities which exist, in the sense of being embodied in the material world, then S can be regarded as true.

The status of the axioms of ZFC is connected in turn with the fact that ZFC is only partially interpreted. The existence of a full Tarskian interpretation of ZFC is only possible if Platonism is correct. In that case there is a Platonic world of sets, and the axioms of ZFC would be true or false depending on whether they held for this Platonic world or not. On an Aristotelian position, however, there is not a full interpretation of all the sets of ZFC. Only those which can be shown to be embodied in the material world can be regarded as having a genuine reference. There are, however, many sets which are definable within ZFC, but which cannot at present be shown to have a material embodiment. The symbols for such sets, e.g. $\aleph_4$, do not have a reference, that is to say they do not have a Tarskian interpretation, since Tarski's semantics are referential in character.

It is useful to consider this account of ZFC in relation to two well-known philosophies of mathematics, namely fictionalism and if, then-ism. Let us start with fictionalism. According to fictionalism, mathematical entities do not exist, and so mathematical writings are similar to literary fiction, which can take the form of novels, plays, etc. Literary fiction describes characters and their doings, but these characters are purely imaginary and do not exist in the real world. For example, consider Shakespeare's play: Hamlet. This presents the actions of a number of interesting characters such as Hamlet himself, Polonius, Ophelia, Laertes, etc. However these characters are purely imaginary and never existed in the real world. Similarly a fictionalist would claim, mathematics deals with entities such as numbers, sets, etc., and describes their properties, but these entities are purely imaginary and do not exist in the real world.

---

[12]I had a discussion with Lakatos on this point in the late 1960s when I was doing my PhD with him. Lakatos thought that the continuum hypothesis would be decided one day, and, when I asked him why, he replied that this was because of the growth of mathematics. Assuming human civilisation continues, mathematics will undoubtedly continue to grow and to develop new concepts, which will be successfully applied to the material world. However, the direction of this future development is uncertain, and it may not consist of further development and successful application of the theory of transfinite cardinals, but rather of the development and successful application of some completely new concepts and theories. Thus the continuum hypothesis may never be decided.

Now it is obvious that fictionalism, as a general philosophy of mathematics, must be rejected from the present Aristotelian point of view. According to this Aristotelian approach, many mathematical entities such as numbers, sets and geometrical figures do exist because they are embodied in the material world. Moreover, many mathematical propositions such as $4 + 5 = 9$ are true because they correspond to what is the case in the material world. However, from our present point of view, fictionalism, though not correct in general, does hold for some mathematical entities, such as, notably, the transfinite cardinals $\aleph_1$, $\aleph_2$, ... , $\aleph_n$, ... , $\aleph_\omega$, ... These really are fictions like Hamlet, Polonius, Ophelia, etc. We could perhaps refer to them as metaphysical fictions rather than literary fictions.

Fictionalism has close connections with if, then-ism. Let us take an arbitrary mathematical statement S. According to if-then-ism, we can never claim that S on its own is true. The most we can claim is that there is a set of axioms A, such that 'If A, then S' is logically true. Once again it is clear that if, then-ism cannot be accepted from the present Aristotelian point of view. As we remarked, the approach leads to the conclusion that many mathematical statements, e.g. $4 + 5 = 9$ can be accepted as true without deducing them from any set of axioms. On the other hand, from the present point of view, if, then-ism is correct for some mathematical statements. Statements such as $\aleph_4 + \aleph_5 = \aleph_5$ cannot be accepted on their own as being true. The most we can say of such statements is that they can be deduced from the axioms of ZFC, or some similar version of axiomatic set theory.

If, then-ism can be applied not just to the metaphysical fictions of mathematics, but to literary fictions as well. To do so, we simply take any work of literary fiction, and regard all the statements in the work as our axioms. We then accept any conclusion which can be deduced from these axioms. So, for example, we can take all the statements in a standard edition of Hamlet, and see what further statements can be deduced from them. In this way we can conclude that Ophelia is the sister of Laertes, and the daughter of Polonius, but that it would be incorrect to claim that Ophelia is the wife of Hamlet. Of course some statements about Ophelia are undecidable by this method. We cannot establish, for example, whether or not Ophelia had measles as a baby. Now some famous literary critics have indeed applied this if, then-ist method. For example, Bradley drew a surprising conclusion regarding Hamlet. It is usually supposed that Hamlet, being a young student, is about 20 years old. However, Bradley deduces from the text of the play that Hamlet is exactly 30 years old (see [9], pp. 407–9). The evidence for this is in Act V, Scene i, where Hamlet has a conversation with the gravedigger. The gravedigger says that he started his present occupation on the day when Hamlet was born, and has been doing the job for 30 years. Moreover, he digs ups the skull of Yorick and says it has been lying in the earth for three and twenty years. Hamlet remembers Yorick carrying him on his back when he was a boy, which confirms that Hamlet must at least be in his late twenties. However, other literary critics have felt that Bradley went too far in these if, then-ist deductions. No doubt Shakespeare did try to create a consistent fictional world, but perhaps he really did intend Hamlet to be a young student of about twenty and did not notice the inconsistency of this with the statements made by the gravedigger. After all, a few inconsistencies can creep into a literary work without the average reader or member of the audience noticing and feeling perturbed.

It is interesting to compare the situation here with the metaphysical fictions generated by ZFC. Of course we cannot be sure that ZFC is consistent, though there is some evidence that it is. However, if an inconsistency were discovered in ZFC, we could no longer accept this axiomatic theory. Because it is based on classical 1st order logic, if it is inconsistent, then any statement of the theory could be proved as a theorem, which is obviously unsatisfactory. Consequently, if an inconsistency were discovered in ZFC, the mathematical community would immediately set to work to produce a new system, which avoided inconsistencies of that type. Literature, by contrast, is more tolerant of inconsistencies, and, if an inconsistency is discovered in a famous novel, it does not have to be immediately rewritten to remove the flaw.

Still there are definitely some points in common between literary fiction and abstract set theory. Perhaps the closest comparison is not between ZFC and a pure work of fiction, but rather between ZFC and a historical novel. Let us suppose that an author tries to create a historical novel with both characters which really existed (RCs), and other characters which are purely imaginary (ICs). The author is conscientious, and makes sure that any scenes involving just RCs are historically correct and well supported by documents as regards the words and actions attributed to the RCs. However, naturally the scenes involving RCs and ICs, or just ICs, though they might be historically plausible, are not literally true. (These hypothetical rules of practice are quite close to those actually adopted by Sir Walter Scott.) Now, if we take any statements involving just RCs from this novel we can regard it as true, but if a statement involves ICs as well as RCs, we should regard it as a fiction, i.e. on Frege's approach, as neither true nor false. Here the analogy with ZFC is very close. If a theorem of ZFC contains only reference to entities which have been shown to be materially embodied, i.e. actually to exist on the Aristotelian criterion, then the theorem can be regarded as true. If the theorem contains reference to some entity, such as $\aleph_4$ which has not been shown to exist by the Aristotelian criterion, then the theorem should be regarded as neither true nor false. Of course this analogy is not entirely accurate. It is always possible that, through the development of science, $\aleph_4$ is shown to exist in the physical world. However, it is hardly likely that a character imagined by a historical novelist is found by later historical research to have really existed and carried out the actions described in the novel.

I have argued that ZFC generates some metaphysical fictions which are surplus to the requirements of applied mathematics. If this is so, it might be argued, surely we should start a programme for paring down ZFC so that the modified system produced only what was required by applied mathematicians, and no surplus structure of a metaphysical character. Such an argument does not seem to be correct. There are two reasons why we should tolerate the metaphysical structures generated by ZFC. First of all ZFC is a remarkably simple and elegant system. It consists of only 8 axioms and 1 axiom schema. From this, virtually all the mathematics needed by applied mathematicians can be developed with little difficulty. ZFC is a remarkable achievement, carrying out for 20th century mathematics, what Euclidean geometry did for ancient Greek mathematics. Each system provided a simple and elegant framework within which the mathematics of its time could be developed. Now if we attempted to pare down ZFC so that the modified system generated only those sets needed for applied mathematics, and no surplus structure, the result would inevitably be a more complicated and less elegant system. Moreover what would be gained by eliminating the metaphysical structure of the Cantorian alephs? This structure, even if it is not used in practice, does no harm.

One could even argue that these metaphysical fictions might do some good. The view of the Vienna Circle that metaphysics is meaningless and should be eliminated has now largely been abandoned. Popper criticised this view and pointed out that metaphysics could be not only meaningful, but also positively beneficial for science by providing a heuristic guide for the construction of scientific theories. It is still, of course, not to be excluded that the Cantorian alephs will play some role in a future successful theory of physics.

Thus some metaphysical fictions should be tolerated within mathematics. Yet, at the same time, such toleration should perhaps not go too far. Earlier we mentioned suggestions about modifying ZFC by introducing stronger axioms of infinity, such as, for example, an axiom postulating the existence of an inaccessible cardinal. However, as we have seen, ZFC, as well as generating the infinities which are actually useful in physics, namely $\aleph_0$ and c, generates many more infinite numbers which have as yet proved of little use in physics or other branches of science. Thus we have so to speak already got more infinite numbers than we really need, and so to postulate the existence of further such numbers hardly seems a good idea. It would only be adding to the metaphysical and fictitious side of mathematics, without contributing anything of practical use.

Acknowledgements I read an earlier draft of this paper at a seminar in La Sapienza, Roma on 16 February 2015, and I am very grateful for the comments I received on that occasion. Particularly useful were some comments from three experts on Aristotle (Silvio Maracchia, Diana Quarantotto, and Monica Ugaglia) who helped me with a number of points. After the seminar, I had a very useful discussion with Carlo Cellucci concerned mainly with the problem of applying the Aristotelian approach to ZFC, and the possible use of ideas of fictionalism and if, then-ism. This led to many improvements in section 7 of the paper. In addition I was fortunate to receive extensive comments from quite a number of people to whom I sent a copy of an earlier draft. These included: Ernie Davis, James Franklin, Andrew Gregory, Ladislav Kvasz, Penelope Maddy, Anne Newstead, Alex Paseau, and Brian Simboli. Their input was very helpful in preparing the final version of the paper.

References

1. Annas, J. (1976) Introduction to her English translation of Aristotle's *Metaphysics* Books M and N, Oxford University Press, Paperback Edition, 1988.
2. Avigad, J. (2015) Mathematics and Language. *This volume*.
3. Azzouni, J. (2015) Nominalism, the nonexistence of mathematical objects. *This volume*.
4. Archimedes *The Method treating of Mechanical Problems*. English translation by Sir Thomas L. Heath in *Great Books of the Western World*, Vol. 11, Encyclopaedia Britannica, 1952, pp. 569–592.

5. Aristotle *Metaphysics*. English translation by W.D.Ross, except for Books M (XIII) and N (XIV), where the J. Annas translation in (1976) is used.
6. Aristotle *Physics*. English translation by R.P.Hardie and R.K.Gaye.
7. Benacerraf, P. (1973) Mathematical Truth. Reprinted in *Philosophy of Mathematics. Selected Readings*. Edited by P.Benacerraf and H. Putnam, 2nd edition, Cambridge University Press, 1983, pp. 403–420.
8. Birkhoff, G. and MacLane, S. (1941) *A Survey of Modern Algebra*. Second Revised Edition, Macmillan, 1961.
9. Bradley, A.C. (1904) *Shakespearean Tragedy*. Reprinting of 2nd Edition, Macmillan, 1964.
10. Cantor, G. (1895 & 1897) *Contributions to the Founding of the Theory of Transfinite Numbers*. English translation by P.E.B.Jourdain, Dover, 1955.
11. Chauvin, R. (1963) *Animal Societies*. English translation, Sphere books, 1971.
12. Chihara, C. (1982) A Gödelian thesis regarding mathematical objects: Do they exist? And can we perceive them? *The Philosophical Review*, XCI, pp. 211–227.
13. Cohen, P. J. (1966) *Set Theory and the Continuum Hypothesis*, Benjamin.
14. Dedekind, R. (1872) *Continuity and Irrational Numbers*. English translation in *Essays on the Theory of Numbers*, Dover, 1963.
15. Dedekind, R. (1888) *Was sind und was sollen die Zahlen?* English translation in *Essays on the Theory of Numbers*, Dover, 1963.
16. Franklin, J. (2014) *An Aristotelian Realist Philosophy of Mathematics*. Palgrave Macmillan.
17. Franklin, J. (2015) Uninstantiated Properties and Semi-Platonist Aristotelianism, *Review of Metaphysics, Forthcoming*.
18. Frege, G. (1892) On sense and reference. In P. Geach and M. Black (eds.) *Translations from the philosophical writings of Gottlob Frege*, Blackwell, 1960, pp. 56–78.
19. Gödel, K. (1947) What is Cantor's continuum problem? Reprinted in *Philosophy of Mathematics. Selected Readings*. Edited by P.Benacerraf and H. Putnam, 2nd edition, Cambridge University Press, pp. 470–485.
20. Gray, J. (2015) Mathematics at infinity. *This volume*.
21. Kolmogorov, A. N. (1933) *Foundations of the Theory of Probability*. Second English Edition, Chelsea, 1956.
22. Maddy, P. (1980) Perception and Mathematical Intuition, *The Philosophical Review*, LXXXIX, 2, pp. 163–196.
23. Maddy, P. (1981) Sets and Numbers, *Nous*, 15, pp. 495–511.
24. Maddy, P. (1990) *Realism in Mathematics*, Oxford University Press.
25. Maracchia, S. (2011) Infinito potenziale e infinito attuale nella matematica greca, *Lettera Matematica Pristem*, **78**, pp. 38–47.
26. Newstead, A.G.J. (2001) Aristotle and Modern Mathematical Theories of the Continuum. In D.Sfendoni-Mentzou, J. Hattiangadi, and D.M.Johnson (eds.) *Aristotle and Contemporary Science*, Peter Lang, pp. 113–129.
27. Plato *The Republic*. English translation by F.M.Cornford.
28. Poincaré, H. (1908) *Science and Method*. English translation by F. Maitland, Dover.
29. Stillwell, J. (2015) From the Continuum to Large Cardinals. *This volume*.
30. Wittgenstein, L. (1921) *Tractatus Logico-Philosophicus*. English translation by D.F. Pears and B.F. McGuinness, Routledge & Kegan Paul, 1963.
31. Wittgenstein, L. (1953) *Philosophical Investigations*. English translation by G.E.M. Anscombe, Blackwell, 1963.
32. Wittgenstein, L. (1966) *Lectures and Conversations on Aesthetics, Psychology and Religious Belief*. Edited by Cyril Barrett. Basil Blackwell.

Let *G* be a group

Jesper Lützen

Abstract Traditional philosophy of mathematics has dealt extensively with the question of the nature of mathematical objects such as number, point, and line. Considerations of this question have a great interest from a historical point of view, but they have become largely outdated in light of the development of modern mathematics. The prevalent view of mathematics in the 20th (and the early part of the 21st) century has been some variant of the so-called formalistic view, according to which mathematics is a study of axiomatic systems or structures. In such structures the objects have exactly the properties set out in the axioms and nothing else can be said about their nature. The question of the nature of mathematical objects has become a non-issue. Similarly, the structure itself is completely determined by its set of axioms and one can say nothing meaningful about its nature other than that. In particular, in the formalistic understanding of mathematics a structure exists if its axioms are consistent among each other (i.e., non-contradicting).

In this chapter we shall first outline this formalistic view of mathematics, then we shall follow the historical developments that led to it, and finally we shall critically analyze the formalistic view of mathematics, not from a purely philosophical point of view but from the view of the practice and history of mathematics. Instead of asking what can in principle exist and what can in principle be proved, we shall ask what is studied in practice, i.e., which mathematical structures are singled out by mathematicians and which statements in the structure are found interesting enough to be worthy of investigation. This question is not answered by the formalistic view of mathematics. In order to investigate the question one must study what drives mathematicians when they develop new mathematics. Here the history of mathematics points to intriguing *problems* as a particularly forceful driving force. Many (perhaps most) mathematical theories have been developed as an aid to solving problems both inside and outside of mathematics.

Aristotle considered mathematical objects as abstractions of phenomena in nature. However, many other generating processes such as generalization, rigorization, and axiomatization have played an important role in the development of the

J. Lützen (✉)
Department of Mathematical Sciences, University of Copenhagen, Universitetsparken 5, DK-2100 København Ø, Denmark
e-mail: lutzen@math.ku.dk

© Springer International Publishing Switzerland 2015
E. Davis, P.J. Davis (eds.), *Mathematics, Substance and Surmise*,
DOI 10.1007/978-3-319-21473-3_9

mathematical structures and theories that are currently being investigated. Thus, for a historian like me, it is tempting to replace an Aristotelian ontology with a richer historical narrative taking all the driving forces and generating processes into account. Although such a narrative may not explain the ontology of mathematical structures and objects it gives an account of how actually studied mathematical structures and objects came into being and to some degree also why they were singled out among the many potentially possible structures and objects.

1 Mathematical structures

A mathematical structure is a set of things called elements of this set, equipped with some relation or operation. It could be an order relation $<$ or a binary operation that to two elements of the set assigns a third element of the set. An example of the latter is addition $+$ which to two numbers a,b assigns a third number called $a + b$. The relation or operation must satisfy certain rules or axioms like the usual commutative law of addition: $a + b = b + a$. So a mathematical structure is a set with one or more relations or operations satisfying a number of axioms. One of the most studied structures is a group that is defined in Box 1. From the axioms mathematicians then deduce theorems, i.e., logical consequences of the axioms.

Box 1: Groups

In mathematics a group is a finite or infinite set of elements $a,b,c,\ldots$ and an operation denoted $*$ which to two elements in the group a,b attaches a third element denoted $a * b$. A simple example of such an operation could be addition $+$ on the set of integers. In order to be a group the operation $*$ must satisfy three axioms.

Axiom 1 called the associative law: for all elements a,b,c in the set,

$$a * (b * c) = (a * b) * c.$$

In other words if one first finds $b * c$ and then operates on this element from the left with a, one gets the same element as when one first finds $a * b$ and then operates with c on the right.

Axiom 2: There exists an element called a neutral element e with the property that for all elements a of the set

$$e * a = a * e = a.$$

In other words, nothing happens when one operates on either side with the neutral element.

Axom 3: For every element a in the set there exists a so-called inverse element denoted a^{-1} such that

$$a * a^{-1} = a^{-1} * a = e$$

In other words if one operates on an element on the left or the right with its inverse element one gets the neutral element.

In the example mentioned above of the integers with the operation $+$, the associative law holds, the neutral element is 0 and the inverse element of a is $-a$.

Another example is the set of displacements of a plane or of space where the $a * b$ means the displacement one gets by first using the displacement b and then the displacement a. Here the neutral element is the "displacement" that does not move anything, and the inverse of a particular displacement is the reverse displacement back to the original position.

Also the set of displacements of a plane or space, that move a certain figure onto itself form a group. For example displacements of a plane that map a given line onto itself make up a group. A reflection in the given line or a parallel displacement in the direction of the line are examples of such motions.

If a is any object the set consisting of this object can be made into a group by stipulating that $a * a = a$. Of course, this is a trivial group, but it shows that any object whatsoever can play the role of an object in a group.

But in what sense do such structures exist, what kind of objects are their elements, and what kind of knowledge do the theorems represent? Let us start with the first question: How do mathematicians argue for the existence of a structure? The philosophically minded reader will probably be disappointed if he/she looks in a modern mathematics book for an answer. For example in an introductory text book in algebra one will meet a definition of a group and then the following theorems will typically start with the phrase: "Let G be a group." In books about mathematical analysis theorems often begin with the words: "Let H be a Hilbert space," etc. Such declarations may remind us of God's creation of the world about which Genesis reports: "And God said: "let there be light and there was light." Do mathematicians play God when they do mathematics? Do they think that they can create what they speak? In a sense they do. When they define a mathematical structure like a group or a Hilbert space, it is thereby created in a mathematical sense. In order for a mathematical structure to exist it is not required that one can point to an instantiation of it in the physical world; the only thing a mathematician will require of such a structure is that it is free of contradiction. That is, it must be impossible to find a statement in the theory that can be shown to be a consequence of the axioms and whose negation is also a consequence of the axioms. This seems to be a minimal requirement. In fact if there were such a statement in a structure, ordinary logic would allow one to prove that all statements in the structure be both true and false. Such a structure is clearly uninteresting.

But what is then the nature of such a mathematical structure and in particular what is the nature of the elements (objects) in the structure? A mathematician will in general refuse to answer this question, not because he does not know the answer or is philosophically naïve, but because he has a good reason to leave the question unanswered. For example it is a central characteristic of the definition of a group that we do not explain what the elements of the group are. In fact that allows us to deal with groups of many different types of objects at once. The elements can be numbers or transformations or matrices or vectors or .... In this way a proof about groups kills many birds with one stone.

Thus, saying that a structure is a group does not tell us anything about what kind of objects the elements are, nor the precise way the operation works. It only tells us that the operation must satisfy certain rules: the three axioms.

A modern formalistically oriented mathematician will have a similar view of the nature of more classical mathematical concepts such as number and geometry. Indeed one can define the ordinary real numbers as a set with two binary operations called addition and multiplication and an ordering which satisfies a long list of axioms. Similarly, plane Euclidean geometry can be defined as two sets of elements called points and lines, and a list of relations between them such as "lie on" (a point can be said to lie on a line) "between" (a point on a line can be said to be in between two other points on the line), etc. that satisfy an even longer list of axioms. In a way, these definitions *create* the real numbers and plane geometry with its points and lines. But they do not tell us anything about the ontology of a real number or a point and a line. They only tell us how their elements (numbers or lines and points, respectively) relate to each other.

In this way the whole question of the ontology of mathematical theories and objects have been side stepped. Of course one can ask if something in nature behaves like a particular mathematical system, but the answer cannot be found by mathematical means. For example it is a physical question to investigate if physical space can conveniently or correctly be described by the mathematical structure we call Euclidean geometry. And the answer does not affect the mathematical validity of Euclidean geometry.

This formalistic or structuralist view of mathematical theories and their objects became widely accepted around 1900. We shall now consider the historical development that led to it.

2   From idealized (or abstract) nature to arbitrary structures

It was the ancient Greek mathematicians and philosophers who developed the idea of mathematics as a deductive science in which one *proves* theorems from a system of axioms. Aristotle clearly explains the need for an axiomatic basis for a deductive science: Indeed, if you want to prove a theorem, you must base the proof on some prior knowledge, i.e., some theorems that you already have proved to be true. But a proof of these more basic theorems must be based on even more fundamental theorems, etc. In order not to end in an infinite regress, one must begin the deductive process with some statements (the axioms) that one assumes to be true. The other theorems of the theory must then be deduced from the axioms. This description of the role of the axioms in a mathematical theory is shared by a modern mathematician. However, the view of the nature of the axioms has changed (For a fuller survey of the historical developments analyzed in this paper see [4]).

For the ancient Greek mathematicians and philosophers the axioms were considered (evident) truths of the world. Not the crude physical world of phenomena that we sense, but an ideal or abstract world. For Plato the real world was the world of ideas, and it was this world that the axioms dealt with. For Aristotle they dealt with abstractions. For example, when a line in Euclid's Elements (c. 300 BC) is defined as having no width and the first postulate (axiom) claims that one can draw a straight line between two points, this is of course not true of the physical world. No mathematician can draw a breadthless line. Similarly, the third postulate claims that it is possible to draw a circle with a given center and a given radius. But a circle is defined as a curve (line) of no width that lies equally far from one given point (the center). Of course we know from experience, that we cannot draw an exact circle. Many years of experience with a compass can teach us to draw very accurate circles, but they will never be perfect. However, that does not matter because the circles and the lines that the mathematician deals with are not the physically imperfectly drawn circles and lines. It is the ideal circles and lines (or the abstracted circles and lines). Mathematicians can prove things about the ideal or abstract circles by basing their arguments only on the definition of the circle and line and the axioms. For Plato this makes mathematics an important discipline for philosophers because it can teach them how to get insight into the world of ideas.

So, for the ancient Greeks the objects of mathematics were idealized or abstracted objects, and axioms were (self-evident) truths about them. Thus all consequences (theorems) deduced from them were also true, although not self-evident truths. This view of mathematics was generally accepted until the 19th century when it was challenged from several directions. We shall now turn to these challenges.

2.1   Symbolic algebra

The natural numbers 1,2,3, ... have been handled by humans as far back as we have written sources. In fact writing began as a way to denote numbers. Archeologists have unearthed 20,000 years old bones with engraved marks that are usually interpreted as representing a number of something. One can question whether such representations can be called writing, but also one of the first real written languages the Mesopotamian cuneiform writing grew out of an earlier way to represent numbers.

Negative numbers can be traced back to about year zero when Chinese mathematicians represented numbers as bamboo sticks on a counting board. One color was used for the positive numbers and another color for the negative numbers. Special rules of calculation were invented for handling the new numbers, for example, the rule saying that negative times negative is positive. Other cultures like the Indians (c. 500 AD) developed similar ideas later but other equally advanced cultures, for example, the ancient Greeks, did not introduce negative numbers.

As late as the 18th century, where the negative numbers were freely used by leading mathematicians like Euler, other mathematicians voiced objections. The problem with negative numbers was not primarily how to work with them (although the rule for negative times negative was problematized) but rather the question of their nature or ontology. For the ancient Greeks a number represented a collection of units; the number of things so to speak. They also developed an advanced theory of continuous quantities such as line segments, but they did not consider such quantities as numbers. The medieval Arab and the early modern European mathematicians gradually erased the dichotomy between number and quantity by extending the notion of number to include all the positive real numbers. But they continued to consider them as a measure of a quantity, such as a length of a line or an area of a surface. Negative numbers are not numbers in this sense because they do not represent the number of something or the measure of a quantity.

There are other ways to interpret negative numbers: In ancient China buying -5 goats just meant selling 5 goats. Later (at least in the 17th century) a directional geometric interpretation was suggested: To advance -5 feet in a particular direction on a straight line simply means to advance 5 feet in the opposite direction.

Despite such interpretations, negative numbers were often shunned. For example in Medieval and Renaissance algebra (Arab as well as Latin) negative solutions to equations were in general left out, and for good reasons. The equations were often set up as a means to find the length of a line segment or an area of a figure, and it is of course unclear what a line of a negative length or a figure with a negative area could mean.

At the beginning of the 19th century, a group of Cambridge mathematicians called the Analytical Society tried to circumvent the problem of negative numbers by dividing algebra into an arithmetical and a symbolical algebra. In arithmetical algebra the objects are (positive) numbers and the operations $+ .-, \cdot$ and: are the usual ones. However, the operations cannot be used unlimited. For example a-b only makes sense if $a > b$. In symbolical algebra on the other hand, the objects are just symbols $a, b, \ldots$ and the operations have unlimited validity. Subtracting a larger symbol from a smaller symbol is allowed and the result is a negative symbol $-c$.

One of the members of the group, George Peacock, defined algebra as "the science which treats of the combinations of arbitrary signs and symbols by means of defined though arbitrary laws." So, according to Peacock the nature of the objects were not important in algebra; only the laws of combination such as the commutative laws: $a + b = b + a$ and $ab = ba$ were essential. However, neither he nor his follower Augustus De Morgan made use of the arbitrariness of the laws of combination. They both assumed that they were the same as in arithmetical algebra. This changed in 1853 when William Rowan Hamilton constructed the so-called quaternions which are a kind of four dimensional number. And here one of the usual laws could no longer be maintained: For quaternions the commutative law of multiplication: $ab = ba$ no longer hold true.

The undefined or symbolic nature of the objects and the arbitrariness of the laws (axioms) in 19th century British algebra was soon taken over to other areas of mathematics.

2.2   Groups

The development of the concept of a group was also an important ingredient in the structuralist turn in mathematics. The word group in the mathematical sense was introduced around 1830 by the young revolutionary French mathematician Evariste Galois in connection with his ground breaking investigations of the question: Which n^{th} degree equations can be solved by a formula in which enters only the operations: $+,-,\cdot,:$ and the extraction of pth roots. Following the lead of Joseph-Louis Lagrange, Galois showed how this problem could be elucidated by studying those permutations of the roots of the equations that keep the expressions of the coefficients fixed. He pointed out that such a set of permutations have the property that if one first permute with one of the permutations and then with another, the combined permutation of the roots will be in the set. He called such a set of permutations a group.

Galois died in a duel when he was only 20 years old and so his ideas only became publically known after their publication in 1846. Gradually the idea of a group found its way into algebra books and it was noticed, that similar structures were found elsewhere in pure and applied mathematics [6]. In fact as early as 1801 Carl Friedrich Gauss had studied transformations of so-called quadratic forms that exhibit a similar structure as Galois' permutations, and around 1870 Camille Jordan, Felix Klein and Sophus Lie showed how the group concept could be used in connection with geometric transformations as well. In this way it became clear that the properties of groups did not depend on the nature of the elements in the group: they can be permutations of the roots of an equation, transformations of space, or numbers. The one distinguishing feature of the concept of a group is the axioms satisfied by the binary operation involved. Arthur Cayley (1854) was the first who explicitly wrote down an abstract set of axioms for a (finite) group similar, but not identical to those in Box 1. However, in the middle of the 19th century it was still unheard of to define a mathematical structure solely by its axioms, and the definition was ignored. It took another half century before the axiomatic introduction of groups and other algebraic structures such as rings and fields became commonplace. The first algebra book in which algebra was introduced from the outset as the study of algebraic structures was Van der Waerden's *Moderne Algebra* (1930-31). Until then, algebra had been introduced as the study of equations.

2.3   Non-Euclidean Geometry

The philosophically deepest root of the structural view of axioms as arbitrary starting points of our deductions came from geometry [2]. Until around 1800 geometry had been viewed as the paradigm of an exact mathematical theory. However, one particular axiom of the theory had been up for debate from the beginning, namely the so-called parallel postulate. From the other postulates, it can be shown that if two straight lines have a common normal they are parallel, in the Euclidean sense, i.e., they do not intersect. The parallel postulate says that if one of the lines is turned even the smallest angle around its point of intersection with the normal then the two lines will intersect. This postulate was considered far less self-evident than the other postulates. Ancient Greek, medieval Arab, and early modern European mathematicians tried to prove the postulate from the other postulates. If they had succeeded, one could have left out the parallel postulate from the list of postulates. It would have become a theorem. However, although many mathematicians believed they had found a proof, the majority of mathematicians were never convinced.

With the Italian Jesuit Giovanni Girolamo Saccheri (1733) and the Swiss mathematician Johann Heinrich Lambert (1766) the approach to the question changed. They tried to prove the parallel postulate indirectly. They assumed that the parallel postulate was invalid and tried to reach a contradiction. If the parallel postulate is false there are infinitely many lines through a given point parallel to a given line (not through the given point). From this assumption they deduced a lot of consequences and at the end they reached what at least Saccheri considered a contradiction.

Around 1820 three mathematicians Gauss in Göttingen, Nikolai Ivanovich Lobachevsky in Kazan, and Janos Bolyai in Hungary came to the conviction that there was no contradiction in Saccheri's system and that therefore the parallel postulate could not be derived from the other postulates. Moreover, at least Gauss and Lobachevsky were convinced that physical space could very well turn out to be a space in which the parallel postulate does not hold. This kind of geometry was called non-Euclidean by Gauss. The three mathematicians derived many weird properties of such a space, for example: The angle sum of a triangle is less than 180° and it decreases when the area of the triangle grows. This means that there is an upper bound for the areas of triangles and that one cannot reduce the size of a figure without distorting it (an architect's nightmare). But despite these unfamiliar properties the three mathematicians decided, that they were not absurd. Both Gauss and Lobachevsky tried if they could measure a discrepancy from 180° in the angle sum of a triangle in physical space (a large one for which such an effect would be measurable), but they could not detect a deviation from Euclidean space.

At first the investigations of non-Euclidean geometry were overlooked. There are many reasons for that:

1. Two of the discoverers, Lobachevsky and Bolyai were otherwise unknown mathematicians who worked far from Paris, the center of mathematics of the time. Gauss was the most famous mathematician of the time, but he chose not to publish his ideas.
2. The conclusions of Gauss, Lobachevsky, and Bolyai were in sharp contrast to the very influential philosophical ideas of Immanuel Kant. Kant had argued that the geometry of space is an a priori synthetic intuition, and the geometry he had in mind was of course Euclidean geometry. As an example of how we argue

geometrically he had in fact shown how one can prove in an a priori way that the angle sum of a triangle is 180°. Gauss explicitly contradicted this point of view and claimed that the nature of physical space had to be determined empirically.

3. None of the three mathematicians could prove that there was no inconsistency in non-Euclidean geometry.

All this changed around 1870. In 1866 Gauss' correspondence was published showing that the prince of mathematics (Gauss's nickname) had entertained such unconventional ideas. This led to re-publications and translations of the works of Lobachevsky and Bolyai. And finally it was proved that non-Euclidean geometry was as consistent as Euclidean geometry. The first mathematician to give such a proof was the Italian mathematician Eugenio Beltrami (1868). He accomplished this by presenting what we would call a model of non-Euclidean geometry. He built his model on Gauss' differential geometry of surfaces (1827). Gauss had introduced a notion of curvature of a surface, and Beltrami could show that the geometry on a surface with constant negative Gauss curvature was non-Euclidean, at least if one considers geodesics (locally shortest curves) as straight lines. Felix Klein (1872) discovered the same model from the point of view of projective geometry and 10 years later the leading French mathematician Henri Poincaré hit upon another model.

As pointed out most clearly by Poincaré, such models show that if there is a contradiction in non-Euclidean geometry, it would, in the model, appear as a contradiction in Euclidean geometry. Thus if Euclidean geometry is free of contradictions then so is non-Euclidean geometry.

This left mathematicians at the end of the 19th century in the following situation. There are two equally consistent geometries. In Euclidean geometry the parallel postulate is an axiom of the theory, in the other the negation of the parallel postulate is an axiom. How should one decide between them? One way to do that was to find out if the parallel postulate was true of physical space or not. However, mathematicians of the late 19th century were reluctant to use this way out. First, during the 19th century, mathematics had emancipated itself from physics. To be sure, physics was still a great source of inspiration for mathematicians, but to let the foundations of mathematics depend on physical measurements was more than many late 19th mathematicians would allow. Second, the measurement of the (non-) Euclidean nature of space turned out to be a non-trivial matter. Indeed, if space is non-Euclidean it differs very little from Euclidean space in the sense that angle sums in measurable triangles are very close to 180°. So it is very likely that measurements will not be able to reveal a discrepancy from Euclidean geometry. But of course that does not exclude that space is non-Euclidean, it only shows that the deviation from Euclidean geometry is below our accuracy of measurement. So, leaving the decision to the physicists would delay the decision indefinitely, perhaps infinitely.

Moreover, Poincaré pointed out that the nature of space cannot be investigated empirically in isolation independently of other physical theories of nature. He argued as follows: assume that some day one can measure, that a large enough triangle has in fact an angle sum less than 180°. In that case we have two alternatives:

We can conclude that physical space is non-Euclidean, or we can conclude that the triangle was not a real triangle after all. For example if the measurement is conducted by optical sighting between the vertices of the triangle we might conclude that the light rays that compose the sides of the triangle are not straight lines. That would force us to revise the laws of physics (Maxwell's laws) but we can uphold Euclidean geometry. For Poincaré this shows that the nature of physical space is a convention. He actually argued that since Euclidean geometry is simpler than non-Euclidean geometry we will always stick to the convention that physical space is Euclidean. In case of a discrepancy as the one explained above, we will always choose to revise the laws of physics rather than the more basic laws of geometry.

Poincaré's prediction turned out to be wrong. In fact, in 1905 and 1917, based on new empirical evidence, Albert Einstein suggested that space (or rather space-time) is best modelled on a kind of non-Euclidean geometry.

But let us return to the situation in the late 19th century. Mathematicians had developed two equally consistent geometric theories and there was no good way to decide which one to choose. In that situation mathematicians decided not to choose. Both alternatives were left as proper mathematical theories. One could choose the parallel postulate as an axiom or one could choose its negation. It is up to the mathematician, and one choice is as good as another. Two different but equally valid geometries will result from the choices. This situation illustrates Georg Cantor's often quoted dictum (1896): "The essence of mathematics lies in its freedom." But of course this freedom does not allow the traditional view of axioms as self-evident truths. One cannot claim that both the parallel postulate and its negation are self-evident. Instead axioms began to be considered as arbitrary starting points for our deductions.

2.4 Hilbert's Foundations of Geometry

Toward the end of the 19th century it was obvious that Euclid's geometry had to be rewritten. To be sure, the millennium long debate concerning the parallel postulate had revealed, that this postulate is indeed necessary, if one wants a geometry as the one described by Euclid. But it had also been revealed that Euclid in his proofs used many geometric assumptions that were neither stated as axioms nor deducible from the explicitly formulated axioms in Euclid's Elements. For example, already in the first theorem, where Euclid constructs an equilateral triangle, he postulates that two specific circles intersect each other despite the fact that no axiom guarantees the existence of an intersection point. For this and other reasons, several mathematicians tried to come up with complete axiom systems for geometry. The most famous of them was David Hilbert's Grundlagen der Geometrie (1899) [1], [3]. Hilbert needed 21 axioms instead of Euclid's 5 in order to be able to deduce the usual Euclidean theorems. One of them was a slight reformulation of the parallel postulate.

Hilbert changed Euclid's presentation in another and even more fundamental way namely with regard to the definitions. Euclid's Elements begins with a definition of the basic objects of geometry.

"A point is that which has no part"
"A line is a breadthless length"
"A straight line is a line which lies evenly with the points on itself."

And so on. It is obvious that these definitions leave something to be desired. For example, one can ask: What are a width and a length, and in particular what does it mean that points lie evenly with other points? One could be tempted to answer the last question with: when they lie on a straight line; but in that case the definition is circular. We are here up against the same problem that we met when we discussed the beginning of the process of deduction. There we argued that we need to begin with axioms that are just stated without requiring proofs. When analyzing the problem of definitions we meet with same problem that was pointed out by Aristotle: If we want to define the meaning of a certain word as "straight line" we must do so in terms of other words. But in order to define these more basic words, we must use even more basic ones, etc. Thus, in order to begin the process of definition we must begin with words that we do not define.

Hilbert took the consequence of this analysis and left words as point, straight line, lie on, between, etc., as undefined. He pointed out that the meaning of the words indirectly appear from the axioms. Such implicit definitions were sharply criticized by the philosopher-mathematician Gottlob Frege. He pointed out to Hilbert that such definitions did not allow one to decide if a particular object is a point or a line. For example he asked Hilbert if his pocket watch could not be considered as a point if one followed Hilbert's approach. Hilbert answered in the positive, but did not consider this as a problem. If one can find a set including the pocket watch that can play the role of points and another set that can play the role of lines, and a suitable interpretation of the words lie on, lie between, etc., then this is indeed a Euclidean geometry in Hilbert's sense and the pocket watch is a point of the theory.

He had another reason to let the objects remain undefined. In projective geometry Joseph Diaz Gergonne (early 19th century France) and other mathematicians had discovered a property called duality. If a theorem about points and lines in a plane is true, then one can find another true theorem by interchanging the words point and line. For example it is true that through two different points there is precisely one line (this is an axiom both in Euclid and in Hilbert). By interchanging the words point and line we get the statement: Through two different lines there is exactly one point. As formulated here, the statement sounds a bit strange, but if we interpret "through" as "on" (which we can because the word "through" is undefined) the statement says that given two different lines there is exactly one intersection point. This still sounds wrong, because two parallel lines do not intersect in Euclidean geometry, but in projective geometry one has adjoined points at infinity that play the role of the missing intersection points. So in projective geometry the dual statement is indeed true.

In view of the duality of projective geometry it is irrelevant whether we consider points and lines as we use to do or whether the long thing we usually call a straight line is considered a point, and the dot we usually call a point is considered a line. In this way the question of the nature of the point and the straight line or any other object in a mathematical theory becomes meaningless.

Thus for Hilbert an axiomatic mathematical system is a system of arbitrary axioms from which one can deduce theorems about undefined objects. However he stressed that one need to check one thing: The consistency of the axioms. If the axioms are consistent then the system exists in Hilbert's sense. Thus Hilbert reduced the problem of existence in mathematics to the problem of consistency.

Existence of objects in a theory is determined by the rules of existence laid down in the axioms. A particular geometric configuration exists in Euclidean geometry, if it is postulated in the axioms of geometry (for example one of Hilbert's axioms posits that there are at least four points not lying in a plane) or can be deduced from the existential claims in the axioms. For example one can prove that there exists a square on a given line segment.

In 1901 the mathematician and philosopher Bertrand Russell described this new view of mathematics as follows:

> "Pure mathematics consists entirely of assertions to the effect that, if such and such a proposition is true of anything, then such and such another proposition is true of that thing. It is essential not to discuss whether the first proposition is really true, and not to mention what the anything is, of which it is supposed to be true. Both these points would belong to applied mathematics. We start, in pure mathematics, from certain rules of inference, by which we can infer that if one proposition is true, then so is some other proposition. These rules of inference constitute the major part of the principles of formal logic. We then take any hypothesis that seems amusing, and deduce its consequences. If our hypothesis is about anything, and not about some one or more particular things, then our deductions constitute mathematics. Thus mathematics may be defined as the subject in which we never know what we are talking about, nor whether what we are saying is true. People who have been puzzled by the beginnings of mathematics will, I hope, find comfort in this definition, and will probably agree that it is accurate".

In fact, since Russell, this definition has been relativized even more. In Russel's definition of mathematics the one absolute is the principles of formal logic. Since then mathematicians and logicians have debated what these laws are, and today there are several competing schools favoring different rules of logical inference (e.g., formalist and intuitionist logic).

Another unsettling discovery concerning mathematical structures emerged in the 1930s when Kurt Gödel proved that consistency of a (suitably forceful system) cannot be proved within the system itself. This has the unpleasant consequence, that we cannot be sure that our mathematical theories are consistent. This in a sense puts an end to the long search for absolute certainty in mathematics.

In this section we have given a historical account of a very important change in the philosophy of mathematics: the change from thinking about mathematics as an idealized or abstract physical reality to thinking about it as a study of mathematical structures whose elements are undefined and whose axioms are in principle arbitrary. This change is often considered as one of the main revolutions of mathematics.

3   How arbitrary are mathematical axioms in practice?

According to Russell one begins a mathematical theory by picking "any hypothesis that seems amusing" and then one starts deducing its consequences. In this view mathematics is just a game the rules of which are chosen entirely arbitrarily. At first this may seem to be in complete accordance with Hilbert's ideas about axiomatic systems. However, in a lecture of 1919–20 about his axiomatic method Hilbert flatly disagreed with Russell emphasizing that "We are not speaking here of arbitrariness in any sense. Mathematics is not like a game whose tasks are determined by arbitrarily stipulated rules." The difference between the two great thinkers may be described as a difference between a philosopher and a working mathematician. From a philosophical point of view it may be defendable to view axioms as arbitrary points of departure of the mathematical deductive game. However, a mathematician like Hilbert would know, that in a mathematical sense the axioms are suggested by earlier mathematical developments. In fact Hilbert's view is revealed in the term he often used to describe his method. He talked about axiomatization. Of course, in order to axiomatize a theory it is necessary that such a theory is already developed in a non-axiomatic way. As Hilbert pointed out (see [1]), mathematics is not build like a house where one starts by laying the foundations. Instead, a mathematical theory is often begun without proper foundation, and only when the lack of a proper foundation becomes urgent, it is laid. According to Hilbert, axiomatization is a way to provide a foundation for already rather mature mathematical theories. Hilbert axiomatized geometry, a discipline that had been around in an imperfectly axiomatized version for more than 2000 years. He did not pick his axioms of geometry just because he thought they were amusing, as Russell put it, but because they were needed to support the structure we call Euclidean geometry. In a sense the axioms can be said to be proof generated in the sense that they are chosen to allow certain deductions that we want to make in order to arrive at the important theorems of the theory such as Pythagoras' theorem. In the same way Hilbert axiomatized the theory of arithmetic of the real numbers and other previously developed theories.

It is doubtful if any important mathematical theory has ever been developed as Russell suggests, by first choosing an amusing set of axioms and then deducing away. This raises an important question: What is an interesting mathematical theory? Or phrased differently which theories will a good mathematician consider worth studying? Hilbert tried to explain and analyze what a true mathematical system is, but he did not develop any way to distinguish interesting from uninteresting axiomatic systems. Considering how many strange axiom systems one could in principle investigate, it is striking how few are in fact studied by mathematicians. In a sense any game with precisely stated rules can be considered a mathematical system, but they are not studied by mathematicians. Chess and a few similar games have been studied from a mathematical point of view, but groups have been studied much more extensively.

Similarly, within an axiomatic system the theorems are usually not discovered simply by deducing away beginning with the axioms. Considering the huge amount

of mostly uninteresting things one can deduce it is not surprising that mathematicians usually discover interesting theorems by formulating heuristic conjectures and
subsequently trying to find a proof. But again one can ask: what distinguishes the
important theorems from the unimportant ones.

4   What is important? An historical approach

There is no formal mathematical way to measure the degree of interest of a mathematical structure or of a theorem within a structure. Still among mathematicians,
one will find wide agreement about the question, with some minor differences of
opinion on the edges.

   In order to approach the question we need to investigate the driving forces in
mathematics. What makes mathematicians develop this theory rather than that; what
makes them want to prove this theorem rather than that? One answer was given by
Hilbert in a famous public address to the international Congress of Mathematicians
in 1900. According to Hilbert *problems* are the driving force behind the development
of mathematics. In the talk and the subsequent written account of it he formulated
23 problems whose solution he expected would develop mathematics in the 20th
century. The problems mentioned by Hilbert were all problems formulated inside
mathematics. But also problems outside mathematics have been powerful driving
forces for the development of mathematics. In a sense such extra-mathematical
problems constitute the origin of mathematics. Take the concept of number. This
concept was surely developed as a means to deal with problems of size and of
number of physical things or social questions as trade. But once it was established
the number concept gave rise to questions of a more internal mathematical nature:
How many prime numbers are there? (The answer: "infinitely many" was found by
the Ancient Greeks). How many prime twins (i.e., primes which are two apart as 3
and 5) are there? (The answer is unknown). Similarly geometry clearly began as a
way to deal with physical space, but it soon raised questions that have little bearing
on practical applications for example: Can one prove the parallel postulate from the
other axioms?

   So problems in nature and society provided the background for the beginning
of mathematics. And extra-mathematical problems have continued to influence
the development of mathematics from antiquity till this day. In ancient Greece,
music, astronomy, optics and statics provided stimulus to mathematics. These areas
of application were followed in the Renaissance and the early modern period by
kinematics and perspective drawing and in the 19th century by heat, electricity,
and elasticity. Finally in the 20th century almost all aspects of science, technology,
and economics and society were studied and managed by mathematical models.
All through history these areas of application continued to pose new mathematical
questions that necessitated the development of existing theories or the creation of
new areas of mathematics.

The development of mathematics has taken place through a combination of theorem proving, problem solving, generalization, abstraction, rigorization, axiomatization, etc. But in the final end, what has determined the success of a theory, a structure or a theorem is its ability to be applied in reality or in other mathematical theories, or more generally to fit in a fruitful way into the web of already developed mathematics [5]. For example the concept of a group was developed in order to solve equations and it has shown its worth because it fits into many other mathematical theories and helps solving problems in physics and chemistry. Similarly, Pythagoras' theorem is continually being taught not only because it can be proved from the axioms of Euclidean geometry but also because it is useful both in practical applications and (often in a generalized higher dimensional version) in many parts of mathematics.

Thus, Russell may be philosophically justified when he claimed that mathematics is like a game whose rules or axioms are entirely arbitrary and that mathematical theorems are found by successive derivations from the axioms. However, this image does not capture the real driving forces behind the practice of mathematics, and it cannot explain why certain structures and theorems are deemed important and others are not. A historian on the other hand, can tell stories of their genesis, stories that can shed light on this question. Such a story will contain a story of repeated abstraction, which according to Aristotle reveals the ontology of mathematical objects, but it will be much richer. In addition to the process of abstraction it will operate with many other driving forces and processes, and it will ultimately be judged by the question of applicability within or outside of mathematics. The story will not be deterministic as an Aristotelian story of abstraction might suggest. Instead there will enter many contingencies and many influences from all areas of society and nature.

5 Conclusion

The modern formalist structuralist view of mathematics leaves very little to be said concerning the nature of mathematical objects. The whole point of the structuralist approach is that the objects are left undefined. Any object whatsoever can play the role of an element in a group, a point in Euclidean geometry or as the real number 708. The only thing that characterizes an element in a mathematical structure is how it relates to the other elements in the structure. This is laid down in the axioms. From a formalistic point of view there is nothing else one can say.

And what is the ontological status of a mathematical structure? According to the formalistic philosophy consistency is the only requirement of existence. If it is consistent the system exists as a mathematical structure. In this way modern mathematics has in a sense circumvented the question of ontology.

However, from the point of view of the practice and history of mathematics the formalist account of existence is not complete. History of mathematics can provide stories that elucidate the formalistic account in several respects. First the historian can tell the story of how the modern formalist, structuralist point of view came about. This is what I tried to do in the long middle section of this chapter. Second the historian can tell the story of how particular structures came to be. This story is always richer than the standard story told by Russell: The mathematician X found this or that set of axioms amusing and began to deduce theorems from them. In real life, the story mostly begins with a problem, either a problem outside mathematics (be it practical or theoretical) or a problem within mathematics. The story will then unfold through a messy series of new mathematical theorems and more importantly through a series of transformations of the theory through generalizations, abstractions, clarifications, axiomatizations, etc. On the way, new problems inside or outside the theory, will drive the development in different directions, until it reaches the intermediate stage that we call the present. And the story will always contain an essential human element.

The story will also reveal the origin of the objects that in the end turn up as undefined objects in the theory: The lines and points in geometry, the real numbers, the elements in a group. In a sense such stories can be considered an alternative ontology of the modern structures and their objects.

And at least the history of mathematics can explain how certain mathematical structures came to be studied intensely while the majority of possible structures are not considered at all. In this way history can reveal why certain structures, objects, and results exist, not in principle but in the sense that they are part of mathematical practice.

References

1. L. Corry, *Hilbert and the Axiomatization of Physics (1898-1918): From "Grundlagen der Geometrie" to "Grundlagen der Physik"*, Dordrecht: Kluwer (2004)
2. J. Gray, *Ideas of Space. Euclidean, Non-Euclidean, and Relativistic*, Oxford University Press (1989)
3. D. Hilbert, *Natur und mathematisches Erkennen*, Ed. D. Rowe, Birkhäuser (1992)
4. V. Katz, *A History of Mathematics*, 3rd ed. Pearson (2009)
5. J. Lützen, "The Physical Origin of Physically Useful Mathematics", *Interdisciplinary Science Reviews*. Vol. 36 (2011) 229-243
6. H. Wussing, *The Genesis of the Abstract Group Concept: A Contribution to the History of the Origin of Abstract Group Theory*, Dover Publications (2007)

From the continuum to large cardinals

John Stillwell

Abstract Continuity is one of the most important concepts of mathematics and physics. This has been recognized since the time of Euclid and Archimedes, but it has also been understood that continuity raises awkward questions about infinity. To measure certain geometric quantities, such as the diagonal of the unit square, or the area of a parabolic segment, involves infinite processes such as forming an infinite sum of rational numbers. For a long time, it was hoped that "actual" infinity could be avoided, and that the necessary uses of infinity could be reduced to "potential" form, in which infinity is merely approached, but never reached. This hope vanished in the 19th century, with discoveries of Dedekind and Cantor. Dedekind defined the points on the number line via infinite sets of rational numbers, and Cantor proved that these points form an **uncountable** infinity—one that is actual and not merely potential. Cantor's discovery was part of his theory of infinity that we now call **set theory**, at the heart of which is the discovery that most of the theory of real numbers, real functions, and measure takes place in the world of actual infinity. So, actual infinity underlies much of mathematics and physics as we know it today. In the 20th century, as Cantor's set theory was systematically developed, we learned that the concept of measure is in fact entangled with questions about the entire universe of infinite sets. Thus the awkward questions raised by Euclid and Archimedes explode into huge questions about actual infinity.

1 Introduction

The real number system $\mathbb{R}$, traditionally called the *continuum*, has been a mystery and a spur to the development of mathematics since ancient times. The first important event in this development was the discovery that $\sqrt{2}$ is irrational; that is, not a ratio of whole numbers. This discovery led to the realization that *measuring*

J. Stillwell (✉)
Mathematics Department, University of San Francisco, 2130 Fulton Street,
San Francisco, CA 94117, USA
e-mail: stillwell@usfca.edu

© Springer International Publishing Switzerland 2015
E. Davis, P.J. Davis (eds.), *Mathematics, Substance and Surmise*,
DOI 10.1007/978-3-319-21473-3_10

(in this case, measuring the diagonal of the square) is a more general task than *counting*, because there is no unit of length that allows both the side and the diagonal of the square to be counted as integer multiples of the unit.

Thus, by around 300 BCE, when Greek mathematics was systematized in Euclid's *Elements*, there was a clear distinction between the theory of magnitudes (measuring length, area, and volume) and the theory of numbers (whole numbers and their arithmetic). The former was handled mainly by geometric methods; the latter was like the elementary number theory of today, and its characteristic method of proof was *induction*. Euclid did not recognize induction as an axiom, but he used it—in the form we now call *well-ordering*—to prove crucial results. For example, in the *Elements*, Book VII, Proposition 31, he uses induction to prove that any composite number n is divisible by a prime as follows.

Given that $n = ab$, for some positive integers $a, b < n$, if either a or b is prime we are done; if not, repeat the argument with a. This process must eventually end with a prime divisor of n. If not, we get an infinite decreasing sequence of positive integers which, as Euclid says, "is impossible in numbers."

Thus Euclid is using the principle that *any decreasing sequence* (of positive integers) *is finite*, which is what we now call the *well-ordering* property of the positive integers.

Euclid's theory of magnitudes is evidently something more than his theory of numbers, since it covers magnitudes such as $\sqrt{2}$, but it is hard to pinpoint the extra ingredient, because Euclid's axioms are sketchy and incomplete by modern standards. (Nevertheless, they served mathematics well for over two millenia, setting a standard for logical rigor that was surpassed only in the 19th century.) The relationship between the theory of magnitudes and the theory of numbers was developed in Book V of the *Elements*, a book whose subtlety troubled mathematicians until the 19th century. Although the details of Book V are complicated, its basic idea is simple: compare magnitudes (our real numbers) by comparing them with *rational* numbers. That is

$$x < y \quad \text{if and only if} \quad x < \frac{m}{n} < y \quad \text{for some integers } m, n.$$

The power of this idea was not fully realized until 1858, when Dedekind used to give the first rigorous definition of the continuum.

In the meantime, the *intuition* of the continuum had developed and flourished since the mid-17th century. Newton based the concepts of calculus on the idea of *continuous motion*, and the subsequent development of *continuum mechanics* exploited the continuum of real numbers (and its higher-dimensional generalizations) as a model for all kinds of *continuous media*. This made it possible to study the behavior of vibrating strings, fluid motion, heat flow, and elasticity. Almost the only exception to continuum modeling before the 20th century was Daniel Bernoulli's kinetic theory of gases (1738, with extensions by Maxwell and Boltzmann in the 19th century), in which a gas is modeled by a large collection of particles colliding with each other and the walls of a container. Even there, the underlying concept was continuous motion, since each particle was assumed to move continuously between collisions.

Thus, long before the continuum was well understood, it had become indispensable in mathematics.

The understanding of the continuum finally advanced beyond Euclid's Book V in 1858, when Dedekind extended the idea of comparing x and y via rational numbers to the idea of *determining* x (or y) via rational numbers. But while we can separate x from y finding just *one* rational number m/n strictly between them, to determine x we need to separate (or *cut*) the *whole set* of rational numbers into those less than x and those greater than x. Thus the exact determination of an irrational number x requires infinitely many rational numbers, say those greater than x. For example, $\sqrt{2}$ is determined by the set of positive rational numbers m/n such that $m^2 > 2n^2$.

Although Dedekind's concept of a cut in the rational numbers is a logical extension of the ancient method for comparing two lengths, it is *not* a step that the Greeks would have taken, because they did not accept the concept of an infinite set. In determining real numbers by infinite sets of rational numbers, Dedekind brought to light the fundamental role of infinity in mathematics. In a sense we will make precise later, mathematics can be defined as "elementary number theory plus infinity." But before doing so (and before deciding "how much" infinity we need), let us examine more closely what we want from the continuum.

2 What does the Continuum Do for Mathematics?

The immediate effect of Dedekind's construction of the continuum was to unify the ancient geometric theory of magnitudes with the arithmetic theory of numbers in a *number line*. Certainly, mathematicians for centuries had assumed there was such a thing, but Dedekind was the first to define it and to prove that it had the required geometric and arithmetic properties: numbers fill the line without gaps and they can be added and multiplied in a way that is compatible with addition and multiplication of rational numbers.

As a consequence, we get the basic concepts of the theory of measure: length, area, and volume. The *length* of the interval $[a, b]$ from a to b is $b - a$; the *area* of a rectangle of length l and width w is lw; the *volume* of a box of length l, width w, and depth d is lwd. Today, we take it for granted that the area or volume of a geometric figure is simply a number, but this is not possible until there is a number for each point on the line.[1]

The absence of gaps in $\mathbb{R}$, which we now call *completeness*, was called *continuity* by Dedekind. Indeed, Dedekind's continuous line is a precursor to the modern concept of continuous *curve* (in the plane, say) which is defined by a continuous function from the line or an interval of the line into the plane. Such a curve has

---

[1]The ancient Greeks had a much more complicated theory of measure, in which the measure of a rectangle is *the rectangle itself*, and two rectangles have the same measure only if one can be cut up and reassembled to form the other.

no gaps because the line has none. Another consequence of the "no gaps" property is the *intermediate value theorem* for continuous functions: if f is continuous on an interval $[a, b]$ and $f(a)$ and $f(b)$ have opposite signs, then $f(c) = 0$ for some c between a and b. Other basic theorems about continuous functions also depend on the completeness of $\mathbb{R}$. In fact, it was to provide a sound foundation for such theorems of analysis that Dedekind proposed a definition of $\mathbb{R}$ in the first place.

With this foundation established, all the applications of analysis to continuum mechanics, mentioned in the previous section, become soundly based as well. The continuum also meets other needs of analysis, because its completeness implies that various infinite processes on $\mathbb{R}$ lead to a result whenever they ought to. For example, an *infinite sequence* of real numbers

$$x_1, \quad x_2, \quad x_3, \quad \cdots$$

has a *limit* if and only if it satisfies the *Cauchy criterion*: for all $\varepsilon > 0$ there is an N such that

$$|x_m - x_n| < \varepsilon \quad \text{for all} \quad m, n > N.$$

With the help of the limit concept we can make sense of *infinite sums* $a_1 + a_2 + a_3 + \cdots$ of real numbers. Namely, if $a_1 + a_2 + \cdots + a_n = S_n$, then $a_1 + a_2 + a_3 + \cdots$ is the limit of the sequence $S_1, S_2, S_3, \ldots$, if it exists.

Infinite sums are also important in geometry. Special cases were used (implicitly) by Euclid and Archimedes to determine areas and volumes. A wonderful example is the way Archimedes determined the area of a parabolic segment by dividing it into infinitely many triangles, as indicated in Figure 1. In modern language, this is the segment of the parabola $y = x^2$ between $x = -1$ and $x = 1$. The first approximation to its area is the black triangle, which has area 1. The second approximation is obtained by adding the two dark gray triangles, which have total area 1/4. The third is obtained by adding the four light gray triangles, which have total area $(1/4)^2$, and so on.

Fig. 1 Filling the parabolic segment with triangles

It is clear that by taking enough triangles we can approach the area of the parabolic segment arbitrarily closely, and it can also be shown (by the formula for the sum of a geometric series) that

$$1 + \frac{1}{4} + \left(\frac{1}{4}\right)^2 + \left(\frac{1}{4}\right)^3 + \cdots = \frac{4}{3}.$$

Thus the area being approached in the limit—the area of the parabolic segment—is 4/3. It is a happy accident in this case that the area turns out to be expressible by a geometric series, which we know how to sum. In other cases we can exploit the geometric series to prove that certain sets of points have *zero* length or area.

For example, let $S = \{x_1, x_2, x_3, \ldots\}$ be any subset of $\mathbb{R}$ whose members can be written in a sequence. I claim that S has zero length. This is because, for any $\varepsilon > 0$, we can cover S by intervals of total length $\leq \varepsilon$. Namely,

cover x_1 by an interval of length $\varepsilon/2$,
cover x_2 by an interval of length $\varepsilon/4$,
cover x_3 by an interval of length $\varepsilon/8$, and so on.

In this way, each point of S is covered, and the total length of the covering is at most

$$\frac{\varepsilon}{2} + \frac{\varepsilon}{4} + \frac{\varepsilon}{8} + \cdots = \varepsilon.$$

Thus the length of S is $\leq \varepsilon$, for any $\varepsilon > 0$, and so the length of S can only be zero.

This innocuous application of the geometric series has two spectacular consequences:

1. The set of rational numbers between 0 and 1 has total length zero, because these numbers can be written in the sequence

$$\frac{1}{2}, \frac{1}{3}, \frac{2}{3}, \frac{1}{4}, \frac{3}{4}, \frac{1}{5}, \frac{2}{5}, \frac{3}{5}, \frac{4}{5}, \frac{1}{6}, \frac{5}{6}, \ldots$$

   (taking all those with denominator 2, then those with denominator 3, and so on). It follows that *the rational numbers do not include all the numbers between 0 and 1*, which gives another proof that irrational numbers exist.
2. The real numbers between 0 and 1 cannot be written in a sequence. We say that there are *uncountable many* real numbers between 0 and 1. It follows that $\mathbb{R}$ itself is uncountable.[2]

---

[2]This argument was discovered by Harnack in 1885, but he was confused about what it meant. It seemed to him that covering all the rationals in [0,1] by intervals would cover *all* points in [0,1], so an interval of length 1 could be covered intervals of total length $\leq \varepsilon$. Fortunately for measure theory, this is impossible, because of the Heine-Borel theorem that if an infinite collection of open intervals covers [0,1] you can extract a finite subcollection that also covers [0,1]. By unifying

Thus the theory of measure, which $\mathbb{R}$ makes possible, tells us something utterly unexpected: *uncountable sets exist, and $\mathbb{R}$ is one of them.* In fact, the theory of measure brings to light even more unexpected properties of infinity, as we will see, but first we should take stock of the implications of uncountability.

3 Potential and Actual Infinity

When Archimedes found the area of the parabolic segment by dividing it into infinitely many triangles he did not argue precisely as I did in the previous section. He did not sum the infinite series

$$1 + \frac{1}{4} + \left(\frac{1}{4}\right)^2 + \left(\frac{1}{4}\right)^3 + \cdots,$$

but only the finite series

$$1 + \frac{1}{4} + \left(\frac{1}{4}\right)^2 + \cdots + \left(\frac{1}{4}\right)^n,$$

which has sum $\frac{4}{3} - \frac{1}{3}\left(\frac{1}{4}\right)^n$. This makes little practical difference, since the finite collection of triangles being measured approaches the parabolic segment in area, and the numerical value of the area clearly approaches 4/3. However, there was a *conceptual* difference that was important to the Greeks: only *finite* sets need be considered.

The Greeks invariably avoided working with infinite sets when it was possible to work with finite sets instead, as is the case here. The avoidable infinite sets are those we now call *countably* infinite. As the name suggests, an infinite set is countable if we can arrange its members in a list

1st member,   2nd member,   3rd member,   . . .

so that each member occurs at some finite position. The prototype countable set is the set $\mathbb{N} = \{1, 2, 3, \ldots\}$ of positive integers. Indeed, another way to say that an infinite set is countable is to say that there is a one-to-one correspondence between its members and the members of $\mathbb{N}$. Another countably infinite set that we have already seen is the set

---

any overlapping intervals in the finite covering you get a covering of [0,1] by finitely many open *disjoint* intervals of total length $\leq \varepsilon$, which is clearly impossible.

    It was to deal with precisely this problem that Borel in 1898 introduced the Heine-Borel theorem. He called it "the first fundamental theorem of measure theory."

$$\left\{ \frac{1}{2}, \frac{1}{3}, \frac{2}{3}, \frac{1}{4}, \frac{3}{4}, \frac{1}{5}, \frac{2}{5}, \frac{3}{5}, \frac{4}{5}, \frac{1}{6}, \frac{5}{6}, \ldots \right\}$$

of rational numbers between 0 and 1.

The Greeks, and indeed almost all mathematicians until the 19th century, believed that countably infinite sets existed in only a *potential*, rather than *actual*, sense. An infinite set is one whose members arise from a *process* of generation, and one should not imagine the process being completed because it is necessarily *endless*. Likewise, a convergent infinite sequence is a process that grows toward the limiting value, without actually reaching it. Mathematicians issued stern warnings against thinking of infinity in any other way. An eminent example was Gauss, who wrote in a letter to Schumacher (12 July 1831) that:

> I protest against the use of infinite magnitude as something completed, which is never permissible in mathematics. Infinity is merely a way of speaking, the true meaning being a limit which certain ratios approach indefinitely closely, while others are permitted to increase without restriction.

As long as it was possible to avoid completed infinities, then of course it seemed safer to reject them. But it ceased to be possible when Cantor discovered the uncountability of $\mathbb{R}$ in 1874. Since $\mathbb{R}$ is uncountable, it is impossible to view it as the unfolding of a step-by-step process, as $\mathbb{N}$ unfolds from 1 by repeatedly adding 1. For better or for worse, we are forced to view $\mathbb{R}$ as a completed whole.

Cantor's first uncountability proof was not the measure argument given in the previous section, but one related to Dedekind's idea of separating sets of rationals into upper and lower parts. In 1891 Cantor discovered a new, simpler, and more general argument that leads to an unending sequence of larger and larger infinities. This is the famous *diagonal argument*, which shows that any set X has more subsets than members.

Suppose that, for each $x \in X$, we have a subset $S_x \subseteq X$. Thus we have "as many" subsets as there are members of X. It suffices to show that the sets S_x do not include all subsets of X. Indeed, they do not include the subset

$$D = \{x : x \notin S_x\},$$

because D differs from each S_x with respect to the element x.

An important consequence of the diagonal argument is that *there is no largest set*. This is because, for any set X, there is the larger set $\mathscr{P}(X)$ whose members are the subsets of X.

The reason for calling Cantor's argument "diagonal" becomes apparent if we take the case $X = \mathbb{N}$. Then subsets $S_1, S_2, S_3, \ldots$ of $\mathbb{N}$ may then be encoded as rows of 0s and 1s in an infinite table like that shown in Figure 2, where a 1 in column n says that n is a member, and 0 in column n says that n is not a member. Thus, in the table shown, S_1 is the set of even numbers, S_2 is the set of squares, and S_3 is the set of primes. The diagonal digits (shown in bold) say whether or not $n \in S_n$, and we construct D by switching each of these digits, so $n \in D$ if and only if $n \notin S_n$.

subset	1	2	3	4	5	6	7	8	9	10	11	
S_1	**0**	1	0	1	0	1	0	1	0	1	0	...
S_2	1	**0**	0	1	0	0	0	0	1	0	0	...
S_3	0	1	**1**	0	1	0	1	0	0	0	1	...
S_4	1	0	1	**0**	1	0	1	0	1	0	1	...
S_5	0	0	1	0	**0**	1	0	0	1	0	0	...
S_6	1	1	0	1	1	**0**	1	1	0	1	1	...
S_7	1	1	1	1	1	1	**1**	1	1	1	1	...
S_8	0	0	0	0	0	0	0	**0**	0	0	0	...
S_9	0	0	0	0	0	0	0	0	**1**	0	0	...
S_{10}	1	0	0	1	0	0	1	0	0	**1**	0	...
S_{11}	0	1	0	0	1	0	0	1	0	0	**0**	...
$\vdots$												
D	**1**	**1**	**0**	**1**	**1**	**1**	**0**	**1**	**0**	**0**	**1**	...

Fig. 2 The diagonal argument

The set $\mathscr{P}(\mathbb{N})$ is in fact "of the same size" as $\mathbb{R}$, because there is a one-to-one correspondence between infinite sequences of 0s and 1s and real numbers. Such a correspondence can be set up using binary expansions of real numbers, but we skip the details, which are somewhat messy. $\mathscr{P}(\mathbb{N})$ is just one of several interesting sets of the same size (or *same cardinality*, to use the technical term) as $\mathbb{R}$. Another is the set $\mathbb{R}^2$ of points in the plane, and an even more remarkable example is the set of *continuous functions* from $\mathbb{R}$ to $\mathbb{R}$. Thus all of these sets are uncountable, and hence not comprehensible as merely "potential" infinities.

4   The Unreasonable Effectiveness of Infinity

In 1949, Carl Sigman and Bob Russell wrote the jazz standard "Crazy he calls me," soon to be made famous by Billie Holliday. Its lyrics include the lines

> The difficult I'll do right now
> The impossible will take a little while

A mathematical echo of these lines is a saying attributed to Stan Ulam:

> The infinite we shall do right away.
> The finite may take a little longer.

I do not know a direct source for this quote, or what Ulam was referring to, but it captures a common experience in mathematics. Infinite structures often arise more naturally, and are easier to work with than their finite analogues.

The continuum is a good example of an infinite structure that is easier to work with than any finite counterpart, perhaps because there *is* no good finite counterpart.

Even when one is forced to work with a finite instrument, such as a digital computer, it is easier to have an underlying continuum model, such as $\mathbb{R}^3$ in the case of robotics, as Ernest Davis has explained in his chapter.

In the history of mathematics there are many cases where $\mathbb{R}$ helps, in a mysterious way, to solve problems about finite objects. There are theorems about natural numbers first proved with the help of $\mathbb{R}$, and theorems about $\mathbb{R}$ that paved the way for analogous (but more difficult) theorems about finite objects.

An example of the former type is Dirichlet's 1837 theorem on primes in arithmetic progressions: *if a and b are relatively prime, then the sequence*

$$a, \quad a+b, \quad a+2b, \quad a+3b, \quad \ldots$$

contains infinitely many primes. Dirichlet proved this theorem with the help of analysis, and it was over a century before an "elementary" proof was found, by Selberg in 1948. Even now, analytic proofs of Dirichlet's theorem are preferred, because they are shorter and seemingly more insightful.

An example of the latter type is the classification of finite simple groups, the most difficult theorem yet known, whose proof is still not completely published. An infinite analogue of this theorem, the *classification of simple Lie groups*, was discovered by Killing in 1890. It is likely that the classification of finite simple groups would not have been discovered without Killing's work, because it required the insight that many finite groups are analogues of Lie groups, obtained by replacing $\mathbb{R}$ by a finite field.

Thus, for reasons we do not really understand, $\mathbb{R}$ gives insight and brings order into the world of finite objects. Is it also possible that some higher infinity brings insight into the world of $\mathbb{R}$? The answer is yes, and in fact we understand better how higher infinities influence the world of $\mathbb{R}$ than how $\mathbb{R}$ influences the world of natural numbers. The surprise is just how "large" the higher infinities need to be. To explain what we mean by "large," we first need to say something about axioms of set theory.

5   Axioms of Set Theory

The idea that sets could be the foundation of mathematics developed in the 19th century. Originally their role was thought to be quite simple: there were numbers, and there were sets of numbers, and that was sufficient to define everything else (this was when "everything" was based on $\mathbb{R}$). But then in 1891 the set concept expanded alarmingly when Cantor discovered the endless sequence of cardinalities produced by the power set operation $\mathscr{P}$. It follows that there is no largest set, as we saw in Section 3, and hence *there is no set of all sets*. For some mathematicians (Dedekind among them) this was alarming and paradoxical because it implies that *not every property defines a set*. In particular, the objects X with the property "X is a set" do not form a set.

However, there is a commonsense way to escape the paradox, first proposed by Zermelo in 1904. Just as the natural numbers are generated from 0 by the operation of successor, and there is no "number of all numbers" because every number has a successor, sets might be generated from the empty set $\emptyset$ by certain operations, such as the power set operation. With this approach to the set concept there is no "set of all sets" because every set X has a power set $\mathscr{P}(X)$, which is larger.

Thus Zermelo began with the following two axioms, which state the existence of $\emptyset$ and the definition of equality for sets:

Empty Set.   There exists a set $\emptyset$ with no members.

Extensionality.   Sets are equal if and only if they have the same members.

All other sets are generated by the following *set construction axioms*.

Pairing.   For any sets A, B there is a set $\{A, B\}$ whose members are A and B. (If $A = B$, then $\{A, B\} = \{A, A\}$, which equals $\{A\}$ by Extensionality.)

Union.   For any set X there is a set $\bigcup_{x \in X} x$ whose members are the members of members of X. (For example, if $X = \{A, B\}$, then $\bigcup_{x \in X} x = A \cup B$.)

Definable Subset.   For any set X, and any property $\varphi(x)$ definable in the language of set theory, there is a set $\{x \in X : \varphi(x)\}$ whose members are the members of X with property φ.

Power Set.   For any set X there is a set $\mathscr{P}(X)$ whose members are the subsets of X.

These axioms produce enough sets to play the role of all mathematical objects in general use. In particular, the natural numbers $0, 1, 2, 3, \ldots$ may be elegantly realized as follows:

$$0 = \emptyset,$$

$$1 = \{0\},$$

$$2 = \{0, 1\},$$

$$3 = \{0, 1, 2\},$$

$$\vdots$$

This definition, proposed by von Neumann in 1923, has the bonus features that the $<$ relation between numbers is simply the membership relation $\in$, and the successor function is $S(n) = n \cup \{n\}$. With the help of an "induction" axiom, which we postpone until the next section, it is possible to develop all of the elementary theory of numbers from these axioms. However, we cannot yet prove the existence of infinite sets. Infinity has to be introduced by another axiom.

Infinity.   There is a set I whose members include $\emptyset$ and, along with each member x, the member $x \cup \{x\}$.

It follows that I has all the natural numbers $0, 1, 2, 3, \ldots$ as members, and the property of being a natural number is definable, so we can obtain the set $\mathbb{N}$ as a definable subset of I. Finally, we can use Power Set and Pairing to obtain the real numbers from $\mathbb{N}$ via Dedekind cuts, and hence obtain the set $\mathbb{R}$.

A useful trick here and elsewhere is to define the *ordered pair* (a, b) as $\{\{a\}, \{a, b\}\}$. Among their other uses, ordered pairs are crucial for the definition of functions, which can be viewed as sets of ordered pairs.

The universe of sets obtained by the Zermelo axioms is big enough to include all the objects of classical mathematics; in fact, most classical objects are probably included in $\mathscr{P}(\mathscr{P}(\mathscr{P}(\mathscr{P}(\mathbb{N}))))$. Nevertheless, it is easy to imagine sets that fall outside the Zermelo universe. One such set is

$$U = \mathbb{N} \cup \mathscr{P}(\mathbb{N}) \cup \mathscr{P}(\mathscr{P}(\mathbb{N})) \cup \mathscr{P}(\mathscr{P}(\mathscr{P}(\mathbb{N}))) \cup \cdots,$$

obtained by repeating the power set operation indefinitely and taking the union of all the resulting sets. U cannot be proved to exist from the Zermelo axioms. For this reason, Fraenkel in 1922 strengthened Definable Subset to the following axiom.

Replacement. If X is a set and if $\varphi(x, y)$ is a formula in the language of set theory that defines a function $y = f(x)$ with domain X, then the range of f,

$$Y = \{y : x \in X \text{ and } \varphi(x, y)\},$$

is also a set.

With this axiom, the set U can be proved to exist, because we can define

$$Y = \{\mathbb{N}, \mathscr{P}(\mathbb{N}), \mathscr{P}(\mathscr{P}(\mathbb{N})), \mathscr{P}(\mathscr{P}(\mathscr{P}(\mathbb{N}))), \ldots\}$$

as the range of a function $f(n) = \mathscr{P}^n(\mathbb{N})$ on $\mathbb{N}$, and then obtain U as $\bigcup_{y \in Y} y$.

The above axioms, together with the "induction" axiom to be discussed in the next section, form what is now called *Zermelo-Fraenkel* set theory, or ZF for short. Its *language*, which was presupposed in the Definable Subset and Replacement axioms can very easily be made precise. It contains the symbols $\in$ and $=$ for membership and equality, letters $x, y, z, \ldots$ for variables, parentheses, and logic symbols. ZF is a plausible candidate for a system that embraces all of mathematics, because there are no readily imagined sets that cannot be proved to exist in ZF. Nevertheless, we find ourselves drawn beyond ZF when we try to resolve certain natural questions about $\mathbb{R}$.

6   Set Theory and Arithmetic

Before investigating what ZF can prove about $\mathbb{R}$, it is worth checking whether ZF can handle elementary number theory. We have already seen that, without even using the axiom of Infinity, we can define the natural numbers $0, 1, 2, 3, \ldots$, the $<$ relation, and the successor function $S(n) = n \cup \{n\}$. This paves the way for inductive definitions of addition and multiplication:

$$m + 0 = m, \quad m + S(n) = S(m + n); \qquad m \cdot 0 = 0, \quad m \cdot S(n) = m \cdot n + m.$$

And we can prove all the theorems of elementary number theory with the help of induction too—*if* we have an induction axiom. Since $<$ is simply the membership relation, induction is provided by the following axiom, saying (as Euclid did) that "infinite descent is impossible":

Foundation.   Any set X has an "$\in$-least" member; that is, there is an $x \in X$ such that $y \notin x$ for any $y \in X$.

Thus, when Foundation is included in ZF and Infinity is omitted, all of elementary number theory is obtainable. Since the language of ZF allows us to talk about sets of numbers, like $\{3, 5, 11\}$, sets of sets of numbers like $\{\{5\}, \{7, 8, 9\}\}$, and so on, ZF−Infinity is somewhat more than the usual elementary number theory. However, it is not *essentially* more, because we can encode sets of numbers, sets of sets, and so on, by numbers themselves, and thereby reduce statements about finite sets to statements about numbers.

A simple way to do this is to interpret set expressions as *numerals* in base-13 notation, where $0,1,2,\ldots,9$ have their usual meaning and $\{$, $\}$, and the comma are the digits for 10, 11, and 12, respectively. For example, $\{3, 4\}$ is the numeral for

$$10 \cdot 13^4 + 3 \cdot 13^3 + 12 \cdot 13^2 + 4 \cdot 13 + 11.$$

Since elementary number theory has the resources to handle statements about numerals, it essentially includes all of ZF−Infinity. This was observed by Ackermann in 1937, and it justifies the seemingly flippant claim in Section 1 that mathematics equals "elementary number theory plus infinity." At least, ZF equals elementary number theory plus Infinity.

When we return to ZF by adding the axiom of Infinity to ZF−Infinity, the axiom of Foundation takes on the larger role of enabling *transfinite induction* and the theory of *ordinal numbers*, which extend the theory of natural numbers into the infinite. We need such an extension when studying $\mathbb{R}$, because it is often the case that processes involving real numbers can be repeated more than a finite number of times. Consequently, one needs ordinal numbers to measure "how long" the process continues.

Cantor was the first to study such a process in the 1870s. His example was *removing isolated points* for a subset of $\mathbb{R}$. If we have the set

$$S_1 = \left\{ \frac{1}{2}, \frac{3}{4}, \frac{7}{8}, \frac{15}{16}, \ldots \right\},$$

all points of which are isolated, then all points are removed in one step. It takes two steps to remove all points from

$$S_2 = \left\{ \frac{1}{2}, \frac{3}{4}, \frac{7}{8}, \frac{15}{16}, \ldots, 1 \right\},$$

because all points except 1 are isolated, hence removed in the first step. After that, 1 is isolated, so it is removed in the second step. By cramming sets like S_2 into the intervals $[1, 3/2], [3/2, 7/4], [7/4, 15/8], \ldots$ and finally adding the point 2, we get a set S_3 for which it takes three steps to remove all points: the first step leaves $\{1, \frac{3}{2}, \frac{7}{4}, \ldots, 2\}$, and the second step leaves 2, which is removed at the third step.

It is easy to iterate this idea to obtain a set S_n that is emptied only by n removals of isolated points. Then, by suitably compressing the sets $S_1, S_2, S_3, \ldots$ into a convergent sequence of adjacent intervals, we obtain a set S that is not exhausted by $1, 2, 3, \ldots$ removals of isolated points. However, we can arrange that just one point of S remains after this infinite sequence of removals, so it makes sense to say that S is emptied at *the first step after* steps $1, 2, 3, \ldots$. The first step after $1, 2, 3, \ldots$ is called step ω, and the *ordinal number* ω is naturally defined to be the set of all its predecessors, namely

$$\omega = \{0, 1, 2, 3, \ldots\}.$$

The successor function $S(x) = x \cup \{x\}$ applies to ordinals just as well as it does to natural numbers, so we get another increasing sequence of numbers

$$\omega + 1 = \{0, 1, 2, 3, \ldots, \omega\},$$
$$\omega + 2 = \{0, 1, 2, 3, \ldots, \omega, \omega + 1\},$$
$$\omega + 3 = \{0, 1, 2, 3, \ldots, \omega, \omega + 1, \omega + 2\},$$

$$\vdots$$

There is likewise a first ordinal number after all of these, namely

$$\omega \cdot 2 = \{0, 1, 2, 3, \ldots, \omega, \omega + 1, \omega + 2, \ldots\}.$$

In general, for any set A of ordinal numbers, the set $\lambda = \bigcup_{\alpha \in A} \alpha$ is an ordinal number. If λ is not a successor, as is the case when A has no greatest member, then λ is called a *limit* ordinal and λ is the *limit of* the set A. This powerful limit operation produces ordinals big enough to measure not merely "how long" a process on real numbers can run, but "how long" it takes to construct any set by means of the ZF set construction operations.

The construction of sets is hitched to the construction of ordinals as follows. If

$$V_0 = \emptyset,$$
$$V_{\alpha+1} = V_\alpha \cup \mathscr{P}(V_\alpha),$$
$$V_\lambda = \bigcup_{\beta < \lambda} V_\beta \quad \text{for limit } \lambda,$$

then the subscripts α run through all the ordinal numbers and every set X occurs in some V_α. It is actually the same to let $V_{\alpha+1} = \mathscr{P}(V_\alpha)$, but the above definition makes it clearer that a member of V_α also belongs to all higher V_β. The least α such that $X \in V_{\alpha+1}$ ("how long" it takes to construct X) is called the *rank* of X. In particular, the rank of each ordinal α is α.

7 Large Cardinals

We mentioned in Section 5 that the Infinity axiom is not provable from the other axioms of ZF. It is now worth explaining in more detail why this is so, since similar details occur in showing that certain *large cardinal axioms* are not provable in ZF.

The first thing to notice is that Infinity does not hold in V_ω, the union of the sets $V_0, V_1, V_2, \ldots$. This is because $V_0 = \emptyset$ and $V_{n+1} = V_n \cup \mathscr{P}(V_n)$, whence it easily follows by induction that all sets in each V_n are finite.

The second thing is that V_ω satisfies all the *other* axioms of ZF. The only ones that are not quite obvious are the set construction axioms, and these hold for the following reasons.

Pairing. If $A, B \in V_n$ then $\{A, B\} \in V_{n+1}$, because all subsets of V_n belong to V_{n+1}.

Union. Suppose $X \in V_{n+2}$, and $y \in x \in X$. Then $x \in V_{n+1}$ and $y \in V_n$. Thus $\bigcup_{x \in X} x$ is a subset of V_n and hence a member of V_{n+1}.

Power Set. Suppose $X \in V_{n+1}$, so any members of X are in V_n. Then each subset of X is a subset of V_n, hence a member of V_{n+1}. $\mathscr{P}(X)$ is therefore a subset of V_{n+1}, hence a member of V_{n+2}.

Replacement. Suppose f is a definable function whose domain is some $X \in V_n$ and whose values lie in V_ω. Since all members of V_n are finite, the range of f is finite, hence all of its members lie in some V_m. But then the range of f is a member of V_{m+1}.

Thus V_ω is a model of all the axioms of ZF except the Infinity axiom. Since the Infinity axiom does *not* hold in V_ω, it cannot be a logical consequence of the other axioms.

Now we are ready to consider a large cardinal axiom; namely, one asserting the existence of an *inaccessible* V_α. V_α is called inaccessible if it has $\mathbb{N}$ as a member and is closed under the power set and replacement operations. That is, if $X \in V_\alpha$, then $\mathscr{P}(X) \in V_\alpha$, and if f is a function whose domain belongs to V_α then the range of f also belongs to V_α.

Since V_α has $\mathbb{N}$ as a member, V_α satisfies the Infinity axiom, and it satisfies Power Set and Replacement because of its closure under the power set and replacement operations. The remaining axioms of ZF hold for the same reasons that they hold in V_ω (bearing in mind that α must be a limit ordinal for closure under the power set operation). Thus an inaccessible V_α satisfies all the axioms of ZF. No inaccessible V_α is actually *known*, however—it is not called "inaccessible" for nothing! This is due to the remarkable fact, essentially noticed by Kuratowski in 1924: *if an inaccessible V_α exists, then its existence is not provable in* ZF.

The explanation of this fact is quite simple. If an inaccessible V_α exists, take the one with smallest possible α (which exists by the Foundation axiom). Then the V_β belonging to V_α are *not* inaccessible because $\beta < \alpha$. Thus V_α satisfies not only the ZF axioms but also the sentence "There is no inaccessible V_α." This means that the existence of an inaccessible V_α is not a logical consequence of ZF.[3]

8 Measuring Subsets of $\mathbb{R}$

In Section 2 we observed the power of infinite operations (specifically, countable unions) to determine lengths and areas. The example of the parabolic segment dramatically shows how a complicated region can be measured by cutting it into a countable infinity of parts (triangles) whose measure is known. The completeness of $\mathbb{R}$ also plays a role, by allowing the formation of infinite sums of measures.

So, given that we need countable unions to determine measures, the question arises: *which regions are measurable when we allow countable unions?* To make the question as simple and definite as possible, let us confine attention to subsets of the unit interval [0,1], and impose the following conditions on the measure μ.

Measure of an interval. $\mu([a, b]) = b - a.$
Subtractivity. If S, T are measurable and $S \subseteq T$, then

$$\mu(T - S) = \mu(T) - \mu(S).$$

Countable additivity. If $S_1, S_2, S_3, \ldots$ are disjoint measurable sets, then

$$\mu(S_1 \cup S_2 \cup S_3 \cup \cdots) = \mu(S_1) + \mu(S_2) + \mu(S_3) + \cdots.$$

Notice that the measure of an interval implies that $\mu(\{a\}) = \mu([a, a]) = a - a = 0$. Subtractivity then implies that $[a, b], (a, b], [a, b)$, and (a, b) all have measure $b - a$. It also follows from countable additivity that any countable set has measure zero, as we already observed in Section 2.

Another consequence of subtractivity is that the complement $[0, 1] - S$ of any measurable set S is measurable. It also follows (not so obviously) from countable additivity that the countable union $S_1 \cup S_2 \cup S_3 \cup \cdots$ of *not necessarily disjoint* measurable sets $S_1, S_2, S_3, \ldots$ is measurable. Thus the class of measurable sets

---

[3]Unless of course ZF is inconsistent, in which case any sentence is a logical consequence of ZF. Thus whenever we show that something is not provable in ZF we are implicitly assuming that ZF is consistent. This is an unavoidable assumption because the consistency of ZF *cannot* be proved from the ZF axioms. The above argument shows the futility of one approach to proving consistency—by finding a V_α that satisfies the ZF axioms—and in fact any approach must fail by Gödel's famous *second incompleteness theorem.*

includes all sets obtainable from open intervals by the operations of complement and countable union. These are called the *Borel* sets after Émile Borel, who in 1898 outlined essentially the approach to measure I have just described.

However, by adding some plausible extra assumptions to ZF (see the next section) we can prove that there are only continuum-many Borel sets, so not all subsets of [0,1] are Borel. Fortunately, there is a natural extension of Borel's measure concept that covers many more sets (as many as there are subsets of $\mathbb{R}$ or [0,1]). This is the concept of *Lebesgue measure*, introduced by Lebesgue in 1902.

Lebesgue assigns measure zero to any subset of a Borel set of measure zero, thereby assigning a measure to any set that differs from a Borel set by a set of measure zero. Under the same assumptions that support the theory of Borel sets it then becomes possible that *all subsets of* [0, 1] *are Lebesgue measurable.* "Possible" does not mean provable, but *consistent*, in the following sense.

If ZF+"an inaccessible V_α exists" is consistent,

then so is ZF+"all subsets of [0,1] are Lebesgue measurable."

This result was proved by Solovay in 1965, and in 1984 Shelah showed that the assumption of inaccessibility cannot be dropped. So, a problem about $\mathbb{R}$ is settled only by assuming the existence of a set large enough to model the whole of set theory. Somehow, a question about $\mathbb{R}$ "blows up" to a question about the whole universe.

This is not an isolated example, as we will see in the next section.

9 Nonmeasurable Sets and the Axiom of Choice

Since it is consistent with ZF for all subsets of [0,1] to be measurable, only a new axiom can give us a nonmeasurable set. Do we want such an axiom? The general consensus is that we do, and in fact such an axiom has been around for over 100 years: the *axiom of choice* (AC for short).

In the late 19th century certain forms of AC were assumed, or used unconsciously, to prove results such as the countability of a countable union of countable sets.[4] But AC escaped attention until 1904, when Zermelo used it to prove the surprising theorem that *every set can be well-ordered* (that is, linearly ordered in such a way that every subset has a least element). AC states that every set X of

---

[4]Not only is this result unprovable without some form of AC, it is even consistent with ZF for $\mathbb{R}$ to be a countable union of countable sets. This amazing result was proved by Feferman and Levy in 1963. It implies that every subset of $\mathbb{R}$ is a Borel set, but at the same time wrecks the theory of measure, because countable sets have measure zero and a countable union of measure zero sets is also supposed to have zero measure.

Thus for both the theory of Borel sets and for measure theory we want a countable union of countable sets to be countable. One axiom that implies this is the so-called *countable axiom of choice*, whose statement is the same as AC but with the extra condition that X is countable.

nonempty sets x has a *choice function*: a function f such that $f(x) \in x$ for each $x \in X$. The converse theorem, that well-ordering implies AC, is almost obvious because $f(x)$ can be defined as the *least* member of x in a suitable well-ordering.

Now well-ordering obviously simplifies the world of sets—for example, it means that every set has the same cardinality as an ordinal—so one would have expected AC to be welcome. In fact, it received a hostile reception from Borel, Lebesgue, and many other eminent mathematicians. They objected to the apparent undefinability of the choice function f, which is indeed a mystery in the important case where the sets x are nonempty subsets of $\mathbb{R}$.

Also, AC had some unwelcome consequences, such as *nonmeasurable subsets of* $[0,1]$. This was proved by Vitali in 1905.

To explain Vitali's proof it is convenient to imagine $[0,1]$ made into a circle of circumference 1 by joining its two ends. Then we partition $[0,1]$ into equivalence classes, where a and b declared to be equivalent if $b - a$ is rational. Thus the rational numbers in $[0,1]$ form one equivalence class R. Every other class is the translate $R + \delta$ of R through an irrational distance δ. (This is where it is convenient to view $[0,1]$ as a circle: translation through distance δ means moving each point around the circumference through distance δ.)

Now AC says that we can form a set S with exactly one member from each distinct equivalence class $R + \delta$. It follows that the translate $S + r$ of S through a rational distance $r \neq 0$ in $[0,1]$ is disjoint from S, because the representative of each equivalence class is translated to a different representative. Moreover, as r runs through all the nonzero rationals in $[0,1]$, the points in $S + r$ run through all representatives of all equivalence classes.

Thus the countably many disjoint sets $S + r$ *fill* the circle of circumference 1. Finally, notice that if S is measurable then all the sets $S + r$ have the same measure, because *Lebesgue measure is translation-invariant* (due to the fact that translates of any interval have the same measure). But now there are only two possibilities: $\mu(S) = 0$ or $\mu(S) > 0$. By countable additivity, the first implies that $\mu([0, 1]) = 0$ and the second implies that $\mu([0, 1]) = \infty$. Both of these conclusions are false, so S is not measurable.

This is a painful predicament. To have measure theory at all, we need some form of AC—in particular we need a countable union of countable sets to be countable— but AC tells us that we cannot measure all subsets of $[0,1]$. Solovay's model offers a way out of this predicament, because it satisfies a weaker axiom of choice, called *dependent choice*, which is strong enough to support measure theory. (In particular, dependent choice implies that a countable union of countable sets is countable.) However, mathematicians today are hooked on AC. Algebraists like it as much as set theorists do, because the equivalents of AC include some desirable theorems of algebra. Among them are the existence of a basis for any vector space, and the existence of a maximal ideal for any nontrivial ring.

Thus, in the universe now favored by most mathematicians, there are nonmeasurable sets of real numbers. The question up for negotiation is: *how complex are the nonmeasurable sets?* The answer to this question depends on large cardinals, and in fact on cardinals much larger than the inaccessible required for Solovay's model.

We already know that the measurable sets include all Borel sets. The simplest class that includes non-Borel sets is that class called the *analytic*, or Σ_1^1, sets, which are the projections of Borel sets in $\mathbb{R}^2$ onto $\mathbb{R}$. They, and their complements the Π_1^1 sets, were shown to be measurable by Luzin in 1917. These two classes form the first level of the so-called *projective* sets, which are obtained from Borel sets in $\mathbb{R}^n$ by alternate operations of projection and complementation. In 1925, Luzin declared prophetically that "one does not know and *one will never know*" whether all projective sets are measurable. Maybe he was right.

In 1938 Gödel showed that it is consistent with ZF+AC to have nonmeasurable sets at the second level of the projective hierarchy, $\Sigma_2^1 \cup \Pi_2^1$. This result is a corollary of his momentous proof that:

If ZF is consistent then ZF + AC is consistent.

Gödel proved this result by defining a class L of what he called *constructible* sets, each of which has a definition in a language obtained from the ZF language by adding symbols for all the ordinals. The definitions of this language can be well-ordered, so L satisfies the well-ordering theorem and hence AC. Moreover, L satisfies all the axioms of ZF, because there are enough constructible sets to satisfy the set construction axioms. It turns out that the well-ordering of constructible real numbers is in $\Sigma_2^1 \cup \Pi_2^1$, and this gives a nonmeasurable set at the same level.

Thus the assumption that all sets are constructible—an assumption that the universe is in some sense "small"—gives nonmeasurable sets that are as simple as they can be, in terms of the projective hierarchy. At the other extreme, it is possible that *all projective sets are measurable*, under the assumption that sufficiently large cardinals exist. This was proved in 1985 by Martin and Steel, and the necessary large cardinals are called Woodin cardinals. They are considerably larger than the inaccessible cardinal required for Solovay's model, so their existence is even more problematic.

Once again, a question about the real numbers "blows up" to a question about the size of the whole universe of sets.

10   Summing Up

The aim of this chapter has been to give a glimpse of our ignorance about $\mathbb{R}$ and to emphasize that some questions about $\mathbb{R}$ call for further axioms about the whole universe of sets. Readers with some knowledge of mathematical logic will know that the same is true of $\mathbb{N}$ because of Gödel's incompleteness theorem. However, there seems to be a difference between $\mathbb{N}$ and $\mathbb{R}$. The only known questions about $\mathbb{N}$ that depend on large cardinals are ones concocted by logicians; in the case of $\mathbb{R}$, the questions that depend on large cardinals include some that arose naturally, before large cardinals had even been imagined.

Thus $\mathbb{R}$, much more so than $\mathbb{N}$, is the true gateway to infinity. $\mathbb{N}$ is the largely avoidable "potential" infinity of the Greeks; $\mathbb{R}$ is an actual infinity forced upon us by the apparent continuity of the natural world. If we are committed to understanding $\mathbb{R}$, then we are committed to understanding the whole universe of sets.

Yet, admittedly, it is a mystery how we can understand anything about $\mathbb{R}$. What are we thinking when we think about $\mathbb{R}$?

11   Further Reading

There is a dearth of literature for students, or even the ordinary mathematician, about $\mathbb{R}$ and its relations to set theory and logic. I attempted to remedy this situation with a semi-popular book, *Roads to Infinity* (AK Peters 2010) and an undergraduate text *The Real Numbers* (Springer 2013).

Beyond that, you will have to consult quite serious set theory books for results such as Solovay's model in which all sets of reals are Lebesgue measurable. The standard one is Jech's *Set Theory* (3rd edition, Springer 2003). A more recent, and less bulky, book that also covers Solovay's model and Woodin cardinals is Schindler's *Set Theory* (Springer 2014). Finally, for the general theory of large cardinals, with many interesting historical remarks, see Kanamori *The Higher Infinite* (Springer 2003).

Mathematics at infinity

Jeremy Gray

Abstract Points just beyond the plane, or points that have been lost by a transformation, were often rationalised as points at infinity, but this term meant different
things to different people at different times. It was a figure of speech, a purely formal
claim, and a claim that there are new real objects in geometry, before it became
assimilated into the modern language concerning mathematical existence.

This essay looks for points, lines, and circles at infinity, as presented by the likes of
Desargues, Newton, Euler, Poncelet, Plücker, Cayley, and others.

The existence of mathematical objects

Naively, mathematical objects have been taken to exist because they are somehow
out there in the world, or (which is not exactly the same thing) because what we
say or know about them helps us do things successfully that we could otherwise
not do at all. This is discussed much more fully in the essays by Avigad and Gillies
in this volume, and can only be taken as background here. But it is worth noting
historically that explicit attention to mathematical ontology has been much more
the concern of philosophers (Aristotle, Plato) than of mathematicians, with some
notable exceptions (Leibniz) until the late 19th and early 20th centuries, when
the arrival of modern mathematics generated the interest of the likes of Poincaré,
Hilbert, Brouwer, Weyl and Gödel. We have, for example, no words from Euclid
himself on the relationship between his *Elements* and the natural world. Nor, on the
other hand, do we have any reason to believe that the statements of mathematics
were regarded as other than true and applicable to the world before, it would seem,
the discovery of a plausible non-Euclidean geometry early in the 19th century.

J. Gray (✉)
Department of Mathematics and Statistics, The Open University, Milton Keynes, UK

Mathematics Institute, University of Warwick, Coventry, CV4 7AL, UK
e-mail: j.j.gray@open.ac.uk

© Springer International Publishing Switzerland 2015
E. Davis, P.J. Davis (eds.), *Mathematics, Substance and Surmise*,
DOI 10.1007/978-3-319-21473-3_11

We lack a clear set of historical opinions about mathematical existence, and must make do in the main with naive analyses. If it is not altogether clear what a point was taken to be, either in Euclid's *Elements* or in the world, it will not be clear what was meant by the term a 'point at infinity'; if the nature of the concept of a natural number is not much discussed by mathematicians it is hard to know what to make of their remarks about the negative integers, still harder to interpret their remarks about $\sqrt{-1}$, the square root of negative one.

To give just one indication of the problem, it is quite possible that mathematical objects may have been used but said not to exist, if they were taken to be terms in a theory that could be replaced by other, more elaborate and cumbersome objects that do exist. The vexed question of infinitesimals in the foundations of the calculus is a case in point; we shall see that points at infinity and the 'number' $\sqrt{-1}$ are others. Indeed, it is quite likely that the whole concept of existence, loaded as it is with a bias towards medium-sized spatio-temporal objects, may be the wrong concept with which to discuss what mathematicians have been talking about.

Informal infinities

Despite their obvious differences the various conic sections—the circle, ellipse, parabola, and hyperbola—have been regarded as forming a family of curves at least since their description by Apollonius [2]. Typically, problems arose when a complicated construction applied, say, to study an ellipse did not apply automatically to a hyperbola, and the case at interest here is when the construction produces a tangent to an ellipse but an asymptote to a hyperbola. In these cases, Apollonius seems to have regarded asymptotes as analogous to tangents, but nonetheless different; for example, in Book III of his *Conics* he presented theorems about tangents (such as theorems 37–40) after getting their exceptional cases that deal with asymptotes (theorems 30–36) out of the way. At no stage did Apollonius, or any other Greek geometer, generalise a construction by speaking of points at infinity: lines and curves either meet or they do not meet, they were never said to meet 'at infinity'.

A particular case of this problem concerns the foci of a conic. The foci of the ellipse and the hyperbola were studied by Apollonius in Book III of his *Conics*, theorems 45–52, but only his contemporary Diocles considered the focus of the parabola and it was he who proved that the sun's rays are reflected to a point by a parabolic mirror (see [18] 4.C1, 2). Thereafter the matter seems well-understood and in need of no further analysis until Kepler wrote his *Ad Vitellionem paralipomena* [26] in 1604, the year before he decided that Mars travelled in an elliptical orbit, as described his *Astronomia nova* [27] of 1609. Kepler was interested in geometrical and physiological optics in this work, and wanted to give a unified account of the conic sections. He began by briefly describing the curves, which he depicted as plane sections of a cone and presented in a single plane. Kepler seems to have considered that the hyperbola consisted of two separate curves, not branches of one curve, and that the limiting case of the hyperbola was a single straight line, to judge by the way he dealt with plane sections of the conic that pass through the vertex of the cone. He then introduced their foci. These are the points F, F' with the property, established by Apollonius, that the lines FP and $F'P$ joining F and F' to a point P on the conic make equal angles with the tangent to the conic at P. Kepler noted that the circle has a single focus, its centre, and the ellipse and hyperbola each have two. As for the parabola:[1]

> In the Parabola one focus, D, is inside the conic section, the other is to be imagined (fingendus) either inside or outside, lying on the axis [of the curve] at an infinite distance from the first (infinito interuallo a priore remotus), so that if we draw the straight line HG or IG from this blind focus (ex illo caeco foco) to any point G on the conic section, the line will be parallel to the axis DK.

Low key though this may be, it seems to be the first time that a point was said to lie at infinity, and it would be interesting to know how forcefully Kepler wished to press his approach. However, it is quite possible to read it as making no ontological commitment at all; indeed, what can one say about the focus of the parabola that lies 'at infinity' that cannot be said immediately in the language of parallel lines? The idea that a family of parallel lines has a lot in common with a family of lines passing through a point was, as Andersen has very thoroughly explained in [1], familiar to both renaissance painters and those who offered mathematical accounts of their work.

The man who introduced points at infinity as more than a figure of speech was the mathematician and architect Girard Desargues, who was born in 1591 in Lyons. The *Brouillon project* or *Rough Draft on Conics* of 1639 [13, 14] was his second publication, the first being his twelve-page *Perspective* that was devoted to perspective constructions. It opens with a clear statement that points at infinity are to be accepted into geometry and treated on a par with other points in the plane or in three dimensions [13, p. 70].

> In this work every straight line is, if necessary, taken to be produced to infinity in both directions. ...
> To convey that several straight lines are either parallel to one another or are all directed towards the same point we say that these straight lines belong to the same ordinance, which will indicate that in the one case as well as in the other it is as if they all converged to the same place.
> The place to which several lines are thus taken to converge, in the one case as well as in the other, we call the butt of the ordinance of the lines.
> To convey that we are considering the case in which several lines are parallel to one another we often say that the lines belong to the same ordinance, whose butt is at an infinite distance along each of them in both directions.
> To convey that we are considering the case in which all the lines are directed to the same point we say that all the lines belong to the same ordinance, whose butt is at a finite distance along each of them.
> Thus any two lines in the same plane belong to the same ordinance, whose butt is at a finite or infinite distance.
> In this work every Plane is similarly taken to extend to infinity in all directions.

---

[1]See [13, p. 187]. Kepler added that the straight line has two coincident foci that lie on the line itself.

But if Desargues was happy in principle that a family of parallel lines meet at a point at infinity, he nonetheless found some occurrences of it difficult to accept, and they came up in the situations that Apollonius had preferred to keep separate. One topic that had occupied both Apollonius and, later in Hellenistic times Pappus, was the construction of the so-called fourth harmonic point. Given three points A, B, C on a line their fourth harmonic point D is the point D on that line such that

$$\frac{AD}{DB} = -\frac{AC}{CB}.$$

Such a set of four points has been given various names, such as Pappus's 'harmonic division' or, Desargues's term, 'four points in involution'. The construction of the fourth harmonic point is as follows: one takes an arbitrary point G not on the line ABC and joins it to A and B, then one draws an arbitrary line through C that meets the line GA at the point E and GB at the point F, then one draws the lines AF and BE which meet at H, and then one draws the line G—it meets the line ABC at the required point D. Remarkably, it does not depend on any of the arbitrary choices. In the case where C is the midpoint of AB the point D will lie 'at infinity' because the line EF is now parallel to the line ABC. The Greek geometers had treated this as a special case, and noted that what plays the role of the above ratio is the simpler one $AC : CB$.

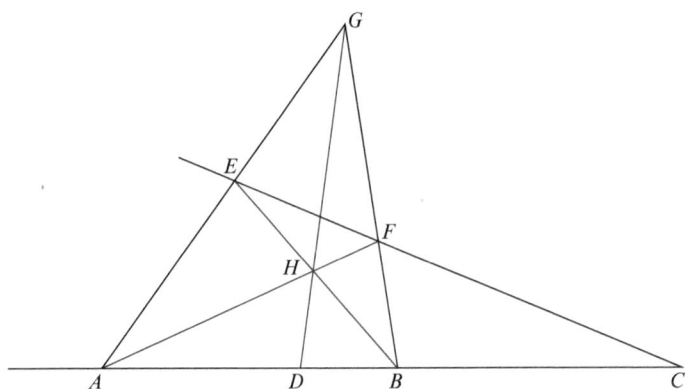

When Desargues came to this issue he remarked that it was incomprehensible [13, p. 82] but a few pages later [13, p. 85] he explained that a set of four points such as A, B, C, D for which $AD : DB = -AC : CB$ occur in "as it were, two species of the same genus", those in which the four points all lie at a finite distance and those in which the fourth point is at an in finite distance.

> So we shall take good note that a straight line divided into two equal parts by a point and understood to be extended to infinity is one of the forms of an involution of four points.

Desargues's *Rough Draft on Conics* was published in an edition of fifty copies, and it quickly met with success. It inspired Blaise Pascal, then only 16, to produce his theorem on six points on a conic in 1642, and in 1648 the engraver Abraham Bosse published the result that we know as Desargues's theorem.[2] However, the limited number of copies and the almost unreadable style of the *Rough Draft on Conics* with its many needless and obscure neologisms then combined to make the work disappear, until by 1679 only one copy was known, that, happily, was copied out by the mathematician Philippe de la Hire.

The difficult, but ultimately much more successful use of algebra in geometry developed by Descartes in his *La géométrie* (1637) surely also diminished the impact of Desargues's approach. Descartes's opinion of the *Rough Draft on Conics* was that the new terminology was unnecessary if Desargues was writing for experts and would only make the bookmark difficult; he also suggested using "the terminology and style of calculation of Arithmetic, as I did in my *Geometry*". As for parallel lines meeting in a butt at an infinite distance, Descartes' commented that[3]

> this is very good, provided you use it, as I am sure you do, as an aid to understanding what
> is difficult to see in one of the types, by comparing it with the other, where it is clear, and
> not conversely."

As for Philippe de la Hire, it is very likely that he had received this copy from his father Laurent de la Hire, a painter and pupil of Desargues who was also a friend and colleague of Bosse. In any case, it was long thought that only de la Hire's copy that survived until a sole copy of the original was found in the Bibliothèqual Nationale and published by René Taton in 1951.[4]

Philippe de la Hire gave several accounts of the geometry of conic sections in the course of his life, of which his [29] is the most important. It can be regarded as a projective treatment of the subject, including a theory of poles and polars of conics, organised around the harmonic division of four points, and it is held together by an appreciation that any two conic sections are projectively equivalent, and the harmonic division is projectively invariant. From there the subject passed to Isaac Newton, who published what became called his 'organic construction' of conic sections in his *Principia Mathematica* in connection with the determination of orbits, and went on to give a largely projective classification of conic sections in [31], published as an appendix to his *Opticks* in 1704.

It is well known that Newton gave an essentially projective treatment of conic and cubic curves in this work, and the details can not be given here. In particular he gave examples of what he called the five divergent parabolas; these are curves that have a branch going to infinity but have no asymptote, unlike the hyperbolic branches that are tangent at infinity to their asymptotes. We would say that the tangent to

---

[2]See [6]. This theorem may have appeared in Desargues's third publication, the *Leçons de ténèbres* (1640), but that book is lost.

[3]See Descartes's letter to Desargues in [18, 11.D4].

[4]De la Hire's copy was edited and published by Poudra in 1864.

a parabolic branch is the line at infinity; Newton said rather [31, p. 10] that "the tangent of the parabolic branch being at an infinite distance, vanishes, and is not to be found". He then indicated 72 species of cubic curve (forgetting the other six, which he had in fact known about) and that every cubic curve is the projection of precisely one of five types of cubic curve (curves of the second genus, as he called them), writing "so the five divergent parabolas, by their shadows, generate all other curves of the second genus". It was the work of several mathematicians to make sense of Newton's account.

It is relevant here to note that Newton may not have accepted a line at infinity, but he did accept points. He noted [31, p. 25] that the straight lines in one direction may all meet a cubic curve in just one point, and so

> we must conceive that those straight lines pass through two other points in the curve at an infinite distance. Two intersections of this sort, when they coincide, whether at a finite or an infinite distance, we shall call a double point.

He then repeated what he had said in the *Principia* about how double points can arise by applying his organic construction (a form of birational transformation) to a conic section.

Newton's contributions might have stimulated a century of work, but it was not to be. The projective approach to geometry was developed by de Gua de Malves in his [12], but not by Leonhard Euler or Gabriel Cramer, and it declined in significance as the 18th century proceeded, until it was revived partly by Gaspard Monge and much more influentially by Jean Victor Poncelet.

We have reached a convenient point to ask about the ontological implications of the 17th-century work, but there can be no simple answer. The authors did not address the issue explicitly, beyond making it clear that they spoke of points at infinity as if they were as good as finite points. They did not treat them as a mere manner of speech that could easily be replaced with statements about parallel lines, and they did not treat theorems involving them as special cases of other theorems. But they did not discuss what could be meant by talking about new points 'at infinity', however, much this might seem to strain the idea of a geometry based on a concept of distance to breaking point. Indeed, the whole question of the relation of mathematical points to points in the world was little discussed. David Hume did discuss how the foundations of geometry can be obtained from sense experience in his *A Treatise on Human Nature* (1739), Part II "On the ideas of space and time", but he did not raise the issue of points 'at infinity'; rather, he was concerned with infinite divisibility and the notion of 'point' in geometry.

Instead, an ontological debate about points at infinity was more apparent in issues to do with the use of algebraic methods in geometry, and here Euler did have things to say. Some concerned points 'at infinity', others the appearance of points with complex coordinates.[5] Euler had nothing to add to the idea that some curves may

---

[5] There was also the issue of how to count multiple points of intersection, but that cannot be pursued here.

have points 'at infinity', which he accepted, but he had interesting views about points with complex coordinates, to which we turn after briefly reviewing attitudes to complex numbers and complex roots of equations.

Complex numbers, complex points

For many centuries, questions such as 'what is the square root of negative one?' or 'what are the roots of $x^2 + x + 2 = 0$?' were answered in various ways by mathematicians, but until the 16th century it was always possible to dismiss them. One could reply that there is no such number, and the equation has no roots. What made the question unavoidable was the discovery in the 16th century of methods for solving cubic equations, and specifically the growing realisation that the formula for the solution of a cubic equation with three real roots necessarily involved the square roots of negative numbers. The 'casus irreducibilis' or irreducible case, as it came to be called, was known to Cardano, who could not find a way round it in [7], his *Ars magna* of 1545, and in his *L'Algebra* [4] Bombelli produced a formalism for handling complex quantities.

Thereafter many different explanations of the question of 'what is a complex number?' were proposed. The ontological context was unpropitious. It was generally agreed that there were numbers, that is, the counting numbers or non-negative integers, and there were magnitudes, such as lengths. In this sense, numbers were properties of objects: one could own three sheep or purchase five yards of cloth. Negative numbers, and still more negative magnitudes, posed further problems, but might be associated with debts or directions along a line. What united all these ideas was that numbers and magnitudes were attributes, and accordingly the ontological question became 'what objects have complex numbers or magnitudes as attributes?'. No answer was to be given for many decades that satisfied its audience, but Euler's proposal in [17] was the simplest.

Euler stepped outside the existing ontological framework, after showing that there was no answer within it, and offered a different definition of number, as a formal quantity that we can handle algebraically exactly as we handle ordinary numbers and that can, indeed, produce ordinary numbers. It is a concept that is free of contradiction and it embraces the usual numbers.

143. And, since all numbers which it is possible to conceive are either greater or less than 0, or are 0 itself, it is evident that we cannot rank the square root of a negative number amongst possible numbers, and we must therefore say that it is an impossible quantity. In this manner we are led to the idea of numbers, which from their nature are impossible; and therefore they are usually called *imaginary quantities*, because they exist merely in the imagination.

144. All such expressions as $\sqrt{-1}$, $\sqrt{-2}$, $\sqrt{-3}$, $\sqrt{-4}$, &c. are consequently impossible, or imaginary numbers, since they represent roots of negative quantities; and of such numbers we may truly assert that they are neither nothing, nor greater than nothing, nor less than nothing; which necessarily constitutes them imaginary, or impossible.

145. But notwithstanding this, these numbers present themselves to the mind; they exist in our imagination, and we still have a sufficient idea of them ; since we know that by $\sqrt{-4}$ is meant a number which, multiplied by itself, produces -4; for this reason also, nothing prevents us from making use of these imaginary numbers, and employing them in calculation.

The new numbers can be called 'imaginary', but they are not impossible per se— they are simply not measuring numbers. Indeed:

149. It remains for us to remove any doubt which may be entertained concerning the utility of the numbers of which we have been speaking; for those numbers being impossible, it would not be surprising if they were thought entirely useless, and the object only of an idle speculation. This, however, would be a mistake; for the calculation of imaginary quantities is of the greatest importance, as questions frequently arise, of which we cannot immediately say whether they include any thing real and possible, or not; but when the solution of such a question leads to imaginary numbers, we are certain that what is required is impossible.

This is another valuable insight. The price of declaring complex numbers to be logically impossible is high: the calculation of unknown quantities requires that complex numbers can be accepted as answers, even though in a problem concerning quantities the valid conclusion would be that 'what is required is impossible'. A similar view was held by Johann Heinrich Lambert, who wrote to the philosopher Immanuel Kant in 1770 that "The sign $\sqrt{-1}$ represents an unthinkable non-thing. And yet it can be used very well in finding theorems".[6]

These views should not be taken as typical of the period, and nor as they incisive as they may seem. Euler was given to many observations and remarks of a foundational kind that resolve nothing; his remarks about infinitesimals and the interpretations of $0/0$ in his *Introductio in analysis infinitorum* (1748) sit alongside the remark that there is no number smaller than one part in ten million and some inspired manipulation of infinite and infinitesimal numbers that cannot properly justify the correct results they are presented as supporting. In the case of imaginary numbers, as is well known, Euler promptly went on to make sign errors in his calculations with these 'numbers' that we apparently know how to use, treating $\sqrt{n}$ as a single-valued expression. And, as we shall shortly see, there are graver problems with his attitude to the 'numbers' that have to do with their status as answers.

Algebra was, of course, a generator of complex numbers. Whereas mathematicians of the 17th century had usually defined a curve to be of order n if it meets a straight line in at most n points, mathematicians of the 18th century preferred to define the order of a curve as the degree of its defining equation, and to prove a theorem to the effect that a curve of degree n meets a straight line in at most n points. This includes the case where the curve is defined by a polynomial equation of the form

$$y = a_n x^n + a_{n-1} x^{n-1} + \cdots + a_0$$

---

[6] See [25, nr. 61] and [18, 16.A3].

and the other curve is the straight line $y = 0$. These meet in the points whose x-coordinates are found by solving the equation that is found by eliminating y from these two equations:

$$a_n x^n + a_{n-1} x^{n-1} + \cdots + a_0 = 0.$$

What became called the fundamental theorem of algebra is the claim that the polynomial equation has n roots, where, as always in the 18th century, the coefficients $a_n, a_{n-1}, \ldots, a_0$ are all real. Or rather, and more commonly, in a form that Lagrange endorsed, it was claimed that a polynomial of degree n can be factored into a certain number, k, of linear terms (corresponding to the real roots of the equation) and a number j of quadratic terms (corresponding to the $2j$ complex conjugate roots) where $k + 2j = n$. This indicates a degree of reservation about the status of complex roots.

The more general claim was that two curves of degrees m and n meet in at most mn points. This is true, for example, of two ellipses meeting in four points, but it is much harder to prove. Euler published a short paper [16] in 1750 that illustrated some of the difficulties involved. As he pointed out, for the theorem to be true one must be able to say that parallel lines meet in a point, and that a parabola meets a line parallel to its axis in two points—which requires the invention of points at infinity. One must be able to explain why two circles meet in at most two points but two ellipses can meet in four by accounting for the 'missing points', which will have complex coordinates. As he put it:[7]

> the meaning of our proposition is that the number of the intersections can never be greater than mn, although it is very often smaller; and thus we may consider either that some intersections extend towards infinity, or that they become imaginary. So that in counting the intersections to infinity, the imaginary ones as well as the real ones, one may say, that the number of intersections is always $= mn$.

In November 1751 Euler wrote to Cramer (who had also considered the problem) that it led to 'such tangled formulas that one completely loses patience in pursuing the calculations'.[8]

The method of eliminating a variable defied every challenge, and the Académie des Marines, the scientific school attached to the French Navy, was provoked to remark in 1770 that:[9]

> The elimination of unknowns is one of the most important parts [of mathematics] to perfect, both because the extreme length of ordinary methods makes it so repugnant and because the general resolution of equations depends on it.

On this occasion they dismissed attempts by Euler and another mathematician, Étienne Bezout, because they had not shown that elimination lead to an equation

---

[7]Transl. Winifried Marshall, Euler Archive.

[8]Euler–Cramer correspondence, OO474 not yet published, quoted in [39, p. 168]).

[9]Quoted in [39, p. 168]).

of the right degree, mn, but finally in 1779 Bezout announced a satisfactory solution. The result has been called Bezout's theorem ever since, although proofs of increasing rigour continued to be produced until the 20th century.

A curious aspect of complex points of intersection of two curves should be noted that does not apply to the case of the fundamental theorem of algebra. Let us fix one curve, of degree m say, and suppose that it meets a curve of degree n in k points with real coordinates and $mn - k$ points with complex coordinates. Now let us vary the curve of degree n, and consider the case where the varying curve continues to cut the fixed curve in k points with real coordinates that depend on the varying curve. We may, if we like, suppose that each of the k points moves around on the fixed curve. But what of the $mn - k$ points with complex coordinates? In what sense does the fixed curve meet the varying curve in complex points, unless one is prepared to say that both curves have complex points and they meet in a set of $mn - k$ complex points that vary with the varying curve? It is much easier to talk around the same question in the context of the fundamental theorem of algebra than in the context of curves, but this problem was to remain unaddressed for several decades, as we shall see shortly when we discuss Plücker's work.

A further contradiction appears with the use of complex quantities in analysis. Famously, in [15], his *Introductio in analysin infinitorum* of 1748, Euler showed how to unify the theory of the exponential, logarithmic, and trigonometric functions, and went on to resolve the question of what the logarithm of a negative and a complex number could be that had confused both Johann Bernoulli and Jean le Rond d'Alembert. His answer was that the logarithm is an infinitely many valued expression, corresponding to the fact that its inverse, the exponential function, is periodic with period $2\pi i$ (an answer d'Alembert refused to accept). It is odd, therefore, to see that the idea that the logarithm of a number is anyone of a set of values of the form $a + 2k\pi i$, where k is an integer was something some people could accept, and yet to find that complex quantities were only admitted into geometry and analysis on condition that they do not appear in the answer—but such was the case. There are many occasions in the writings of d'Alembert, Euler, and Lagrange where such quantities are admitted only because at the end of the calculation they 'destroy themselves', as the phrase had it. Or, as we would say, they occur in complex conjugate pairs and the imaginary parts cancel each other out. This is true in d'Alembert's work on flows in the plane, in the work of Euler and Lagrange on cartography, and in Euler's work on partial differential equations.

For example, d'Alembert used formal complex methods in his [11] of 1752. In Chapter IV § 45 he considered a two-dimensional flow in the (x, z)-plane, and in §§ 57–60 he looked for conditions on functions M and N of x and z under which the differentials

$$Mdx + Ndz, \quad Ndx - Mdz \tag{1}$$

are exact. He argued that when they are exact there are functions p and q such that

$$Mdx + Ndz = dq, \quad Ndx - Mdz = dp,$$

and so the differentials

$$(M + iN)(dx - idz) \quad \text{and} \quad (M - iN)(dx + idz).$$

are also exact. This gave him an answer to his question in only a few further lines of work, but he then gave a second, simpler argument to the same effect. The definitions of p and q imply that

$$p_z = -q_x \text{ and } q_z = p_x,$$

so $qdx + pdz$ and $pdx - qdz$ are exact differentials and therefore $q + ip$ is a function of $x - iz$ and $q - ip$ is a function of $x + iz$. He now remarked that if one wants p and q to be real then q must be a function of the form

$$\xi(x - iz) + i\zeta(x + iz) + \xi(x + iz) - i\zeta(x - iz),$$

where the functions ξ and ζ have real coefficients (with a similar result for p that he omitted). This gave him expressions for p and q in which, in his phrase, the imaginary quantities destroy themselves.

Lagrange made a similarly formal use of imaginary quantities in cartography in his memoir [30] of 1781. This led him to the equation

$$dx^2 + dy^2 = n^2(du^2 + dt^2)$$

in which u and t are obtained from coordinates on the Earth (taken to be a spheroid) and x and y are coordinates on the plane. He factorised each side formally into factors of the form

$$dx \pm idy = (\alpha \pm i\beta)(du \pm idt),$$

by introducing a new variable ω, such that $n \sin \omega = \alpha$, $n \cos \omega = \beta$. This gave him the equations

$$dx = \alpha du - \beta dt, \quad dy = \beta du + \alpha dt,$$

and for these equations to be integrable, he said that he would follow d'Alembert's method, and so these conditions must hold:

$$\frac{d\alpha}{dt} = \frac{d\beta}{du}; \quad \frac{d\beta}{dt} = \frac{d\alpha}{dt}.$$

from which he deduced that

$$\alpha \pm i\beta = f(u \pm it),$$

and $\alpha \pm i\beta$ must be a function of $u \pm it$.

He then defined the prime meridian (given by $t = 0$) and used it to obtain expressions for x and y that, he said, "have the advantage that the imaginaries destroy themselves". Once again the formal complex quantities act as a catalyst and are made to vanish in the end.

These examples from the 18th century firmly suggest that complex numbers in whatever context were acceptable as means to an end, but much less acceptable in the end itself.

The arrival of projective space

We have seen that Euler, and for that matter, Cramer, were willing to regard curves as having points at infinity, but most likely regarded them as a special kind of point, not to be confused with finite points. Such a position became inadequate when transformations were introduced into geometry that map points at infinity to finite points and finite points to ones at infinity.

The decisive figure in this context is Poncelet, but his presentation was so entangled with other, less acceptable ideas that other mathematicians—Michel Chasles in France, August Ferdinand Möbius, Jakob Steiner, Julius Plücker and ultimately von Staudt in Germany—made equally important contributions in the 1820s and 1830s.

Poncelet's paradoxical name in his [38] for his discovery—'non-metrical geometry'—highlights its novelty. It is a study of the properties of plane figures that are preserved under central projection from one plane to another. Straightness of the straight line is such a property, so is the fact that conic sections are curves that meet a straight line in at most two points, but length and angle are not such properties, and Poncelet was able to show that non-metrical geometry was a rich subject with clear connections to the metrical geometry of conic sections. But he did so in a way that won little support, because he urged a very general interpretation of figures through what he called a principle of continuity.

He introduced the principle to give geometry the generality of algebra, and he used it to talk indifferently of configurations in which a line meets a circle and in which a line does not meet a circle. He argued that when the line is moved from the first position to the second the points of intersection become imaginary, but many things can be said about them and the segment they define, and so it was appropriate to treat real and imaginary intersections on a par. He extended this principle greatly, to include imaginary centres of projection. From an ontological standpoint, as is argued in [5], he may not have been introducing new objects so much as changing the meaning of terms such as 'intersect'. However, in so doing he made few converts, and the enduring part of his legacy was the elucidation of projective properties and projective transformations.

But Poncelet's ideas were the source of a peculiar debate in the 19th century on the nature of complex points on real curves. Various authors put some effort into giving a real interpretation for the pairs of conjugate imaginary points that satisfy an algebraic equation with real coefficients. This is the subject of [10], one of the better, but less well-known, books by the historian of mathematics Julian Lowell Coolidge.

Typically, complex points on real curves were interpreted as corresponding to an elliptic involution, which is a map of period two having conjugate imaginary fixed points. This theory was extensively developed by Laguerre, as Coolidge described.

Julius Plücker is an interesting case, and a more relevant one here, because the subject of the first half of his [33] deals with the affine coordinate geometry of algebraic curves; the larger second half is a rich study of cubic curves.

In this treatment a point is determined by its coordinates and therefore by the intersection of two lines, and Plücker remarked (1835, 19) that the treatment would be incomplete without an account of imaginary coordinates,

> that is, those values that, based on given points, receive given imaginary linear functions. Such coordinates, like the corresponding functions, arise in pairs. Each coordinate, taken singly, corresponds to an imaginary straight line, two connected coordinates correspond to two imaginary straight lines that meet in a real point. But we can also determine geometrically the coordinates through the use of two real straight lines that meet in the point.

A long discussion of the algebra led Plücker to system of four lines that he called harmonical; two real lines with equations $p = 0, q = 0$ and two lines with equation $p + \lambda q = 0, p - \lambda q = 0$, where $\lambda \bar{\lambda} = 1$ that are real or imaginary according as λ is either real or imaginary. He called this system an involution of line pairs, because there is a linear map that fixes the lines of one pair and exchanges the other two, and distinguished the cases when the involution has no imaginary lines and when it has two.

He included in his treatment of cubic curves their points at infinity, and extended this study in his [34] to the geometry of the infinite branches of algebraic curves, in particular curves defined by equations of degree four. He did not study this by using homogeneous coordinates and changing axes. Instead, he worked with the Cartesian equation, which he then sought to write in a helpful form. Given curve defined by an equation $F_n(x, y) = 0$ of degree n he took the equation of line in the form $p(x, y) = ax + by + c = 0$ and looked at the expression

$$F_n = pG_{n-1} + \varphi,$$

where G_{n-1} is a polynomial in x and y of degree $n - 1$ and in general φ will also be of degree n. However, the points where the line meets the curve are given by the zeros of φ, and if any of them lie at infinity then the degree of φ will drop. In particular, if the line is an asymptote of the curve then it is a tangent at infinity and the degree of φ must drop by two (or more, if the contact at infinity is of higher order). So his method was to work with suitably chosen lines $p = 0$ that meet, and preferably touch, the curve $F = 0$ at infinity.

As an example, suppose

$$F(x, y) = x^3 - x^2 y + xy^2 - y^3 - x + 2y = 0.$$

This can be written as

$$(x - y)(x^2 + y^2 - 1) + y = 0,$$

with $p = x - y$ and $\varphi = y$, showing that the curve $F = 0$ has the line $x = y$ as an asymptote.

Plücker developed this theory for real asymptotes and then considered imaginary asymptotes, but the study was nonetheless about the shape of real curves in the real plane. One can therefore wonder when the idea came in that a complex curves has complex points in the same way that a real curve has real points, and in so doing one is in the company of Cayley, who wrote [8, p. 316] that "I was under the impression that the theory as a known one; but I have not found it anywhere set out in detail".

The obvious, and correct answer, is that this was first set out by Riemann in his [40, 41], but his treatment of complex algebraic curves as branched coverings of the complex sphere (although not exactly what Cayley had in mind, as he observed) makes no substantial use of points at infinity and need not be described here.

Projective transformations offered a way of evading the full implications of non-metrical geometry, in that they could always be used to concentrate attention on a finite part of the plane. A projective transformation can always be found to reduce a given quadrilateral to another, any conic to a circle, to bring in almost all the points at infinity in a given figure to the finite part of the plane. Chasles's view in [9] seems to have been that one was study the properties of figures in the finite plane that are invariant under projective transformations. These properties were largely Euclidean, so although he showed that given five points A, B, C, D, P on a conic any other point Q on the conic has the property that the cross-ratio of the lines PA, PB, PC, PD is equal to the cross-ratio of the four lines QA, QB, QC, QD he did not prove the converse, and so he came up with a projective property of a conic but not a projective characterisation of conics.

After Poncelet, mathematicians in France and Germany gradually took up non-metrical geometry, and slowly disentangled a number of complicated questions—such as the projective definition of the basic figures—until it was possible to give entirely projective definitions of the basic concepts, a process that is usually said to conclude with the work of von Staudt [44, 45] in the 1850s or, perhaps, of Felix Klein in the early 1870s. All of this work had to discuss what was meant by points 'at infinity' because now a transformation can map a finite point to one at infinity and vice versa. So to say what a point at infinity might be, it became necessary to say what projective space might be.

A word should be said here about Pasch's approach to geometry in his [32] of 1882. It is well known that he aimed to start from Helmholtz's ideas about the empirical origin of the geometrical axioms and arrive at a coherent system of geometry by specifying at each stage how the empirical ideas were codified into geometrical statements. Once a family of empirical claims about geometrical objects was stated in the form of Grundsätze, every statement about these objects had to be proved on the basis of the Grundsätze alone. The hope was that in this way a logically impeccable geometrical system would be established with a clear relationship to physical space.

Pasch began with bounded regions of the plane and space, and codified the behaviour of points and line segments; the eponymous Pasch axioms belongs here. To deal with anomalous intersection properties, such as lines in a plane that do not meet, Pasch then extended the meaning of the terms 'point', 'line' and 'plane' so that he could say, for example [32, p. 41] that "Two lines in a plane always have a point in common". In so doing he was following the lead of von Staudt and Klein, and he compared what he was doing to the way that the concept of 'number' is stretched in going from meaning 'positive integer' to number in general.

Pasch left it unclear how the experience of a finite but arbitrary part of a plane could ever lead unambiguously to the Euclidean theory of parallels, but that was what he had in mind. So his geometry did not have points or lines 'at infinity', only families of lines no two of which had points in common and families of planes no two of which had a line in common.

Plane projective space was a novelty. It was not metrical, so it could not be said to be the space we inhabit. Also, because it was not metrical it was not at first very clear how considerations of continuity could apply to it. Also, it has an unexpected global property of being non-orientable that is invisible in any finite part of the plane. Two definitions of it co-existed. One added points to the Euclidean plane that were the 'intersection points' of families of parallel lines, and then allowed points to belong indifferently to families of either convergent or parallel lines. The other turned to coordinate methods, pioneered by Möbius, Plücker, and Hesse in Germany.

If the second approach proved more powerful this was because the synthetic methods associated with the first approach seldom dealt successfully with curves other than conic sections, but the algebraic approach was fertile. Plane projective space could be regarded as the collection of all homogeneous triples of numbers, the triple $[0, 0, 0]$ excluded. Mathematicians were at liberty to let the entries in the triples be real or complex numbers, and the ontological question was now clear: in what sense does a seemingly consistent set of mathematical definitions and theorems constitute a description of something that exists but does not correspond to an object in the world?

This question joined with several others in the second half of the 19th century. In what sense do the three dimensions of space and the fourth dimension of time constitute a four-dimensional *space*? In what sense do a set of four variables, such as two of position and two of momentum, constitute a four-dimensional *space*? If, following Grassmann in [19], we work with extensive magnitudes of order n, can we go further and speak of n-dimensional *space*? Is there any good reason to do so? Is there any logical reason *not* to? At what point does a theory go from being figures of speech to being something like a naturalistic account of an object that just happens not to be in the world? And indeed, at what point does a mathematician start to lose confidence in the comfortable connection between Euclidean geometry and the world around him or her?

The existence of Non-Euclidean geometry

The discovery of non-Euclidean geometry has been told in many places and will not be told here.[10] It is well known that Bolyai and Lobachevskii independently and around 1830 published accounts of a geometry that differed from Euclid's in only one respect—the definition of parallel lines—and that was a plausible, and above all, metrical account of space that, however, failed to convince any important mathematician except Gauss. He, however, did essentially nothing to advance their cause. But it is worth noting that what Bolyai and Lobachevskii did, and Gauss had not done in the 1830s, was to describe a three-dimensional space, not a two-dimensional one. Intuition is a poor guide in such work; so accustomed are we to reading geometry from the world around us that we have to struggle to imagine ourselves in another world entirely. It is true that neither Bolyai nor Lobachevskii were able to provide an entirely convincing account, inasmuch as a remote possibility of a self-contradiction in their work could not be ruled out, but the deeper reason that their work met with such strong rejections when it was discussed at all was that it required a wholesale reassessment of our relationship with the world, and surely implied that there were two possible, but incompatible, geometrical theories of it. Even if one was correct the other would be wrong, and the only way to decide which was correct was to introduce an empirical element. Gone would be centuries of naive confidence in the truth of (Euclidean) geometry.

Non-Euclidean geometry was first accepted, and first advocated successfully, by mathematicians who were long past worrying about the Euclidean doctrine of parallel lines, and for whom the whole question of what geometry was about had deepened greatly. The inspirational figure here was Bernhard Riemann, and his view, in his Habilitation lecture of 1854 (published posthumously in 1867 as [42]) was that geometry was the study of any collection of points (most likely themselves described by coordinates) in which it was possible to talk about distance. This included, but was not limited to, sets of points that were n-tuples of numbers and a distance formula that resembled the Pythagorean distance formula of Euclidean differential geometry. Above all, it rendered Euclidean two- and three-dimensional geometries just two among infinitely many and denied them any foundational role.

Geometry was now talk involving distance and angle however they might be defined. In particular, the non-Euclidean geometry of Bolyai and Lobachevskii was, Riemann implied, the geometry of points in an n-dimensional ball of (Euclidean) radius r in which the distance between two points is obtained from the formula (see [42], III, 4)

$$ds = \frac{1}{1 + \frac{\alpha}{4} \sum_j x_j^2} \sqrt{\sum (dx_j)^2},$$

where $\alpha = \frac{-4}{r^2}$ is the curvature.

---

[10]See [20], [21] or [22].

Inspired by Riemann's ideas Beltrami in [3] gave a similar account, but with a different metric in which (non-Euclidean) geodesics appeared Euclidean straight lines. This allowed Klein in [28] to find a way of including non-Euclidean geometry as a special case of projective geometry, and Poincaré in 1881 to give further interpretations of the subject that mark its full acceptance into the world of mathematics.

But it was also clear that the existence of these new geometries was the product of a novel ontology. What the various metrical descriptions provided was a map of non-Euclidean space in almost the same sense that a map of Europe in an atlas provides an account of Europe. The difference is that Europe exists, it is patiently surveyed, and the results of those surveys provide the data for the map; all that exists in the world of non-Euclidean space is the map. The claim that the space exists is based on an argument that the map is internally self-consistent, and that nothing else is required for mathematical existence other than consistency.

It is well known that it took time for these ideas to be accepted, especially by philosophers. Kantians had to reckon with the clear suggestion from Kant that Euclidean geometry was the inevitable consequence of a simple construction; Frege simply refused to accept it because there could only be one set of truths about the world. Their reluctance says a lot about how geometry was taken to apply to the world, and it hints at a possible reason for the ultimate success of Bolyai and Lobachevskii. They set up a complicated system of points, lines, curves, and surfaces and then moved quickly to describe them in formulae. They could not give the ultimately convincing reason for placing trust in their formulae—that was the specific contribution of Riemann and Beltrami—but by displaying a new set of consistent formulae they opened up a path to an argument that would run in precisely the opposite direction to theirs: from formulae to interpretation.

Points at infinity in non-Euclidean geometry have a particular significance in non-Euclidean geometry. They were introduced by Klein in [28] in order to show how to locate Beltrami's account within projective geometry. Klein interpreted Beltrami's work as being about the space of points interior to a fixed conic Ω together with the group of projective transformations that maps this space to itself. He obtained the distance between two points A and B in this space as a suitable multiple of the logarithm of the cross-ratio of the four points A, B, C, D, where C and D are the points where the line AB meets the conic Ω. On this definition, points on Ω are infinitely far away.

Points at infinity were also used by Poincaré in his work on non-Euclidean geometry (see, for example, [35]), but in a different way. What he called a Kleinian group acts on three-dimensional non-Euclidean space and a fundamental domain for the action can cut out a region of the two-dimensional boundary, which of course consists of points at infinity in the geometry. When this happens the study of the group is intimately tied to the study of this boundary under the action of the group.

In these cases it can be fairly argued that the points at infinity are not mysterious at all; they are simply points on the boundary of a finite region drawn in the Euclidean plane or Euclidean three-dimensional space. They nonetheless have an interpretation in the geometry under study that would be difficult to understand if that geometry was taken to be real.

Mathematical existence

On the other hand, it was beginning to be clear to mathematicians that all they could mean by mathematical existence was a consistent theory: if a theory of a class of objects was free of self-contradiction that those objects existed, even if they had no counterparts in the physical world. This was the explicit view of Poincaré and Hilbert by 1900.

The freedom from self-contradiction was not necessarily easy to establish, and Euclidean geometry is a case in point. Once the naive identification of Euclidean geometry and physical space was broken, it had to be asked of Euclidean and non-Euclidean geometry alike: do they exist (are they each self-consistent)? Famously, as Poincaré explained in [36, 37], if one of these two geometries has a consistent mathematical account, then so does the other, and therefore, for as long as they were taken to be the only possible geometries of space, if one must exist then both do. As for which would correctly describe space, however, his position was that it was a matter of convention and we could never decide on logical or empirical grounds alone. But what actually establishes the *mathematical* existence of the objects described in Euclidean geometry?

The naive answer could not be the definitions in Euclid's *Elements*, which by the standards of 1900 were no definitions at all. A better answer was the space of Cartesian geometry, which replaced Euclid's objects with objects described by algebra and the real numbers. Hilbert's *Grundlagen der Geometrie* [23] went further, and offered the use of models of various axiom systems to establish the consistency of those axioms systems, but this ultimately led back to the question of the existence of the real numbers.

As Hilbert saw clearly by about 1900 existence questions in geometry were coming down to existence questions in arithmetic, and they led directly to questions in the new theory of sets and ongoing investigations into logic. Hilbert seems never to have doubted that infinite sets can be satisfactorily handled in mathematics—he proclaimed in [24] that mathematicians would never be banished from the paradise that Cantor had created.

But influential voices spoke out against this view: those of Kronecker, Poincaré, Brouwer, and Weyl, and others. What concerned them was the loose way in which statements were being made about infinite sets of objects. They inclined in various degrees to the view that in principle a statement about all members of a set of objects could only be said to be true if every member of the set had been examined. To be sure, this was an idealisation of the human mind, but it seemed clear that few statements about uncountably infinite sets could be strictly defended. One could certainly say that such a set contained an element of such-and-such a kind if that element could be produced, but arguments by contradiction that could not even in principle produce such an element or, worse, that purported to show that no such element existed, seemed doubtful at best.

French mathematicians around Borel raised other objections. It is one thing, they said, to contemplate an infinite sequence of, say, numbers that is given by a rule for producing either each number from the one before or for specifying the nth number, for each n, but is another thing to speak of an arbitrary sequence given by no rule at all. They also objected, for example, to the axiom of choice.

All these objections were critiques of the human mind. The concern was that if the objects of mathematics were not physical objects but conceptual, mental objects, then the (idealised) mind that produced these objects had to be analysed. In what sense could it be legitimately said to produce objects of all kinds except those that fell into self-contradiction?

The contrary view, held by Hadamard in France and by Hilbert, was that all that was needed was a consistent way of handling these infinite objects. Understanding, in the sense of mental grasp, was not required; all that was needed was valid use.

At this stage, points at infinity no longer posed special problems. Their ontology had become part of a general question about the existence of mathematical objects, particularly those created from classes of objects whose existence was granted (for example, ideal numbers as sets of ordinary numbers).

Concluding remarks

This essay has concentrated on problematic entities for mathematicians in the period before mathematics was given its modern, largely set-theoretic, form, and merely hinted that these difficulties were dissolved in the larger questions about an ontology for modern mathematics. In this volume those questions are addressed most directly by Avigad and Gillies, and they both offer a somewhat Wittgensteinian perspective. They argue that coherent use of a term, in a language that governs its use, does the bulk of the work in establishing the existence of a mathematical object. Gillies, borrowing a distinction from Frege, suggests that consistent use of a term gives it is sense, and allows that some objects (transfinite cardinals) may lack a reference. Avigad, with a nod to a Quinean pragmatism about how we operate in the world, likewise sees mathematics primarily as a linguistic activity.

This linguistic standpoint is supported by the topics discussed in this essay. At one point I asked, rhetorically, "At what point does a theory go from being figures of speech to being something like a naturalistic account of an object that just happens not to be in the world?" One answer to this question is to say that it rests on a false antithesis. It suggests that there is loose talk about things we make up, and hard talk about things around us. But this is inadequate for discussing mathematics, and results in the contortions of some philosophies of mathematics. It is surely better to ask: what more could we want from talk about mathematical objects than that it be consistent with naive mathematics and self-consistent? To put the point another way, the supposed distinction rests on an unduly naive idea of how we know about the world, and a naive epistemology has resulted in an impoverished ontology.

One thing mathematicians want from any discursive analysis of mathematics is that it respects their sense that what they validly discover is true or correct in some straight-forward way. It may be that in this system a certain theorem holds that is false in some other system (compare Euclidean and non-Euclidean geometry), but in each system taken separately certain theorems hold and others do not. What they find disturbing in linguistic analyses, I believe, is the word 'game'. Dickens could have made Oliver Twist throw in his lot with Fagin and become a master criminal and no linguistic rule or rule for writing fiction, however broadly defined, would be violated; a mathematician cannot similarly decide that a well-defined space shall, after all, have negative curvature. The mathematicians' hands are tied.

This much the linguistic analysis of mathematics respects: it is the formalised rules of a system that tie the mathematicians' hands. The remaining puzzle is why this suggests so strongly to many mathematicians that they are dealing with objects when trying to formulate their ideas. In this they are like physicists, who it seems have a tendency to attribute homely properties to the most arcane particles that must otherwise be dealt with using highly abstract, and not always rigorous, mathematics, and we can speculate as to why this is.

As Stillwell suggests in another essay in this volume, a great deal of mathematics can be written in the formal language of ZFC and its underlying logic, and in this system a certain hyperbolic seven-dimensional manifold (or any other mathematical object) has a definition, albeit one that would take up some space to write out fully. Mathematicians working with this manifold habitually use some of its defining features (let us suppose that it is compact and has negative curvature of such-and-such a kind), and these will generally be the ones that distinguish it from other manifolds, or that it has because it is a manifold rather than merely a metric space. They also have a sense of how the theory ought to go by analogy with related fields, and they have quite naive intuitions, often of a spatial kind (say, about two-dimensional manifolds). These factors combine in talk about the properties of the hyperbolic seven-dimensional manifold, rather than in talk about the permissible deductions within the system that defines it, and the manifold becomes an object with properties that determine what can be said about it.

This happens, it is productive, and one day psychologists and neuro-anatomists may have something valuable to say about it. The relevant question here is whether this feeling should drive a philosophy. At this point, the parallel between the issue as posed for the mathematics of the period before the mid-19th century and today is quite close. Whenever mathematicians discuss objects that they agree they cannot kick (in the fashion of Doctor Johnson kicking a stone to refute Bishop Berkeley's idealism) they have only talk about it. If this talk is sufficiently constrained to be coherent it seems to be a matter of time before the object is said to exist, be it a point or a line at infinity, complex points, or a seven-dimensional manifold. Whether there is anything more to this talk by way of ontology may be to ask for a false view of ontology.

References

1. K. Andersen (2007). *The Geometry of an Art: The history of the mathematical theory of perspective from Alberti to Monge*, Springer.
2. Apollonius, Conics *Apollonii Pergaei quae graece exstant cum commentariis antiquis. Edidit et latine interpretatus est I. L. Heiberg*. Gr. & Lat. Leipzig, 1891, 1893, Stuttgart 1974. See also *Treatise on Conic Sections. Edited in modern notation, with introductions, including an essay on the earlier history of the subject*, by T. L. Heath. Eng. Cambridge : University Press, 1896.
3. E. Beltrami (1868). Saggio di interpretazione della Geometria non-euclidea. *Giornale di Matematiche* 6, 284–312. French trans. by J. Hoüel, "Essai d'interprétation de la géométrie non-euclidienne," *Annales ećole normale supérieure* 6, 251–288 (1869).
4. R. Bombelli (1572, 1579). *L'Algebra*, Bologna.
5. H.J.M. Bos, C. Hers, F. Oort, D.W. Raven (1987). Poncelet's closure theorem, *Expositiones mathematicae* 5, 289–354.
6. A. Bosse (1648). *Maniere universelle de M. Desargues*, Paris.
7. G. Cardano (1968). *Ars magna*, T. R. Witmer (transl.), MIT Press, 1968.
8. A. Cayley (1878). On the geometrical representation of imaginary variables by a real correspondence of two planes. *Proc. LMS* 9, 31–39 in *Collected Mathematial Papers* 10, 316–323.
9. M. Chasles (1837). *Aperçu historique sur l'origine et le développement des méthodes en géométrie*, Hayez, Brussels.
10. J.L. Coolidge (1924). *The geometry of the complex domain*. Clarendon Press, Oxford.
11. J. le Rond D'Alembert (1752). *Essai d'une nouvelle théorie de la résistance des fluides*. Chez David l'aîné, Paris. Rep. Culture et Civilisation, Bruxelles 1966.
12. de Gua de Malves, J.P. (1740). *Usages de l?analyse de Descartes, pour découvrir, sans le secours du calcul différéntiel, les propriétés, ou affections principales des lignes géométriques de tous les ordres*. Paris.
13. G. Desargues (1986). *The geometrical work of Girard Desargues*, J.V. Field, J.J. Gray (eds.), Springer.
14. G. Desargues (1988). *L'oeuvre mathématique de G. Desargues: textes publiés et commentés avec une introduction biographique et historique*, 2nd edn., avec une post-face inédit par René Taton, Institut interdisciplinaire d'études epistemologiques.
15. L. Euler (1748). *Introductio in analysin infinitorum*, two vols, *Opera Omnia* (1) vols 8 and 9. English transl. *Introduction to Analysis of the Infinite*, Book I, Springer, 1988, Book II, Springer, 1990 (E101, 102).
16. L. Euler (1750). Demonstration sur le nombre des points, ou deux lignes des ordres quelconques peuvent se couper, *Mémoires de l'académie des sciences de Berlin* 4, 34–248 in *Opera Omnia* (1), 26, 46–59 (E148).
17. L. Euler (1770). *Vollständige Einleitung zur Algebra* in *Opera Omnia* (1) 1, (E387). English transl. *Elements of Algebra*, Rev. J. Hewlett, London 1840, repr. Springer, 1972.
18. J.G. Fauvel and J.J. Gray (1987). *The history of mathematics – a reader*, Macmillan.
19. H.G. Grassmann (1844). *Die Wissenschaft der extensiven Grösse, oder die Ausdehnungslehre, eine neue mathematische Disciplin, etc.*, Leipzig. English transl. *A new branch of mathematics: the "Ausdehnungslehre" of 1844 and other works by Hermann Grassmann*; translated by Lloyd C. Kannenberg. Chicago: Open Court, c1995.
20. J.J. Gray (1989). *Ideas of space: Euclidean, non-Euclidean, and relativistic*. Clarendon Press, Oxford.
21. J.J. Gray (2010). *Worlds Out of Nothing; A Course on the History of Geometry in the 19th Century*, 2nd rev. ed. Springer.
22. M.J. Greenberg (2008). *Euclidean and non-euclidean geometries : development and history*, 3rd ed. New York: W. H. Freeman; Basingstoke: Palgrave.

23. D. Hilbert (1899). *Festschrift zur Feier der Enthüllung des Gauss–Weber–Denkmals in Göttingen [etc.] Grundlagen der Geometrie*. Leipzig.

24. D. Hilbert (1926). Über das Unendlich, *Mathematische Annalen* 95, 161–190.

25. I. Kant (1967). *Philosophical correspondence, 1759–1799*, ed. and transl. A. Zweig.

26. J. Kepler (1604). *Ad Vitellionem paralipomena, quibus Astronomiae pars optica traditur; potissimum de artificiosa observatione et aestimatione diametrorum deliquiorumque solis et lunae. Cum exemplis insignium eclipsium, etc.* Frankfurt.

27. J. Kepler (1609). *Astronomia Nova, seu physica coelestis, tradita commentariis de motibus stellae Martis ex observationibus G.V. Tychonis Brahe*. Prague.

28. C.F. Klein (1871). Ueber die sogenannte nicht-Euklidische Geometrie I, *Mathematische Annalen* 4, 573–615, in *Gesammelte Mathematische Abhandlungen* I, 254–305.

29. P. de la Hire (1685) *Sectiones Conicae, in novem libros distributae*, Paris.

30. J.-L. Lagrange (1781). Sur la construction des cartes géographiques. *Nouv. Mém. Acad. Berlin*, 161–210, in *Oeuvres* 4, 637–691.

31. I. Newton (1704) Enumeration of lines of the third order, Appendix to *Opticks*, transl. C.R.M. Talbot, London, 1860.

32. M. Pasch (1882). *Vorlesungen über neuere Geometrie*. Teubner, Leipzig.

33. J. Plücker (1835). *System der analytischen Geometrie, auf neue Betrachtungsweisen gegründet, und insbesondere eine ausführliche Theorie der Curven dritter Ordnung enthaltend*. Berlin.

34. J. Plücker (1839). *Theorie der algebraischen Curven gegründet auf eine neue Behandlungsweise der analytischen Geometrie*. Bonn.

35. H. Poincaré (1883). Mémoire sur les groupes Kleinéens. *Acta Mathematica* 3, 49–92. In *Oeuvres* 2, 258–299.

36. H. Poincaré (1902a). L'espace et la géométrie. *La Science et l'hypothése* 77–94. Paris, Flammarion. English transl. W. J. Greenstreet, Space and geometry, in Poincaré *Science and hypothesis* 51–71. London, Walter Scott, 1968.

37. H. Poincaré (1902b). L'expérience et la géométrie. *La Science et l'hypothése* 95–110. Paris, Flammarion. English transl. W. J. Greenstreet, Experiment and geometry, in Poincaré *Science and hypothesis* 72–88. London, Walter Scott, 1968.

38. J.V. Poncelet (1822). *Traité des Propriétés projectives des Figures*, Gauthier-Villars.

39. R. Rider (1981). Poisson and algebra: against an eighteenth century background, *Siméon-Denis Poisson et la Science de son Temps*, M. Métivier, P. Costabel, P. Dugac (eds), École Polytechnique.

40. G.F.B. Riemann (1851). *Grundlagen für eine allgemeine Theorie der Functionen einer veränderlichen complexen Grösse*. Inauguraldissertation, Göttingen. In *Werke*, 3rd ed. 35–80. English transl. in Riemann (2004), 1–42.

41. G.F.B. Riemann (1857). Theorie der Abel'schen Functionen. *Journal für Mathematik* 54. In *Gesammelte mathematische Werke*, 120–144. English transl. in Riemann (2004), 79–134.

42. G.F.B. Riemann (1867). Ueber die Hypothesen welche der Geometrie zu Grunde liegen. *K. Ges. Wiss. Göttingen*, 13, 1–20. In *Gesammelte mathematische Werke*, 304–319. English transl. in Riemann (2004), 257–272.

43. G.F.B. Riemann (2004). *Collected Papers*, R. Baker, C. Christenson, H. Orde transl. and ed., Kendrick Press.

44. von Staudt, C.G.C. (1847). *Geometrie der Lage*. Nürnberg.

45. von Staudt, C.G.C. (1856–1860). *Beiträge zur Geometrie der Lage*, 3 vols. Nürnberg.

Mathematics and language

Jeremy Avigad

Abstract This essay considers the special character of mathematical reasoning, and draws on observations from interactive theorem proving and the history of mathematics to clarify the nature of formal and informal mathematical language. It proposes that we view mathematics as a system of conventions and norms that is designed to help us make sense of the world and reason efficiently. Like any designed system, it can perform well or poorly, and the philosophy of mathematics has a role to play in helping us understand the general principles by which it serves its purposes well.

I learned empirically that this came out this time, that it usually does come out; but does the proposition of mathematics say that? ... The mathematical proposition has the dignity of a rule.

So much is true when it's said that mathematics is logic: its moves are from rules of our language to other rules of our language. And this gives it its peculiar solidity, its unassailable position, set apart.

— Ludwig Wittgenstein

... it seemed to me one of the most important tasks of philosophers to investigate the various possible language forms and discover their characteristic properties. While working on problems of this kind, I gradually realized that such an investigation, if it is to go beyond common-sense generalities and to aim at more exact results, must be applied to artificially constructed symbolic languages.... Only after a thorough investigation of the various language forms has been carried through, can a well-founded choice of one of these languages be made, be it as the total language of science or as a partial language for specific purposes.

— Rudolf Carnap

Physical objects, small and large, are not the only posits.... the abstract entities which are the substance of mathematics... are another posit in the same spirit. Epistemologically these are myths on the same footing with physical objects and gods, neither better nor worse except for differences in the degree to which they expedite our dealings with sense experiences.

— W. V. O. Quine

J. Avigad (✉)
Department of Philosophy, Carnegie Mellon University,
Baker Hall 161, Pittsburgh, PA 15213, USA
e-mail: avigad@cmu.edu

© Springer International Publishing Switzerland 2015   235
E. Davis, P.J. Davis (eds.), *Mathematics, Substance and Surmise*,
DOI 10.1007/978-3-319-21473-3_12

"When I use a word," Humpty Dumpty said in rather a scornful tone, "it means just what I choose it to mean — neither more nor less."
"The question is," said Alice, "whether you can make words mean so many different things."
"The question is," said Humpty Dumpty, "which is to be master — that's all."
— Lewis Carroll

1 Introduction

Mathematics holds a special status among the sciences. We expect the results of mathematical calculation to tell us things about the world, such as where to look for Mars or Venus at a given date and time, whether a bridge is strong enough to withstand the loads that will be placed on it, and whether a straight is more likely than a flush in poker. At the same time, we do not look to the world to confirm our mathematical knowledge: we don't rely on empirical observations to justify the claim that two and two make four, and it is hard to imagine a scientific experiment that could disconfirm that belief.

There are other features that set mathematics apart. Unlike physical objects, mathematical objects are not located in space. It is almost grammatically incorrect to speak about mathematical objects in temporal terms, for example, to say that seven was prime yesterday or that it will, in all likelihood, be prime tomorrow. And it seems that, at least ideally, mathematical knowledge admits a kind of certainty that cannot be attained empirically. Our mathematical claims are sometimes mistaken, but that is generally attributed to a failure to understand the proper methods of reasoning or follow them correctly. In contrast, the doubts we have about empirical claims seem to be inherent in the nature of empirical methods, and not in our ability to carry them out.

The philosophy of mathematics tries to explain these features: the *ontology of mathematics* aims to provide an account of mathematical objects and their nature, and the *epistemology of mathematics* aims to provide an account of the means by which we can reliably come to have mathematical knowledge. Whatever we have to say about the nature of mathematical objects, an important part of the game is to explain how it is we can possibly know things about them. And whatever we have to say about that knowledge, an important part of the game is to explain why it is knowledge worth having, and, in particular, how it grounds rational behavior and expectations with respect to empirical phenomena.

Philosophers have played this game since the origins of philosophy itself. Plato relied on his theory of forms, while Aristotle invoked a rival, less Platonic, theory of forms. Descartes' *Meditations* explains why we can be nearly certain that our mathematical claims are true and have something to do with the world, though they are not quite as certain as the fact that God exists. Leibniz classified mathematical truths as analytic truths, Locke classified them as ideas of relation, and Hume classified them as relations of ideas. For Kant, arithmetic and geometric truths were important examples of knowledge that is synthetic (roughly, not "contained" in the definitions of the constituent concepts) and *a priori* (epistemologically independent of our experience of the world). His momentous *Critique of Pure Reason* sought to explain how such knowledge is possible.

Without getting bogged down in knotty metaphysics, modern logic gives us *descriptive* accounts of how mathematics works. Various formal axiomatic foundations—set-theoretic foundations, type-theoretic foundations, categorical foundations—provide us with models of the mathematical language and its general principles of inference. To be sure, there are puzzling features of mathematical language; for example, we sometimes refer to "the seven-element cyclic group" knowing full well that it is only defined up to isomorphism. Different frameworks account for such features in different ways, but they are generally inter-interpretable, and taken together, they give us a pretty good low-level account of the structure of mathematical language and inference.

There are also fairly lowbrow explanations as to why mathematics exhibits the features enumerated above. Mathematical objects are what the norms of mathematical practice say they are, no more, and no less, and we come to have mathematical knowledge by following the norms of reasoning correctly, not by performing experiments and measurements. We learn mathematics by learning the appropriate norms from our parents, our teachers, and the mathematical literature, with additional feedback and correction from graders and journal referees.

But why are the norms of mathematical practice what they are, and how is it that following those norms tells us anything at all about the way the world is? In the first half of the twentieth century, the logical positivists viewed these questions as two sides of the same coin. Mathematics tells us something about the world precisely because it embodies norms and conventions with which we describe the world and make sense of physical experience. And we have settled on these norms just because they provide us with useful ways of making sense of our experiences; if they weren't useful, we would adopt different norms.

On this view, what makes something a piece of mathematics proper is that we hold the rules and conventions relatively fixed in our ordinary patterns of reasoning, blaming any mismatches with experience on the application rather than the rules. In a sense, there are plenty of counterexamples to the mathematical claim that one and one make two: one drop of water and another drop of water make a bigger drop of water, one rabbit and another rabbit make lots of rabbits, and one rock combined with another rock with sufficient force makes dust. But in each case, we reject the counterexample as a misapplication of the mathematical principle. We hold our norms for counting and computing sums fixed because they provide generally useful ways of interpreting our experiences, but in ordinary circumstances we keep experience beholden to the rules, and not the other way around. This point of view has been nicely summed up by William Tait:

> Only empirical explanation is possible for why we have come to accept the basic principles that we do and why we apply them as we do—for why we have mathematics and why it is at it is. But it is only within the framework of mathematics as determined by this practice that we can speak of mathematical necessity. In this sense, which I believe Wittgenstein was first to fully grasp, mathematical necessity rides on the back of empirical contingency.

For the philosopher Rudolf Carnap, adopting a set of mathematical norms and conventions was part of choosing a suitable "language form," a task which he believed could be informed by philosophical reflection and guidance. Carnap and his fellow logical positivists drew a distinction between "analytic" truths, like mathematical truths, that are grounded in the norms of the linguistic framework, and "synthetic" truths, which are grounded by empirical observations. In his influential essay, "Two dogmas of Empiricism," W. V. O. Quine famously denied that there is a sharp distinction between the conventional and empirical parts of language: any empirical claim can be held true in the face of recalcitrant experience by modifying other parts of our language accordingly, and, conversely, even our most deeply held mathematical beliefs are subject to revision in light of future experiences. Thus, for Quine, *all* of science is a matter of maintaining a web of beliefs and making it fit with our experiences. On this view, the allegedly special character of mathematical claims as unassailably fixed is an illusion; at best, we can say that mathematical claims are more stable, or less subject to revision, than their ostensibly empirical counterparts.

In support of Quine's criticisms, one can observe that mathematical concepts do indeed evolve over time, though the subject seems to be remarkably adept at reinterpreting prior developments in a way that preserves their essential correctness. In any case, whether we view the outward features of mathematical practice as reflective of a special status of its claims or simply a provisional stance we adopt towards them, there remains a substantial point of agreement in the views of Carnap and Quine: mathematical and scientific reasoning are shaped by the rules of our language, and these rules are, in turn, adopted for pragmatic scientific reasons.

My goal in this essay is to take this point of view seriously, and argue that it is an important task for philosophy to clarify the rules and norms of our mathematical practices and to help us understand the role they play in our reasoning. For all their talk about the general considerations that influence our choice of language, neither Carnap nor Quine was very specific about how to assess the effects of the choices that we have made historically and continue to make in our everyday activities. Note that I am using the word "language" here in a very broad sense to refer not only to our choice of words and grammar, but also the way we frame, conceptualize, analyze, and reason about the problems before us.

What I am advocating is a view of mathematics as a linguistic artifact, something we have designed, and continue to design, to help us get by in the world. As a community of practitioners, we choose to do mathematics the way we do, and when we do mathematics in the usual ways, we are bound by these choices. Philosophy can help us become aware of the choices we have made, and understand why we have made them, and even, perhaps, why we might want to change them. Various pieces of mathematics are designed to help us achieve various goals, and like any human artifacts, they can serve their purpose well or poorly. The philosophical challenge is to understand the general principles by which they serve our goals well, and to support the development of better artifacts. In that sense, the philosophy of mathematics can be viewed as a kind of language engineering, akin to any other design science.

In this essay, I will try to pursuade you that this view of mathematics makes sense, and provides us with a framework for understanding aspects of mathematics that are important to us. In the next section, I will share some reflections on ordinary mathematical language, using the history of mathematics as a guide. I will then turn to formal mathematical languages, which is to say, symbolic languages whose grammar and rules of use are expressed in precise mathematical terms. Finally, I will explore some of the relationships between the two, and use those observations to bolster my recommendations for the philosophy of mathematics.

2   Informal mathematical language

I will begin with some observations regarding ordinary mathematical concepts and the way they evolve over time. In 2010, a graduate student in my department, Rebecca Morris, asked me to supervise her research on the history and philosophy of mathematics. She was particularly interested in the development of the concept of a "function," and hoped that studying the history would help illuminate the way we reason about functions in everyday mathematics. From a philosophical standpoint, she wanted to understand *why* we talk about functions the way we do, and the senses in which we are justified in doing so.

Today, we think of a function as a correspondence between any two mathematical domains: we can talk about functions from the real numbers to the real numbers, functions that take integers to integers, functions that map each point of the Euclidean plane to an element of a certain group, and, indeed, functions between any two sets. As we began to explore the topic, Morris and I learned that most of the historical literature on the function concept focuses on functions from the real or complex numbers to the real or complex numbers. There is a good reason for that: for most of the history of the function concept, from Euler's 1748 introduction to the analysis of the infinite to the dramatically different approaches to complex function theory by Riemann and Weierstrass in the latter half of the nineteenth century, the word "function" was used exclusively in that sense. Moreover, authors generally tended to assume, implicitly, than any function is given by an analytic expression of some sort. In 1829, in a paper on the Fourier analysis of "arbitrary functions," Dirichlet went out of his way to reason about functions extensionally, which is to say, without making use of any such representation. In that paper, he famously put forth the example of a function that takes one value on rational numbers and another value on irrational numbers as an extreme case, one that was difficult to analyze in conventional terms.

Dirichlet's "arbitrary" functions were, nonetheless, functions from the real numbers to the real numbers. Even the notion of a "number theoretic function," like the factorial function or the Euler function, was nowhere to be found in the literature; authors from Euler to Gauss referred to such entities as "symbols," "characters," or "notations." Morris and I tracked down what may well be the first use of the term "number theoretic function" in a paper by Eisenstein from 1850, which begins with a lengthy explanation as to why it is appropriate to call the Euler phi function a "function." We struggled to parse the old-fashioned German, which translates roughly as follows:

> Once, with the concept of a function, one moved away from the necessity of having an analytic construction and began to take its essence to be a tabular collection of values associated to the values of one or several variables, it became possible to take the concept to include functions which, due to conditions of an arithmetic nature, have a determinate sense only when the variables occurring in them have integral values, or only for certain value-combinations arising from the natural number series. For intermediate values, such functions remain indeterminate and arbitrary, or without any meaning.

When the gist of the passage sank in, we laughed out loud. The opening phrase is a nod to Dirichlet's notion of a function as an arbitrary tabular pairing of input and output values. We had expected Eisenstein to go on to say something like this: "Of course, there is no reason to restrict attention to the real and complex numbers; we can now take a function to be a tabular pairing of values in any two domains." Instead, he went on to argue that we can extend the function concept to functions defined on the integers by considering them to be *partially defined functions on the real numbers*, that is, functions from the real numbers to the real numbers that happen to take values only on integer arguments. Indeed, the first hints of the modern notion of a function as an arbitrary correspondence between any two domains did not appear until the late 1870s, and Eisenstein's introductory paragraph is a nice reminder of the fact that mathematical concepts evolve gradually, and often in surprising ways.

Morris and I decided to focus our attention on another type of function that was studied in the nineteenth century (though they were not labeled as such). In 1837, Dirichlet proved a beautiful theorem that states that there are infinitely many prime numbers in any arithmetic progression in which the first two terms do not have a common factor. For example, there is clearly only one prime number in the progression $2, 12, 22, \ldots$ because the first two terms, and hence all the terms, are multiples of 2. Dirichlet's theorem states that this is essentially the only thing that can go wrong. For example, in the progression $3, 13, 23, 33, \ldots$, the first two terms have no factor in common, and so there will be infinitely many primes. This was conjectured to be true around the turn of the nineteenth century but even the great master, Gauss himself, had been unable to prove it. Dirichlet's proof is a landmark not only because it solved a hard problem, but also because it introduced methods that are now central to algebraic and analytic number theory.

Modern presentations of Dirichlet's proof rely on the notion of a *Dirichlet character*. To deal with the arithmetic progression with first term m and common difference k, we consider the multiplicative group $(\mathbb{Z}/k\mathbb{Z})^*$ of invertible residues modulo k, that is, the numbers modulo k that have no common factor with k. For example, when $k = 10$, we consider the set of numbers $\{1, 3, 7, 9\}$ with multiplication modulo 10. A *character* on this group is a nonzero homomorphism from this group to the complex numbers, that is, a function χ with the property that $\chi(x \cdot y) = \chi(x) \cdot \chi(y)$ for every x and y. For example, there is the trivial character on $(\mathbb{Z}/10\mathbb{Z})^*$ which sends all of the residues to 1, and another character χ which sends

$1, 3, 7, 9$ to $1, i, -i, -1$, respectively. There are four distinct characters modulo 10, and one can show that, for any k, there are only finitely many characters modulo k. To be precise, a *Dirichlet character* is obtained from an ordinary (group) character by "lifting" the values to the integers, i.e., defining $\hat{\chi}(n)$ to be the value of χ at n modulo k if n has no common factor with k, and 0 otherwise. As is common practice, I will blur the distinction between the group character and the associated Dirichlet character, and use the term "character" for both.

What made Dirichlet's theorem interesting to us is that in contemporary proofs, characters are treated as ordinary mathematical objects like the natural numbers. For example, in an ordinary textbook proof:

1. we show that the characters modulo k form a group under pointwise multiplication, just as the integers modulo k form a group under addition;
2. we define functions $L(s, \chi)$, called *Dirichlet L-series*, whose second argument is a character;
3. we write down sums $\sum_{\chi} \chi(m)^{-1} L(s, \chi)$ where the index χ ranges over characters, just as we write down sums like $\sum_{i=1}^{n} i^2$ where the index i ranges over the integers.

These features are distinctly modern: mathematicians from Euler to Cauchy seemed to think of functions as syntactic expressions, like $f(x) = x^2$, that are somewhat different from objects like numbers. In today's terms, functions are "higher-order objects," and what is notable is that in contemporary mathematics their status is no different from more fundamental mathematical entities. Morris and I reasoned that if we could understand how the features above were implemented in Dirichlet's original proof, it would help us understand how functions attained their contemporary status.

When we read Dirichlet's original papers, however, we were surprised to discover that there is no notion of character there at all! Rather, there are certain algebraic expressions which, today, we view as values of the characters. Where we would write $\chi(n)$, Dirichlet wrote expressions

$$\theta^{\alpha} \varphi^{\beta} \omega^{\gamma} \omega'^{\gamma'} \dots,$$

where $\theta, \varphi, \omega, \omega', \dots$ are complex roots of 1 and $\alpha, \beta, \gamma, \gamma', \dots$ are certain integers that implicitly depend on n. Roughly, whereas today we define the characters as the nonzero homomorphisms from the group of invertible residues to the complex numbers, Dirichlet used an explicit symbolic representation of the characters and implicitly relied on the fact that they have that key property.

Dirichlet showed that the relevant symbolic expressions can be parameterized by tuples $\mathfrak{a}, \mathfrak{b}, \mathfrak{c}, \mathfrak{c}', \dots$ of integers, and wrote $L_{\mathfrak{a}, \mathfrak{b}, \mathfrak{c}, \mathfrak{c}', \dots}(s)$ where we would write $L(s, \chi)$. Moreover, where we would write a summation over characters, Dirichlet summed over the representing tuples. For example, where we would write

$$\sum_{\chi\in\widehat{(\mathbb{Z}/k\mathbb{Z})^*}} \overline{\chi(m)}\log L(s,\chi),$$

Dirichlet wrote

$$\sum \Theta^{-\alpha_m \mathfrak{a}}\, \Phi^{-\beta_m \mathfrak{b}}\Omega^{-\gamma_m \mathfrak{c}}\Omega^{-\gamma'_m \mathfrak{c}'}\cdots \log L_{\mathfrak{a},\mathfrak{b},\mathfrak{c},\mathfrak{c}',\dots}$$

Dirichlet's reliance on representing data played out in other ways. In contemporary presentations of Dirichlet's proof, we divide the characters into three groups: first, there is the trivial character, $\chi(m) = 1$ for every residue m; then there are the characters, other than the trivial character, that take only the real values ± 1; and, finally, the remaining characters, each of which takes on at least one non-real complex value. This categorization is *extensional*, which is to say, it is cast in terms of the *values* of the characters χ, and not the data used to represent them. In contrast, Dirichlet's division into cases was cast in terms of the defining data, something we would describe as an *intensional* categorization.

While treating characters as first-class mathematical objects, modern proofs also make them objects of study in their own right. Morris and I were curious to understand how the understanding of Dirichlet characters evolved over time, and so we studied a number of subsequent presentations of Dirichlet's theorem and related results: extensions, by Dirichlet, of his result to Gaussian integers and quadratic forms; an 1863 presentation by Dedekind, in a supplementary appendix to his writeup of Dirichlet's lecture notes on number theory; generalizations by Dedekind, in 1879, and Weber, in 1882, of the notion of a character to an arbitrary group; a constructive proof, due to Kronecker, presented in the 1880s and later written up by his student, Hensel; presentations of Dirichlet's theorem and extensions by Hadamard in 1896 and de la Vallée Poussin in 1897; and presentations by Landau in 1909 and 1927, the latter which reads much like a contemporary textbook proof.

What we found is that the path from Dirichlet's proof to Landau's was long and tortuous, with various "modern" features appearing hesitatingly and tentatively, in fits and starts. Over time, authors gradually identified the expressions defining the characters as key components in the proofs, separated out their central properties, and proved them as independent lemmas. One advantage to doing this is that the fiddly technical details of manipulating representations become isolated and black-boxed, so that readers only have to remember the key properties, and not their proofs. Another advantage is that identifying the notion of a character and their essential properties supports the generalization of the notion to other groups, as well as generalizations of Dirichlet's proof itself. Dirichlet himself made moves in this direction in his later work. Dedekind's 1863 presentation used χ to denote the values of the characters and separated out some of their axiomatic properties, and in 1882, Weber gave the general definition of a character of an abelian group, and proved the general properties. In 1909, Landau enumerated four "key properties" of the characters, and emphasized that these are all that is needed in the proof of Dirichlet's theorem; once we have them, we can forget all the details of the representations of the characters.

The treatment of characters on par with common mathematical objects like numbers came gradually as well. Where we would write sums over the characters, Dedekind's 1863 presentation summed over representing data with expressions like $\sum_{a,b,c,c',\dots}$, like Dirichlet's. In 1897, Hadamard followed a different approach: he introduced an arbitrary numbering of the characters, writing them $\psi_1, \psi_2, \dots, \psi_M$, which meant that he could sum over the characters with conventional notation $\sum_{i=1}^{M}$. In contrast, de la Vallée Poussin introduced special notation S_χ to sum over characters, while at the same time using the notation $\sum_i$ for summing over a finite set of integers. This suggests an abstraction of the summation operation to the set of characters, but marked by a concern that summing over a finite set of characters is nonetheless an operation that is distinct from summing over a finite set of integers. In 1909, Landau followed Hadamard's approach, but in 1927 he finally wrote $\sum_\chi$, as we do today.

The transition to treating the characters as arguments to L-functions was similarly prolonged. For example, Hadamard simply wrote L_v for the series corresponding to ψ_v. In 1897, de la Vallée Poussin wrote $Z(s, \chi)$, and Landau adopted the notation $L(s, \chi)$ that we use today. And the way authors described the classification of L-series, which plays a crucial role in the proof, also changed over time. Most authors we considered favored an intensional characterization like Dirichlet's, though some authors gave both characterizations, that is, specifying the division in terms of the output values as well as in terms of the representing data.

In short, it took almost 90 years for the treatment of Dirichlet characters to evolve to the point where they were identified as objects of interest, seen to have a group structure, passed as arguments to functions, taken to be the indices of a summation, and so on. Since then, from Landau's 1927 proof to the present day, the treatment of Dirichlet characters has remained remarkably stable.

The point I wish to make is that these changes constitute an ontological shift, pure and simple. In 1837, Dirichlet could not say "a character is a nonzero homomorphism from $(\mathbb{Z}/\mathbb{Z}_k)^*$ to $\mathbb{C}$," because there was no general notion of a function between two domains. He certainly could not do all the things that Landau could do, because the linguistic and methodological resources were simply not available to him. Here I am once again using the term "language" in a broad sense, to include not just grammar and vocabulary but also the conceptual and inferential apparatus that determine the way we talk about and reason about a subject.

It should not be surprising that such changes occur only gradually. Whenever we write down a piece of mathematics, it is with the intention that others will read it, understand it, and judge it to be correct and worthwhile. When we introduce new terminology, notation, and patterns of inference, it is usually the case that the intended usage can be explained and justified in conventional terms. But when the changes represent a marked departure from the status quo, they bring with them a host of nagging concerns. What do the new terms and symbols mean? What are the rules governing their use? Are these rules reasonable and justified? Are they appropriate to the subject matter, and do they give us legitimate answers to the questions we have posed? It is not sufficient that a single author becomes comfortable with the changes; for mathematics to work, the entire mathematical community has to function as a coherent body and come to agreement as to what is allowed. Thus, when changes occur, they occur for good reasons. The ultimate goal of mathematics is to get at the truth, and when new methods make it possible to push the boundaries of our knowledge, there is great incentive to make sense of them, get used to them, and incorporate them into the canon.

Our study of the history of Dirichlet's theorem showed that there are, indeed, good reasons to treat characters as ordinary mathematical objects. Simply identifying characters as objects of interest makes it possible to highlight the properties that make then useful. The very act of naming them simplifies the expressions that occur in complex proofs, and developing their properties independently makes those proofs more modular. Dependencies between pieces of a mathematical text are minimized, so that one can verify properties of the characters independently, and then suppress those details later on. This makes it easier to read the proof, understand it, and verify its correctness.

Such a modular structuring has additional benefits, in that the modularized components are adaptable, reusable, and generalizable. Having a theory of characters at one's disposal means having a tool that can be used to solve other problems, and having an abstract description of what makes Dirichlet characters useful enables us to design or discover other objects with similar properties. Recall that Dedekind and Weber transferred the notion of character to arbitrary finite groups. Later generalizations to continuous groups and nonabelian groups are now fundamental to the subjects of representation theory and harmonic analysis.

Moreover, treating characters on par with other mathematical objects allows other bodies of mathematics to be invoked and applied uniformly. Mathematicians were already comfortable summing over finite sets of numbers; if characters have the same status as numbers, conventional notation and methods of calculation transfer wholesale. Similarly, by the 1870s, the mathematical community began to gain facility with the abstract notion of a group. If one can consider a group of characters in the same manner as a group of residues, both are subsumed under a common body of knowledge.

The history of Dirichlet's theorem is only one manifestation of the momentous changes that swept across mathematics in the nineteenth century. The development of the modern theory of functions and the introduction of ideals and quotients in algebra are additional forces that pushed for both algebraic and set-theoretic abstraction—a new "language form," in Carnap's terminology. These are the sorts of changes that we need to study carefully, because understanding them is crucial to understanding how we do mathematics today. In the next two sections, I will argue that we can be more scientific in this study, and that formal languages and logical methods can play an important role.

3   Formal mathematical languages

In 1931, Kurt Gödel, then still a relatively unknown young logician, published his two *incompleteness theorems*. These are now recognized as among the most profound and important results in the history of logic. The first incompleteness theorem shows, roughly speaking, that no formal axiomatic system of mathematics can settle all mathematical questions, even questions about the natural numbers; there will always be relatively straightforward statements about the natural numbers that cannot be proved or disproved on the basis of the axioms. The second incompleteness theorem shows that, moreover, no such system can establish its own consistency. The opening sentences of Gödel's paper are compelling even today, delivering the dramatic news in a calm, matter-of-fact tone:

> The development of mathematics toward greater precision has led, as is well known, to the formalization of large tracts of it, so that one can prove any theorem using nothing but a few mechanical rules. The most comprehensive formal systems that have been set up hitherto are the system of *Principia Mathematica (PM)* on the one hand and the Zermelo-Fraenkel axiom system of set theory (further developed by J. von Neumann) on the other. These two systems are so comprehensive that in them all methods of proof used today in mathematics are formalized, that is, reduced to a few axioms and rules of inference. One might therefore conjecture that these axioms and rules of inference are sufficient to decide *any* mathematical question that can at all be formally expressed in these systems. It will be shown below that this is not the case...

What I would like to focus on here are not Gödel's negative, limitative results, but the positive side of his assessment. Ernst Zermelo presented his system of axiomatic set theory in 1908, and Bertrand Russell and Alfred North Whitehead presented a system of "ramified type theory" in their monumental work, *Principia Mathematica*, the three volumes of which first appeared between 1911 and 1914. It is striking that as early as 1931, Gödel could convey the conventional understanding that, indeed, these systems suffice to formalize the vast majority of ordinary mathematical reasoning, a fact that remains true to this day.

But how do we get to a proof of Fermat's last theorem from a few simple axioms and rules? The phenomenon is analogous to the way that we obtain complex computational systems from primitive components. Conceptually, computer microprocessors and memory chips are built from very simple elements, components that can store a bit of memory or carry out a simple logical operation. These are combined to form components that are capable of carrying out more complex operations, and those operations, in turn, form the basis for low-level programming languages. Working our way up in a modular fashion, we get operating systems and compilers, and ultimately systems that manage airline schedules, control industrial power plants, play chess, and compete in game shows.

The "evidence" Gödel had in mind for the claim that mathematics can be formalized is similar. Starting with a few axioms and rules, we can define basic mathematical objects like numbers, functions, sequences, and so on, and derive their properties. From there, we work our way up through the mathematical canon. In practice, however, the process is rather tedious: we think about mathematics at a higher level of abstraction, and even with a modular approach, there are far too many details to write down in practice. Famously, in the three volumes of formal mathematics in the *Principia*, Russell and Whitehead never got past elementary arithmetic. Explaining how complex objects and inferences can be built up from more simple ones and carrying out a number of examples is enough to make a compelling case that mathematics can be formalized *in principle*, but that is not the same as saying that it can be formalized *in practice*.

Towards the end of the twentieth century, however, computer scientists began to develop "computational proof assistants" that now make it possible to construct axiomatic proofs of substantial theorems, in full detail. The languages with which one interacts which such systems are much like higher-level programming languages, except that the "programs" are now "proofs": they are directives, or pieces of code, that enable the computer to construct a formal axiomatic proof. Proof assistants keep track of definitions and previously proved theorems, as well as notation and abbreviations. They also provide various types of automation to help fill in details or supply information that has been left implicit. Users can, moreover, declare bits of knowledge and expertise as they develop their mathematical libraries, such as rules for simplifying expressions and inference patterns to be used by the automation procedures.

There are a number of proof assistants now in existence, as well as fully verified libraries of number theory, real and complex analysis, measure theory and measure-theoretic probability, linear algebra, abstract algebra, and category theory. Substantial mathematical theorems have been verified in such systems. Proof assistants are also used to verify claims as to the correctness of hardware and software with respect to their specifications, something that is especially valuable when human lives are at stake, as is the case, for example, with control systems for cars and airplanes. Describing the state of the field today would take me too far afield, but there are suggestions for further reading in the notes at the end of this essay.

Designing a theorem prover thus involves designing a language that can express mathematical definitions, theorems, and proofs, providing enough information for the computer to construct the underlying formal mathematical objects, in a manner that is convenient, efficient, and practically manageable. To start with, one has to choose a formal system that constitutes the underlying standard of correctness. Axiomatic set theory, in which every mathematical object is encoded as a set, has proved to be a remarkably flexible and robust foundation, but experience shows that additional structure is useful when it comes to interactive theorem proving. The fact that any expression in set theory can be interpreted as a natural number, a function, or a set makes it hard for the system to infer the user's intent, thus requiring users to provide much more information. In various systems of *type theory*, every object is tagged as an element of an associated *type*. A given expression may denote a natural number, a function from natural numbers to natural numbers, a pair of numbers, or a list of numbers; the syntactic form of the expression, and the manner it is presented, indicates which is the case. Such languages thus provide a closer approximation to ordinary mathematical vernacular.

Mathematical definitions and theorems are then expressed in the chosen foundational language. Here is a statement of the prime number theorem in the Isabelle theorem prover:

```
theorem PrimeNumberTheorem: "(λn. pi n * ln n / n) → 1"
```

Here pi denotes the function $\pi(n)$ that counts the number of primes less than or equal to n:

```
pi n ≡ card {p. p ≤ n ∧ p ∈ prime}
```

The Feit-Thompson theorem is a landmark theorem of finite group theory; it asserts that every finite group of odd order is solvable, or, equivalently, that the only simple groups of odd order are cyclic. These two assertions are expressed as follows in the SSReflect dialect of Coq:

```
Theorem Feit_Thompson (gT : finGroupType) (G : {group gT}) :
    odd #|G| → solvable G.

Theorem simple_odd_group_prime (gT : finGroupType)
    (G : {group gT}) :
    odd #|G| → simple G → prime #|G|.
```

The Kepler conjecture asserts that there is no way of packing equally-sized spheres into space that results in a higher-density than that achieved by the familiar means of arranging them into nested hexagonal layers, the way oranges are commonly stacked in a crate. Here is a statement of this theorem in HOL Light:

```
∀V. packing V => (∃c. ∀r. &1 <= r =>
    &(CARD(V INTER ball(vec 0,r))) <=
        pi * r pow 3 / sqrt(&18) + c * r pow 2))
```

And here is a statement of the Blakers-Massey theorem, a theorem of algebraic topology, formulated in a formal framework known as *homotopy type theory* and rendered in the Agda proof assistant:

```
blakers-massey : ∀ {x₀} {y₀} (r : left x_0 ≡ right y₀) →
    is-connected (n +2+ m) (hfiber glue r)
```

All of the theorems just listed have been fully verified in the systems indicated. To that end, all of the objects they mention, such as natural numbers, real numbers, groups, and homotopy fibers, had to be defined in the relevant foundational frameworks. In the Lean theorem prover, for example, we can define the natural numbers as follows:

```
inductive nat : Type :=
| zero : nat
| succ : nat → nat
```

This declares the natural numbers to be the smallest type containing an element, zero, and closed under a successor function, succ. We can then define, say, addition recursively:

```
definition add : nat → nat → nat
```

```
| add x zero := x,
| add x (succ y) := succ (add x y)
```

We can also define familiar notation:

```
notation 'N' := nat
notation 0 := zero
notation x '+' y := add x y
```

We can then prove that addition is commutative and associative, and go on to define more complex functions and relations, like multiplication, the divisibility relation, and the function which returns the greatest common divisor of two natural numbers.

To serve that purpose, a proof assistant's language has to support not only writing definitions, but also proofs. Such languages vary widely. In Lean's library, the proof that the greatest common divisor function is commutative runs like this:

```
theorem gcd.comm (m n : N) : gcd m n = gcd n m :=
dvd.antisymm
  (dvd_gcd !gcd_dvd_right !gcd_dvd_left)
  (dvd_gcd !gcd_dvd_right !gcd_dvd_left)
```

The proof demonstrates the equality of the left- and right-hand side by showing that each one divides the other, using the abstract characterization of gcd. The proof in Isabelle's library does the same thing, but relies on built-in automation to fill in the details:

```
theorem gcd_comm: "gcd (m::nat) n = gcd n m"
by (auto intro!: dvd.antisym)
```

In contrast, the proof in the SSReflect library is more direct, unfolding the computational definition:

```
Lemma gcdnC : commutative gcdn.
Proof.
move=> m n; wlog lt_nm: m n / n < m.
  by case: (ltngtP n m) => [||-> //]; last symmetry; auto.
by rewrite gcdnE -{1}(ltn_predK lt_nm) modn_small.
Qed.
```

Proof assistants also have to provide general "mathematical knowledge management" and support mathematical conventions. For example, in Lean, the following commands are used to import library developments of the natural numbers and rings, and make information and notation from those libraries readily available:

```
import data.nat algebra.ring
open nat algebra
```

One can globally declare the types of variables used in a particular development:

```
section
  variables m n k : N
  ...
end
```

In ordinary language, this says "in this section, we use the variables m, n, and k to range over the natural numbers." We can also fix parameters:

```
section
  parameter G : Group
  variables g₁ g₂ : G
  ...
end
```

In words, "in this section, we fix a group, G, and let g_1 and g_2 range over elements of G."

These examples provide only a small sampling of the ways that users interact with proof assistants. Developing mathematical theories involves not just defining basic objects and proving their properties, but also establishing complex notation and telling the system how to resolve ambiguous expressions, defining algebraic structures and instantiating them, declaring pieces of information to be used by the system's automated procedures, organizing data and structuring complex theories, and tagging facts and data in order to signal intended usage. There is a substantial distance from a foundational definition of addition to the lemmas and theorems of everyday mathematics, and theorem provers are finely engineered to make the passage as smooth as possible.

The point I wish to make here is simply this: whenever someone communicates with a computational proof assistant, they are speaking a formal mathematical language. This language has been designed for a specific purpose, namely, to enable users to develop formal theories smoothly and efficiently. As such, it should be powerful, expressive, and convenient. It should allow users to write mathematical expressions concisely, carry out reasoning steps in an intuitive way, and manage complex information effectively. At the foundational level, the choice of axiomatic system determines the objects that exist and the appropriate rules for reasoning about them, but the higher-level organizational features of the system are no less important: they determine how we can talk about the objects that are guaranteed to exist by the foundational framework, what sorts of conceptual apparatus we can bring to bear upon then, and the styles and patterns of argumentation we can carry out.

4   The philosophy of mathematics as a design science

In Section 2, I described some changes in informal mathematical language that emerged over the course of the nineteenth century, and their effects on mathematical practice. In Section 3, I described formal mathematical languages, essentially programming languages, that are used to carry out reasoning in proof assistants, and the role those languages play in structuring our interactions with a system. The parallel should be clear: in both cases we are dealing with the use of language to convey mathematical content and support mathematical reasoning. The languages we use can either serve our purposes well or not, and, in both cases, we continue to tinker with the languages to enable them to serve us better.

The analogy is not perfect, and there are important differences between formal and informal mathematics. Computational proof assistants are developed on the scale of years or decades; mathematical language has been evolving for centuries. The features of a proof assistant are determined by a relatively small design team, responsive, though it may be, to a community of users; changes in mathematical language take place organically across a much larger community. A change to the language of a theorem prover is typically the result of an explicit design decision, and has a precise implementation date; changes to mathematical language are sometimes discussed, but often just happen, sometimes gradually and without fanfare. Perhaps most significantly, the language of a theorem prover has a precisely specified grammar and semantics, and a reference manual that tells you what the various commands do and how they are interpreted by the system. We have to look to more nebulous collections of textbooks and speakers to ascertain the meaning of a mathematical claim. It is a remarkable fact that the mathematical community can communicate with such a high degree of agreement and coherence, without such a reference manual.

Here is the central claim of this essay: when it comes to understanding the power of mathematical language to guide our thought and help us reason well, formal mathematical languages like the ones used by interactive proof assistants provide informative *models* of informal mathematical language. The formal languages underlying foundational frameworks such as set theory and type theory were designed to provide an account of the correct rules of mathematical reasoning, and, as Gödel observed, they do a remarkably good job. But correctness isn't everything: we want our mathematical languages to enable us to reason efficiently and effectively as well. To that end, we need not just accounts as to what makes a mathematical argument correct, but also accounts of the structural features of our theorizing that help us manage mathematical complexity.

Let me provide some examples of how formal mathematical languages can help. Consider, for example, some of the virtues that I ascribed to the historical restructuring of Dirichlet's proof in Section 2: dependencies between components of the proof were minimized, it became easier to understand the arguments and ensure their correctness, and the components themselves became more adaptable, reusable, and generalizable. Computer scientists and programmers will recognize immediately that these are all benefits typically associated with the use of modularity in software design. This is not a coincidence; Section 3 made the analogies between computer code and formal proof manifest. A formal understanding of modularity, its benefits, and the means by which it can be achieved should therefore illuminate the informal mechanisms by which it is achieved in ordinary mathematics. In a similar way, the means by which programmers manage abstract interfaces that hide the internal details of a program from the outside world should illuminate the informal practices that make mathematical abstraction such a powerful tool.

For another example, consider the act of solving a mathematical problem. This typically requires us to decide, heuristically, which steps from among a combinatorial explosion of options will plausibly move us closer to the goal, and which facts from among a vast store of background knowledge are relevant to the problem at hand. The general problem of using heuristics and contextual cues to navigate an unruly search space is central to artificial intelligence. In informal mathematics, a well-written proof provides subtle yet effective means to manage information and keep the relevant data ready to hand, and a powerful theory is one that provides cues that trigger the steps needed to solve problems in its intended domain. It seems likely that formal modeling and insights from the field of automated reasoning can help us make sense of how informal mathematics manages its problem-solving contexts, and, conversely, informal mathematical data and examples can inform research in AI.

For yet another example, consider the observations above that informal mathematics functions remarkably well without a formal reference manual, and that mathematics has a remarkable way of reinterpreting its past to preserve the validity of prior insights and bodies of knowledge. Here, mathematics has computer science beat: programmers struggle to share algorithms across different code bases, and to preserve code while surrounding infrastructure—compilers, operating systems, and supporting libraries—continue to evolve. But once again, the formal mechanisms with which computer scientists cope with the problem can shed light on the mechanisms by which informal mathematics preserves meaning through time and across varied environments, and, conversely, these mechanisms can inform the design of computational systems.

To close this essay, I would like to relate these considerations to the central theme of the essays in this collection, namely, the ontology of mathematics. There are various senses in which ordinary mathematical practice might be said to sanction the existence of an object. Some objects exist only in an imprecise, heuristic sense. When we say that a theorem illuminates an important "connection" between algebra and geometry, for example, we do not take "connections" to be *bona-fide* mathematical objects with precise properties. We might assign Kepler's points at infinity, described in Jeremy Gray's essay, to this category, if the notion is only used to motivate some of his mathematics without playing a substantial inferential role.

Next, there are objects that do seem to play a substantial role in our reasoning, but are taken to be *façons de parler*, convenient shorthands that can be eliminated from "proper" mathematical discourse. Some early algebraists seem to have viewed negative magnitudes and nonhomogeneous quantities in this way, so that, for example, an expression like $x^2 + x$ was understood to be the sum of two "areas," x^2 and xy, where y is a unit length. In 1742, Colin MacLauren wrote *A Treatise of Fluxions*, which aimed to defend the methods of Newton's calculus against Bishop Berkeley's attacks by explaining them in more accessible geometric terms:

> This is our design in the following treatise; wherein we do not propose to alter Sir ISAAC NEWTON's notion of a fluxion, but to explain and demonstrate his method, by deducing it at length from a few self-evident truths... and, in treating of it, to abstract from all principles and postulates that may require the imagining any other quantities but such as may be easily conceived to have a real existence.

The message is this: it is convenient to speak of fluxions, but to be fully rigorous we can proceed in more elementary ways.

At the next level, there are objects that are defined in terms of more basic ones, the way we define a complex number to be pair of real numbers and define a group to be a tuple of data satisfying the group axioms. This, of course, is the most straightforward way to introduce new mathematical objects.

Finally, there are mathematical objects that exist in the foundational sense that we do not feel the need to define them or explain them in other terms. For the ancients, these included the points, lines, and circles of Euclid's *Elements*; Euclid's "definitions" serve to motivate the axioms and diagrammatic rules of inference, but they do not play a direct justificatory role. For most of the history of mathematics, numbers, viewed as discrete or continuous magnitudes, enjoyed such an existence. Today, we generally understand sets and/or functions in this way.

The development of foundational systems like set theory was a watershed in mathematics, providing bedrock foundations that can, in principle, resolve all ontological issues. We no longer need to scratch our heads and wonder whether something exists, since it suffices to produce a set-theoretic definition. And we no longer have to search our intuitions to determine what properties our objects have, since the axioms of set theory serve as the final arbiter.

There are nonetheless good reasons to pursue the broader ontological questions considered in this collection of essays. For one thing, there is still the philosophical question as to what justifies set theory, or any other foundation. Choosing such a foundation amounts to sweeping all one's ontological dust into a corner, but that in and of itself doesn't eliminate it. The best justification for using set theory as an ontology is that we have lots of good reasons to talk about the mathematical objects that we do—numbers, functions, points on the projective plane, and so on—and set theory gives us a clean and uniform account of those. In other words, we are justified in talking about sets because we are justified in talking about all those other things, and not the other way around. To round out the story, we need to say something about why we are justified in talking about those other things, and that, in turn, should be based on a meaningful analysis of the inferential and organization role they play in our scientific practices.

Characterizing these roles as "linguistic" may convey the impression that mathematical objects are somehow less objective than things we bump into like rocks and chairs, and that we somehow need to explain the "illusion" that they are really there. This is nonsense; the fact that we don't bump into something doesn't mean it does not exist. Language works in funny ways, and there are plenty of proper nouns in our language that are not neatly situated in time and space. Any of the buildings of my home university, Carnegie Mellon, may be torn down, and students and faculty come and go, but the university is an entity that somehow survives those changes. We can even imagine the board of trustees shutting down the Pittsburgh operation and relocating to Cleveland, in which case, it would be hard to say exactly what had moved. We can speak about the fifth novel in the Harry Potter series, but we can only point to copies of it. We can produce plenty of red things, but it is hard to point to the color red. How many letters are there in the word "book"? I can name three, "b," "o," and "k," but some would insist there are four. Perhaps we can disambiguate the question by distinguishing between letters and instances of a letter, but now you have to deal with objects known as "instances," and you still can't point to a letter. And if you manage to bump into the government of the United States, let me know.

It is one thing to ask how we can regiment our informal language to make our claims more precise, for example, for the purposes of scientific inquiry. We may settle on one way of talking about colors in physics and another for the purposes of psychological experimentation, with suitable ways to connect the two. The fact that mathematical objects play a fundamental role across the sciences does not make them any less real than novels, colors, universities, or governments. It seems to me much easier to justify the claim that numbers exist than the claim that novels do.

Nor does aligning our talk of mathematical objects with linguistic norms mean that these norms are "arbitrary." The rules of mathematical discourse are quite rigid, and the language is remarkably stable over time. But, as we have seen, time can even alter the way we talk about fundamental mathematical objects, and it is important to understand the forces that bear upon such changes. To that end, it helps to take a broad view of mathematics, and understand what the objects of mathematics do for us.

Which brings me to an even better reason to take a more expansive view of the ontology of mathematics: mere existence isn't everything. Of all the "objects" that set theory or type theory or any other foundation allow us to talk about, only a tiny fraction are of any mathematical interest. So the question is: of all the objects we *can* talk about, which ones *should* we talk about? Trying to answer this question forces us once again to consider the roles that our ontological posits play in the mathematics that we actually do.

In short, it is high time we got over the worry that there is anything inherently mysterious or dubious about mathematical objects, and started paying attention to the things that really matter. On the view I have put forward, even traditional ontological concerns are best addressed by asking why we speak about mathematical objects the way we do, and how the rules and methods we adopt to reason about them provide useful ways of making sense of the world. Questions as to how mathematics enables us to communicate and reason effectively are then of central philosophical importance. The history of mathematics has been an enduring struggle to balance concerns about the meaningfulness, appropriateness, and reliability of our mathematical practices against the drive to adopt innovative methods that enable us to think better, push our understanding further, and extend our cognitive reach. We need a science that can help us understand the considerations that bear upon our choices of mathematical norms, and the philosophy of mathematics should be that science.

Notes

The opening quotation by Wittgenstein is from Part I, § 165 of his *Remarks on the Foundations of Mathematics*, G. H. von Wright, R. Rhees, G. E. M. Anscombe, eds.,

MIT Press, 1983. The quotation by Carnap is from his autobiographical chapter in *The Philosophy of Rudolf Carnap*, Paul Schilpp ed., Open Court and Cambridge University Press, 1963. The quotation by Quine is taken from his "Two Dogmas of Empiricism," which first appeared in *The Philosophical Review*, 60: 20–43, 1951 and has been reprinted in many other sources. Alice's exchange with Humpty Dumpty is from Chapter 6 of *Through the Looking Glass*.

The quotation from William Tait is from an essay titled "The Law of the Excluded Middle and the Axiom of Choice," from Alexander George, ed., *Mathematics and Mind*, Oxford University Press, 1994. It is reprinted in William Tait, *The Provenance of Pure Reason: Essays in the Philosophy of Mathematics and Its History*, Oxford University Press, 2005.

The research on the history of Dirichlet's theorem described in Section 2 can be found in Jeremy Avigad and Rebecca Morris, "The concept of 'character' in Dirichlet's theorem on primes in an arithmetic progression," *Archive for History of Exact Sciences*, 68:265–326, 2014.

Gödel's incompleteness theorems were first published in "Über formal unentscheidbare Sätze der Principia Mathematica und verwandter Systeme, I," *Monatshefte für Mathematik und Physik* 38:173–98, 1931. The excerpt in Section 2 is taken from the translation in volume I of Gödel's *Collected Works*, Solomon Feferman et al., eds., Oxford University Press, 1986.

For more on interactive theorem proving and formal verification, see:

- Jeremy Avigad and John Harrison, "Formally verified mathematics," *Communications of the ACM*, 57: 66–75, 2014.
- Thomas C. Hales, "Developments in formal proofs," *Séminaire Bourbaki*, no. 1086, 2014.
- Thomas C. Hales, "Mathematics in the age of the Turing machine," in Rod Downey, ed., *Turing's Legacy: Developments from Turing's Ideas in Logic*, ASL Publications, 2014.
- The December 2008 special issue on formal proof in the *Notices of the American Mathematical Society*.

You can easily find any of the theorem provers mentioned in Section 3 online, where you can browse their libraries and even download the systems and run them on your own computer. Among the snippets of formal text used in Section 3, the statement of the prime number theorem is from a formalization I completed in 2004 with the help of Kevin Donnelly, David Gray, and Paul Raff; the statement of the Feit-Thompson theorem is from the "Mathematical Components" project, led by Georges Gonthier; the statement of the Kepler conjecture is from the "Flyspeck" project, led by Thomas Hales; and the statement of the Blakers-Massey theorem is from a formalization by Kuen-Bang Hou (Favonia), based on a proof by Peter LeFanu Lumsdaine, Eric Finster, and Dan Licata.

Excerpts from MacLauren's *Treatise of Fluxions* can be found in Volume 1 of *From Kant to Hilbert: A Source Book in the Foundations of Mathematics*, edited by William Ewald, Clarendon Press, 1996.

I may be accused of providing an overly simplified view of ontology. My discussion of the difficulties in making sense of everyday natural language borrows examples from Gilbert Ryle's *The Concept of Mind* (University of Chicago Press, 1949), W. V. O. Quine's *Word and Object* (MIT Press, 1960), and John Burgess and Gideon Rosen's *A Subject with No Object: Strategies for Nominalistic Interpretation of Mathematics* (Oxford University Press, 1999). All are well worth reading.

I am grateful to Ernest Davis, Jeremy Gray, and Robert Lewis for discussions and helpful comments, and to Eric Tressler for corrections.

The linguistic status of mathematics

Micah T. Ross

Abstract Mathematicians have often discussed mathematics as a language. Common linguistic categories have analogous mathematical objects. A comparison of linguistic categories and mathematical objects is developed with reference to the early history of mathematics.

1 Introduction

A modern aphorism declares mathematics a universal language. Because this aphorism presents a complicated philosophical proposition which assumes definitions of mathematics, language and universality, the appearance of the aphorism and its ultimate source merit identification. A well-known instance of this aphorism occurs in the 1997 movie *Contact*, when the character Ellie Arroway states "Mathematics is the only true universal language." Carl Sagan did not include the phrase in the 1985 novel [37], but the dialog echoes [34, p. 18], which describes the composition of the Pioneer plaque in "the only language we share with the recipients: science." Before he drafted the Pioneer plaque, Sagan had addressed the question of extraterrestrial communication and explicitly cited Hans Freudenthal [35, p. 150] who had designed the artificial language Lincos to contact alien life.[1] In his work, Freudenthal developed, but did not cite, a brief description of an earlier artificial language, Astraglossa, by Lancelot Hogben.[2] Both Freudenthal

---

[1] The initial proposal of Lincos most closely approximated the aphorism in its claim that "mathematical expressions and formulae belong to a language different from that of the surrounding context... The syntactical structure of 'mathematical language' differs enormously from that of all natural languages" [16, p. 6].

[2] For the initial proposal of Astraglossa, see [22]. Hogben presumed the recipients of the messages to be Martians and Astraglossa depends on two-way communication. Hogben most nearly approximated the aphorism by writing that "[n]umber will initially be our common idiom of

M.T. Ross (✉)
National Tsing Hua University, Hsin Chu, Taiwan
e-mail: micah@micahross.com

© Springer International Publishing Switzerland 2015
E. Davis, P.J. Davis (eds.), *Mathematics, Substance and Surmise*,
DOI 10.1007/978-3-319-21473-3_13

257

and Hogben identified mathematics as the key to extra-terrestrial communication. In contrast, Sagan initially identified science as a language. In so doing, Sagan took up a line of metaphor which leads back to Galileo. In a famous metaphorical passage of *Il Sagiattore*, Galileo described natural philosophy as written in a book, which he identified as the universe, "in mathematical language, and the characters are triangles, circles and other geometrical figures (*in lingua matematica, è i caratteri son triangoli, cerchi, ed altre figure geometriche*)" [19, p. 25]. Semioticians may note, and mathematicians may decry, the exchange of the signifier (mathematics) for the signified (natural philosophy), but Sagan frequently used language as a metaphor. Sagan claimed that "DNA is the language of life," "neurons are the language of the brain," and "science and mathematics" are the language of all technical civilizations [36, p. 34, 277, and 276]. Regardless of whether Sagan recanted the materialist heresy in the screenplay of *Contact*, the construction of artificial languages for the expression of mathematical statements differs from the proposition that mathematics is a language, universal or otherwise.

While Sagan may have been its most powerful publicist, the aphorism does not lack supporters. In 1981, Davis and Hersh confirmed the currency of the aphorism among mathematicians [11, pp. 41–43] and even tentatively proposed an analogy between grammar and mathematics [11, p. 140, 156]. Davis and Hersh proposed that "mathematical adjectives are restrictors or qualifiers" but declined to press the grammatical analogy. Others had taken up that standard before. In 1972, Roger Schofield appended an extended grammatical analogy to a chapter titled "Sampling in Historical Research" [38, pp. 185–188]. Schofield used the analogy as a pedagogical tool to acquaint social scientists with mathematical expressions and identifies the nouns, adjectives, verbs, and adverbs of mathematics. The analogy of subscript indices with adjectives has been adopted in at least one other introduction [33, p. 11]. In these instances, the aphorism appears as a metaphor to encourage mathematical novices.

For some, though, the aphorism ascended to dogma. In 1972, Rolph Schwartzenberger wrote that

> My own attitude, which I share with many of my colleagues, is simply that mathematics is a language. Like English, or Latin, or Chinese, there are certain concepts for which mathematics is particularly well suited: it would be as foolish to attempt to write a love poem in the language of mathematics as to prove the Fundamental Theorem of Algebra using the English language" [39, p. 63].

Despite the clarity of the sentiment, the aphorism appeared alongside appeals to unnamed authorities and a tacit acceptance of the so-called Sapir-Whorf Hypothesis. Perhaps Schwartzenberger numbered Alfred Adler among his colleagues because in the same year, Adler declared that

---

reciprocal recognition; and astronomy will be the topic of our first factual conversations" [22, p. 260].

> Mathematics is pure language—the language of science. It is unique among languages in its ability to provide precise expression for every thought or concept that can be formulated in its terms. (In a spoken language, there exist words, like "happiness," that defy definition.) It is also an art—the most intellectual and classical of the arts" [1, p. 39].

Adler afforded some compromise, but a published conversation between Jean-Paul Changeux and Alain Connes drove the point further. Changeux opened with a bold assertion and followed with a related question: "Mathematical language is plainly an authentic language. But is it therefore the *only* authentic language?" Connes answered "It is unquestionably the only *universal* language" [8, p. 10]. The strength of the sentiment outweighs the clarity of the qualifications.

The aphorism may have entered academic parlance because it neatly juxtaposes unconscious assumptions about mathematics and language. These unconscious assumptions ultimately derive from romantic notions about both mathematics and language. By comparing these unstated assumptions, the aphorism demonstrates the verbal artistry of a poet, a fact confirmed by an earlier popular version of the dictum. In 1939, Rukeyser published the poem "Gibbs," which posthumously inserted the phrase into the mouth of Josiah Gibbs [31]. Poets are not wont to cite sources, but in 1942, Rukeyser developed her interests into a biography and popularized the aphorism by titling a chapter "Mathematics is a language." In this chapter, Rukeyser reports the origin of the phrase:

> At home, in New Haven, Gibbs was preparing for a journey to Buffalo. A story is told of him, the one story that anyone remembers of Willard Gibbs at a faculty meeting. He would come to meetings—these faculty gatherings so full of campus politics, scarcely veiled manoeuvres, and academic obstacle races—and leave without a word, staying politely enough, but never speaking.
> Just this once, he spoke. It was during a long and tiring debate on elective courses, on whether there should be more or less English, more or less classics, more or less mathematics. And suddenly everything he had been doing stood up—and the past behind him, his father's life, and behind that, the long effort and voyage that had been made in many lifetimes—and he stood up, looking down at the upturned faces, astonished to see the silent man talk at last. And he said, with emphasis, once and for all:
> "Mathematics *is* a language." [32, pp. 279–280].

Rukeyser may have held this sentiment, but Lynde Wheeler, Gibbs' student, revoked both poetic license and semiotic insight:

> [H]e was nevertheless a very regular attendant at faculty meetings. It was in connection with his infrequent participation in the discussions at these meetings that those of his colleagues not on the mathematical or physical faculties gained the greater part of their knowledge of him. There is ample evidence that his comments on such occasions were pithy and to the point; but because he always approached every problem, whether of science or of the ordinary affairs of life, from an original and frequently an unconventional point of view, some of his contributions to these discussions were not always appreciated in a sense he intended. Thus his injection of the thought, "mathematics is a language," into a discussion of the revision of the language requirements for the bachelor's degree was in some quarters regarded as an irrelevance; whereas for Gibbs it meant that if language is a means of expressing thought then mathematics is obviously in the same category in the educational scheme as Greek or Latin or German or French, and logically the claims of no one of them can be ignored in planning a balanced course of study [44, p. 173].

Rukeyser may have drawn inspiration from Nathanael West's phrasing of the aphorism in *Miss Lonelyhearts*, "Prayers for the condemned man will be offered on the adding machine. Numbers constitute the only universal language" [43, p. 30]. Because West addressed themes of religion, despair and disillusionment, he may have intended a satire of the Adamic language.

Before West, Joseph Fourier described analytical equations in terms of language. "There may not be a more universal and simple language, more free of errors and obscurities, that is to say, more worthy to express the invariable relations of natural objects" *(Il ne peut y avoir de langage plus universel et plus simple, plus exempt d'erreurs et d'obscurités, c'est-à-dire plus digne d'exprimer les rapports invariables des êtres naturels)* [15, p. xiv]. However, Fourier did not describe mathematics as a language. Fourier praised analytical equations at the expense of other mathematical systems. In the same passage that praised analytical equations for their universality, Fourier limited their domain to the representation of natural objects. Aside from Galilean references to the universal language of nature, earlier authors rarely described mathematics as a language, universal or otherwise. Language was the domain of the arts. The motif of a universal language predates its identification with mathematics. In 1662, John Evelyn claimed that "For *Picture* is a kind of *Universal Language*" [14, p. 140], and in 1736 James Thomson sang in "Liberty" that "Music, again, her universal language of the heart, Renew'd" [41, pt. 4, ll. 246–7]. Perhaps the aphorism reflects a peculiarly modern anxiety. West developed a pessimistic view of numeric language, and Rukeyser inverted the cold connotations of mathematics in support of an unconventional hero. A universal language of mathematics appeared as an element of a post-religious age when alien contact seemed a real possibility. For their predecessors, language remained a religiously charged topic and most references to a universal language recall the linguistic state of man before the *confusio linguarum* at the Tower of Babel described in Genesis 11:2, which may be traced to the Sumerian epic *Enmerkar and the Lord of Aratta*.

An analysis of the validity of the proposition that "mathematics is a language" depends on the definitions of both mathematics and language. Because neither field has a universally accepted definition, the semantic status of either is open to the fallacy of special pleading. Linguists often limit the definition of languages to vocal expression or to potentially descriptive communication or to other constraints; in contrast, semioticians generally consider signs in any medium or mode capable of forming a language. Thus, by some semiotic definitions, mathematics is a language because it forms a system of symbols which communicates to a community. Francophone philosophers distinguish these uses with the division between *langue* and *langage* but the distinction does not survive in English. The validity of the aphorism has ontological ramifications. The status of mathematics as a language could instantly clarify other discussions. For example, languages are not discovered by their speakers; they are developed. If the evaluation of the aphorism is momentarily abandoned, an expansion of the aphorism according to linguistic methods elucidates differences between mathematics and language. These differences, in turn, may be clarified by recourse to the historical development of mathematics.

2   Mathematics as a Language

Without an axiomatic definition of languages or mathematics, a direct proof of the equivalence between these two abstractions cannot attain Euclidean certainty, but an empirical relationship between the two ideas may yet be evaluated. A structuralist method may be adapted from the proofs of congruent triangles. Given the proposition that mathematics is a language, if the two ideas are congruent, then the elements of the two ideas ought to be congruent. In other words, if mathematics is a language, it ought to have categories analogous to linguistic categories. If mathematics lacks an analog to any of these categories, mathematics may yet possibly be a language by some definition, but it would differ from the languages under the domain of linguistics.

Linguists generally hold that every language has open classes of words called syntactic categories which can be identified as verbs, nouns, and adjectives. Because this tenet has empirical support but no rationalistic necessity or justification, challenges to its universality occasionally arise. Thus, claims arise that a given language lacks verbs, nouns, or (most frequently) adjectives. These claims fall into two categories: constructed languages created to challenge linguistic assumptions and descriptions of natural languages. Constructed languages do not relate to the aphorism which depends on a naturalistic understanding of language and the descriptions of verbless, nounless, or adjectiveless languages often permit other descriptions and are not universally accepted. Those intent on establishing a linguistic outlier to justify the aphorism might more soundly insist on a semiotic definition of language.

A language may have categories other than verb, noun, and adjective, but these three categories seem minimally necessary [12, vol. 2, p. 37–8, 62]. Languages assign words to these categories independently of their meaning. Thus, English may express the idea "he is good," but another language may express goodness with a noun. (Even English permits the construction "he is a saint.") Yet another language may express this meaning with a verb. (Here, English struggles. Perhaps, in some contexts, "he behaves" expresses the same sentiment.) Nor must a particular word remain in a single category. One may butter bread, or bread a chicken, or be too chicken to complete a dare. Each change of category creates a new lexeme. The two words may be homophonous, but they are categorically separate. If mathematics is a language, it ought to have categories analogous to verbs, nouns, and adjectives.

3   The Verbs of Mathematics

In the case of verbs, mathematics conforms easily with natural languages. Some grammatical systems divide verbs into stative verbs (which describe states of being) and dynamic verbs (which describe actions). Modern mathematical expressions present a ready analogy to both. Mathematical relations such as "is greater than," "is less than or equal to," or "is a subset of" resemble stative verbs, whereas operations such as addition or subtraction indicate the actions of mathematics.

Languages apply different morphological processes to verbs in order to mark tense, aspect, modality, or evidentiality. In modern mathematical expressions, these morphological processes are largely absent. The rules governing orders of operation encode a local, relativistic tense, but two unrelated operations cannot be compared.[3] Not all languages communicate the same information about verbs, nor did any Adamic language once contain all possible information about verbs. Languages develop within cultures, and collective choices determine which features of verbs are developed. Mathematics, too, develops within cultures, and the aspectless nature of operations may reflect the value which mathematical culture places on a rhetorical preference for abstraction and generalization.

Just as every language has a category of verbs, every axiomatic mathematical system necessitates a small set of fundamental operations. The number of fundamental operations has changed throughout the history of mathematics. Johannes de Sacrobosco reported nine fundamental operations of numeration, addition, subtraction, duplation, mediation, multiplication, division, progression, and the extraction of roots in *De Arte Numerandi* (c. 1225). For the past millennium, the trend has been to reduce the number of fundamental operations. Most modern texts identify the four fundamental operations as addition, subtraction, multiplication and division but limit their range to arithmetic. However, modular expression has recently appeared as a fundamental operation of arithmetic, and current mathematical hearsay attributes the insight to Martin Eichler. Different sub-fields of mathematics specify different fundamental operations, and the number of these operations can vary within a field. For example, as developed by Gödel [20, p. 35], set theory needs eight operations, but later set theorists [24, pp. 178–180] have increased that number (even while retaining the name "Gödel operations"). Other branches of mathematics have clarified that fundamental operations need not resemble algorithmic executions. Numerical analysis includes "table lookup" as a fundamental operation [45, p. 1]. Just as modern linguists risk imposing the grammar and categories of their native language on the languages they study, when modern mathematicians attempt to identify the fundamental operations in ancient mathematics, they hazard projecting modern mathematical structures onto ancient mathematics.

---

[3]For example, the order of operations suggests that in $7 \times (3 + 2)$ the addition happens before the multiplication. Independent mathematical statements exist outside of a temporal framework. Without a context established by natural language, the mathematical statements $6 + 2 = 8$ and $7 \times 5 = 35$ have no relationship to each other. In contrast, natural languages can often easily suggest a temporal relationship, such as "Six was increased by two to make eight; Seven will be multiplied by five to find thirty-five."

3.1 Mesopotamia

Although Mesopotamian scribes relied heavily on lists, no published cuneiform tablet lists the fundamental operations of mathematics. Two approaches have been used to identify the fundamental operations, with different results. In his 1935 landmark study of cuneiform mathematics, Otto Neugebauer undertook a mathematical approach toward the identification of the fundamental operations of Babylonian mathematics [27, pp. 4–8]. Neugebauer recognized that lists and tables formed a large portion of the education of scribes and derived a set of fundamental Babylonian operations from collections of tablets used for pedagogy. Neugebauer noted that the structure of these texts resembled non-mathematical economic texts of supplies, lists, and inventories. Thus, he discounted nearly 100 tablets of metrical conversions as non-mathematical. Then, he sorted the mathematical tablets according to their contents. The resulting categories included tables of multiplicative reciprocals, multiplication, squares, and cubes. A small sample of texts also tabulated geometric progressions, but the orthodoxy of geometric progression as a fundamental operation was not clear. Thus, Babylonian mathematicians had composed reference works for four or five fundamental operations. However, in the course of their computations, Babylonian mathematicians also employed addition and subtraction. Perhaps, these operations were too intuitively obvious to merit the composition of standard reference texts, but their inclusion would increase the number of fundamental operations to six or seven.

Recently, Christine Proust proposed that although the metrical tables do not conform to modern lists of fundamental operations, these tables shared a genre with Babylonian mathematics [29]. Babylonian scribes associated the shape and size of a tablet with the use for the writing on that tablet. Within the Babylonian system of cultural axioms, then, the decision to separate the tables of metro-logical conversion was unwarranted. Objections to the inclusion of metrological conversions as fundamental operations illustrate the seductive power of cultural perspective. A modification of the textual corpus allows modern mathematicians to state the fundamental operations of Babylonian mathematics succinctly, but numerical analysts are not held to the same standard. The exclusion of "clerical" elements by Neugebauer also reveals the rhetorical presumption that mathematics is proscriptive and executive but not descriptive. That is, by removing troubling elements from the history of mathematics, the illusion may be maintained that mathematics embodies certain (potentially universal) rules and the instructions for using these rules but denies evidence of the connection of mathematics to the culture that created it.

Jens Høyrup adopted a philological strategy for identifying the fundamental oper-ations of Babylonian mathematics [23, pp. 18–32]. This approach focused on the distinct vocabulary for eight operations: two types of addition, two corresponding types of subtraction, three types of multiplication, and one geometric construction, of which a special case yields squares, cubes, and their roots. Babylonian math-ematicians separated addition into two operations and used different verbs for the two processes. One type of addition applied to operations which were not considered commutative. The verb (*wasābum*) for this operation took the augend as an object and expressed the addend as the object of a preposition (*ana*). The summation retained the identity of the augend, albeit with a greater magnitude. The other type of addition was considered commutative and merely united two quantities with a conjunction (*u*). The second operation applied to addends which were mathematical abstractions, as was the result. These two types of addition spawned analogous types of subtraction, also marked by word choices. The verb (*nasāhum*) for the inverse of the first type of addition took the subtrahend as the object, marked the minuend with a noun (*libbi*) used adverbially, "from the heart of," and reported the result as the remainder (*šapiltum*). The inverse of the second type of addition compared two magnitudes but seems to have been open to some variation in the Babylonian phrasing: A can "go beyond" (*itter*) B, or B can "be smaller than" (*matûm*) A.

Three words denoted separate operations of multiplication. The development of Babylonian mathematics left traces on the words chosen for calculations. In the first type of multiplication, Babylonian mathematicians used a word from their cultural predecessors, the Sumerians. This Sumerian verb (a.rá) described an operation analogous to modern multiplication. This verb used the multiplicand as the subject and the multiplier as the object. The terminology of this construction seems to have arisen from the measurement of areas. A purely Babylonian category of multiplication seems to have resulted from the determination of volumes. The verb for this operation (*našûm*) took the multiplicand as the subject and denoted the multiplier by a preposition (*ana*). Although it may have begun as a volumetric construction, this type of multiplication extended to constants, metrological conversions, volumes found from a base and height, and areas of pre-existing figures. A third verb (*esēpum*) denoted duplation. Unique among the multiplicative operations, duplation has an inverse, which could be indicated by either the verb (*hepûm*) or the noun (*bāmtum*). However the operation of bisection was denoted, it constituted another fundamental operation even though no generalized division formed a fundamental operation for Babylonians. Instead, the identification of a multiplicative inverse constituted a fundamental operation. Because Babylonians used a Sumerian phrase for reciprocals (igi # gal.bi), reciprocality, like the first type of multiplication, had probably been long recognized as a fundamental operation.

In modern mathematics, geometric constructions form a category independent of arithmetical functions. Since Euclid, geometers have limited themselves to fundamental constructions, often denoted by their tools: the compass and the straight-edge. Euclidean geometers connect two points with a line, draw circles with a given center and a point on the circumference, identify the point at the intersection of two lines, locate the point or points at the intersection of a line with a circle, and specify the point or points at the intersection of two circles. Mathematicians recognize these constructions, but they are not arithmetic operations, fundamental or otherwise. Babylonian mathematicians did not subdivide their mathematics in the same way. They "constructed" rectangles to solve arithmetic problems. To be sure, these geometric constructions were often mental constructs, but given two magnitudes, Babylonian reckoners could construct a rectangle with sides of those magnitudes. The terminology of this operation is complex, but the two magnitudes serve as objects of a causative verb, "to make hold" (*šutakullum*). The expressions for square numbers either use this verb but repeat the same magnitude or insist on the equality (*mitḫartum*) of the magnitudes. Through this construction, the operations for squares and square roots share the same vocabulary. The only question is whether the side or the area of the relation is known. Another magnitude can even be added to the expression to invoke the determination of cubes or cube root, but the fundamental operation remains the same geometric construction.

Neugebauer adopted a mathematical approach which ignored textual genres, and his conclusions conformed to the expectations of modern mathematicians. Whatever criticisms might be raised against this approach, this effort managed to establish that a pre-Greek civilization conceived of mathematics beyond simple arithmetic. Høyrup opted for an etymological method and clarified the mathematical meaning of abstract expressions through reference to natural language. His conclusions depend on the use of Babylonian words in literary contexts which employ natural language. Nonetheless, his careful study of word choice reveals conceptual differences between Babylonian mathematics and modern mathematics. The ability to isolate mathematical differences by reference to Babylonian language calls into question whether Babylonian mathematics existed as an entity ontologically separate from natural Babylonian language. Proust extended the connection beyond language to reunite Babylonian mathematics with Mesopotamian culture. Perhaps the fact that Babylonian mathematicians and modern mathematicians differ in their assessment of the number of fundamental operations challenges the premise that mathematics is universal.

3.2   Egypt

If mathematics constitutes a universal language, similar constructions should appear throughout mathematical texts. In other words, the mathematics of one culture should resemble the mathematics of another. Whereas Babylonian mathematics is known from hundreds of brief texts, Egyptian mathematics is known from a small handful of texts, particularly the *Rhind Mathematical Papyrus* (pBM 10057 and 10058; hereafter, *RMP*). The philological and mathematical approaches of Høyrup and Neugebauer are echoed in the publishing history of the *RMP*. When Peet edited the RMP, he explicitly acknowledged that "[t]here is only one truly fundamental process in arithmetic, that of counting" [28, p. 11], but he focused on an elucidation of the vocabulary of mathematics. When Chace reedited the same papyrus, he declared that Egyptian mathematicians had a "thorough understanding of the four arithmetical operations,—addition, subtraction, multiplication, and division" and proceeded to demonstrate the existence of these operations [7, p. 3]. Like the cuneiform corpus, no Egyptian mathematical text explicitly addresses the question of fundamental operations, but an empirical understanding may be gathered from the *RMP*. Sometimes, Egyptians identified operations similar to those used by Babylonians. In some ways, they differed. In every case, though, the conception of Egyptian mathematics is directed by the expectations of the reader.

The *RMP* contains examples of addition and subtraction. Usually, a preposition (*ḥr*) indicates the operation of addition. The augend stands as the noun to which the preposition phrase is a complement; the addend is the object of the propositional. This expression may be expanded to include a verb (either *wꜣḥ*, *di*, or *dmḏ*), in which case the augend serves as the object of the verb, and the addend again serves as the object of the preposition. The two instances ("Problem 41" and "Problem 42") which use the verb *di* to denote addition also relate to the area of circles. However, not every detail of preservation relates to mathematical conceptions. As Jöran Friberg has hypothesized, the *RMP* may be a "recombination text" which excerpts selections from other texts and recombines them into a new text. The variations in verb choice may reflect geographical or temporal variations in dialect among the source material [17, p. 26]. Whereas the methods of Høyrup work well for many texts which share the same tradition, they do not apply equally well to a single text which draws from varying traditions: the Egyptian choice of verbs for addition has eluded mathematical clarification. Whereas a preposition denotes addition, Egyptians considered subtraction to be a verbal activity (either *ḥbi* or *skm*). The subtrahend is the object of the verb, and the minuend is indicated as the object of a preposition (either *n* or *ḥnt*).

Modern analyses of Egyptian mathematics often focus on the algorithm behind multiplication because it differs from modern techniques. Nevertheless, it can be introduced easily. Babylonian multiplication and modern multiplication work in roughly the same way: a memorized value is assigned to each pair of digits, and those values are combined in the case of multiple digits. Throughout the *RMP*, though, multiplication is accomplished by successively doubling up to half of the multiplier and subsequently summing those doublings which correspond to a binary expression of the multiplier. As in Babylon, doubling formed a fundamental operation of Egyptian mathematics. Unlike Babylon, direct multiplication seems to be a secondary operation. Peet noted the primacy of doubling among Egyptian calculation, but he omitted to mention that the Egyptians themselves seem not to have had a word for doubling [28, p. 15]. A verbal phrase (*wꜣḥ tp*) indicates the generalized, secondary operation of multiplication, but this phrase uses a preposition (*m*) to indicate the multiplicand and prepositional phrase (*r spw*) to indicate the multiplier. Literally translated, the phrase implies counting by a number other than one: "incline the head with the multiplicand for the multiplier's number of times." The mental image behind the phrase seems to be that the reckoner would nod his head with each increment of the multiplicand, as if counting aloud by sixes or elevens. (Against this interpretation, though, *wꜣḥ* had long been connected to addition and *tp* has a wide range of metaphorical uses.)

Despite this descriptive metaphor, the *RMP* never executes multiplications in this way. Scribes are not expected to count by arithmetic progressions; they are expected to double efficiently. The same verb also denotes division, again with a preposition (*m*) to indicate the divisor and the dividend as the object of an infinitive in a prepositional phrase (*r gmi*). (Literally translated, the phrase reads: "incline the head with the divisor to find the dividend.") As it did in the case of multiplication, the *RMP* preserves variations in the vocabulary of division. Another verb (*nis*) takes the dividend as its object and indicates the divisor by a preposition (*ḥnt*). Another fundamental operation of Egyptian multiplication appears in "Problem 69" of the *RMP*: the increase of a multiplier by tenfold. In Egyptian hieroglyphics, each power of 10 necessitated an independent grapheme. By exchanging the graphical expression of a number, Egyptian mathematicians had recourse to multiplication by 10 as a fundamental operation. Although such multiplication is tabulated and occasionally employed, no specific term defines the operation. Because both doubling and multiplication by 10 are repeated throughout the admittedly limited corpus of Egyptian mathematics, they may be accepted as fundamental operations. The acceptance of these operations as fundamental suggests an explanation for "Section 1" and "Section 2" of *RMP*. Rather than merely simplistic tables of division, these "problems" may be explained as tables of inverse operations for fundamental operations of Egyptian mathematics. By interpreting these passages as inverse tables of fundamental operations, Egyptian mathematics shares similarities with Babylonian mathematics.

This interpretation derives from a consideration of tables as the fundamental element of composition in the *RMP*, and it holds a certain well-ordered allure. The *RMP* uses two fundamental operations and begins with two tables of inverses. However, the interpretation depends on an assumption that the composition of the *RMP* exhibits a (pre)-Aristotelian unity within the papyrus. The author of the papyrus does not guarantee a cohesive structure to its composition. In fact, the same internal evidence (variations in verb choice, repeated problems, thematic grouping of contents) which suggests that the papyrus was composed as a recombination text could be used to dismiss this reading. As stated before, the conclusions about the structure of Egyptian mathematics depend on the expectations of the analyst. Difficulty in the reconstruction of the mathematical ideas underlying the composition of the papyrus reveals limitations in the understanding of Egyptian mathematical culture.

Other tabular sections of the *RMP* have passed largely unconsidered. Like the Babylonian tables of metrological conversion, they have been assumed to be pedestrian in nature. However, given that "table lookup" is accepted as a fundamental operation of numerical analysis and similar metrological tables formed a large part of Babylonian mathematics, these portions of the *RMP* may preserve operations fundamental to Egyptian mathematics. "Problem 47" reports subdivisions of 100 quadruple grain measures (4-*ḥqȝt*) into simple measures (*ḥqȝt*). "Problem 80" and "Problem 81" present tables for the conversion of fractions of grain measures (*ḥqȝt*) into volumetric measurements (*hnw*). Oddly, the *RMP* uses idiosyncratic Egyptian units of area but includes no tables of conversion. Whether this absence represents an oversight of the scribe, his limited source material, or an exclusion of area conversion from the fundamental operations again depends on assumptions about the state of Egyptian mathematics.

Although the *RMP* opens with a somewhat grandiose title which promises "the model for inquiring into affairs, for knowing all which is unclear [and deciphering] every mystery," it does not explicitly identify any fundamental operations. The assumption that the fundamental operations remained constant among all Egyptian mathematicians reflects another set of assumptions about mathematics, language, and culture. The Moscow Mathematical Papyrus (pGolnischev, henceforth *MMP*) contains fundamental operations which do not appear in the *RMP*. Specifically, the scribe of the *MMP* assumes that squares and square roots can be found directly. The *RMP* and *MMP* are both documents of the Middle Kingdom but do not share the same fundamental operations.

The eleven operations (addition, subtraction, duplation, mediation, ten-fold multiplication, decimation, identification of multiplicative inverses, two types of metrological conversion, squares, and square roots) identified here as fundamental to Egyptian mathematics derive from a comparison of the Egyptian mathematical texts with Babylonian mathematical texts. To be clear, the assumption of a parallel development between Egypt and Babylon invites as many potential risks as the assumption that mathematics developed in a continuous unbroken chain from the *RMP* to the modern rules of arithmetic. However, at times, the perspective suggested by the Babylonian mathematical texts clarifies the compositional style of the *RMP*. In the case of Babylon, historians of mathematics have access to a wide corpus of texts in several genres. Recent examinations have revealed how modern editors have fit this corpus to support mathematical assumptions drawn from modern mathematics.

3.3 Universality

In order to establish an analogy between language and mathematics, the number of the minimal set of operations is not particularly important, but the variations between Babylonian, Egyptian, and modern mathematics suggest that if mathematics is a language, different mathematical communities communicate in divergent dialects. Moreover, within Babylonian and Egyptian mathematics, mathematical texts must be understood through recourse to the metaphors of natural language. Little surprise, then, that the mathematics of different cultures reflect variations between those cultures. Just as they spoke different languages and wrote in different scripts, they held different fundamental mathematical concepts. Even the fields of modern Western mathematics, in which mathematicians may be assumed to have had similar educations and access to the same publications, develop different mathematical objects. In the case of modern mathematics, at least, the objection may be raised that the mathematicians address different problems. The objection applies less easily to ancient mathematics. As Friberg noted [17, pp. 25–26], the types of problems solved by Egyptians and Babylonians are roughly similar. Nonetheless, the mathematics of these ancient cultures differ no less than their languages and cultures.

Although variations among the fundamental operations erode the notion of universality, the fundamental operations of each mathematical culture fill a category analogous to the syntactic category of verbs. Indeed, the fundamental operations may be repeated, combined, or repeated and combined into more complicated expressions. Beginning in the nineteenth century, mathematicians developed special functions so extensively that operations resemble an open category.[4] Essentially, special functions enabled a morphology of verbal phrases. In this way, every function from the fundamental operations to trigonometric functions may be considered analogous with a lexeme in the syntactic category of verbs. To make an analogy with the field of genetic linguistics, which seeks to compare languages grouped together by similarities, much work remains to be done in "genetic mathematics," particularly among those systems which did not originate from a classical root. Some similarities between Babylon and Egypt might be explained by contact mathematics, but a monogenesis of mathematics remains as elusive as a universal grammar or the monogenesis of language.

4   The Nouns of Mathematics

The category of nouns seems readily congruent with several mathematical categories. Three categories of mathematical objects present an analogy to nouns. Perhaps sets have the strongest claim to being universal nouns but the primacy of sets in mathematics derives from the labors of Bourbaki to establish all of mathematics as a logical development of set theory. A concerted intellectual effort has been applied to describing all mathematics in terms of sets. Notwithstanding the potential universality of set theory, this undertaking is a modern endeavor. Of course, mathematics could have been conducted with no awareness of set theory. After all, Vedic hymns were composed before Pāṇini elucidated Sanskrit grammar. However, an analysis of pre-modern mathematics in terms of set theory invites an attempt to establish continuity retroactively. The descriptions of ancient mathematics from this perspective would have been foreign to people who worked in those paradigms.

The mathematical category of geometrical objects also conforms to the linguistic category of nouns. The geometric objects of Babylon are not so different from the geometric objects of Euclid's *Elements* but the techniques of proof and problem solving differ greatly. Cuneiform mathematics recognizes right angles, can identify similar angles, but does not compare dissimilar angles. Mesopotamian mathematicians could identify the orientation of axes and calculate the areas of triangles, squares, and trapezoids or solve for the lengths of line segments of their constituent parts [23, pp. 227–31]. The geometric systems of Babylonian scribal schools differ from the axiomatic system of Euclid in that Babylonians preferred to solve geometric problem through arithmetic or algebraic techniques. Mesopotamian mathematicians were capable of stating geometric rules but they often invoked

---

[4]By contrast, the category of Babylonian and Egyptian operations is clopen. Both categories are closed because new Babylonian or Egyptian mathematical texts cannot be created but they are also open because new texts might be discovered or the reconstructions might be reassessed. For example, Plimpton 322 has yielded recent surprises.

these rules as arithmetic statements. For example, ancient Babylonians proposed and solved right triangle problems but they relied more on the transformation of Pythagorean triples than geometric constructions. The Nile Valley shared this algebraic Mesopotamian approach. Egyptian mathematicians added a new geometric object, the *sqd*, or the reciprocal of a slope of an inclined surface but they retained the preference for algebraic methods. Moreover, from a linguistic perspective, the terminology of Babylonian and Egyptian geometry and the statement of problems depends heavily on natural language and the terms for ordinary objects.

As mathematical objects which resemble nouns, only numbers would please ancient mathematicians as much as their modern counterparts. After all, mathematics is often described as the study of numbers. Just as the verbs of a language govern nouns, so operations (and functions) govern numbers. Numbers even operate as nouns in natural languages. For example, in English, "four" is a countable noun which refers to the digit four or any grouping of four units (among other specialized uses), but the classification "countable" belies a significant difference between languages and mathematics. In mathematics, a countable set is a set with the same cardinality as the natural numbers. That is, every element in the set may be paired with a natural number. In linguistics, a countable noun can be modified by a number; an uncountable noun cannot. An uncountable noun is sometimes called a mass noun. For example, a farmer may have two cows which produce milk. In order to discuss the two gallons of milk each cow produces, a measure word quantifies the uncountable noun "milk." Some languages (such as Chinese and Japanese) have developed systems of measure words, but English preserves only a few measure words. The previous example may be expanded to introduce such a measure word and illustrate how countability, in the linguistic sense, can modify meaning. Thus, the two head of cattle come from different breeds and produce milks of different quality. In this way, countable and uncountable nouns are grammatical categories which do not depend on the nature of the signified object.

4.1   Countability and Infinity

One obvious difference between nouns and integers is that integers are countably infinite, but no language actually has an infinite number of nouns. Although the number of nouns is theoretically infinite, the phonemes of a language are a closed category, and most languages have only a few thousand nouns. Thus, linguists describe the category of nouns as open. That is, limitless new nouns can be generated. The morphology of language is more complicated than the morphology of mathematics. To generate a new noun, a language must add an affix (a morpheme attached to a word stem to form a new word); mathematics must merely increase the value by 1. The complexity of linguistic morphology, though, can be productive. A simple set of rules can be used (sometimes recursively) to generate as many lexemes as are needed from a limited set of phonemes. Linguistic communities differ somewhat in the upper bound for the length of a lexeme, but eventually noun phrases replace unwieldy lexemes. In the case of mathematics, the patience and memory of the reckoner forms a practical limitation.

The difference between the open set of nouns and the infinite set of integers forms one of the earliest ontological problems of mathematics. Because different cultures answered the challenge differently, the expression of large numbers has become a standard opening among historians of mathematics. In Egypt, each power of ten necessitated a new hieroglyphic symbol. A coiled rope denoted a hundred; a lotus flower stood for a thousand. Far from a primitive notation of tallies, scribes displayed an insight into the intersection of cognition and mathematics by subitizing these graphemes. In other words, they arranged the signs into a pattern which allowed the reader to perceive the number of units quickly rather than tallying up the individual marks [13, pp. 521–522]. As noted in the comparison of verbs and operations, replacing one set of subitized units with another set of similarly subitized units rendered multiplication by 10 a primary operation. Each power of ten took its name from a relatively common noun. Whereas the names for the numbers from 1 through 10 must be reconstructed from later alphabetic writings, rendering the vocalizations of large numbers presents little challenge, and the system seems to have had an upper bound in the millions.[5] Amazingly, a document from the dawn of Egyptian history preserves an instance of a tally in the millions. King Narmer, who may have unified Egypt, counted 1,422,000 goats captured in the course of his wars and inscribed this number on a ceremonial macehead. Egyptians rarely had to count into the millions. The number was rarely used, and as the Egyptian language developed, meiosis weakened the meaning to "many." The millions took their name from the god of multiplicity, and the signs for expressing numbers beyond nine million are unclear (Tables 1 and 2).

This problem of rendering mathematical infinity in a natural language is not a feature of hieroglyphic writing or some "primitive" state of Egyptian mathematics. Every culture faces this problem. Although modern English speakers might offer a "googol" or "googolplex" as an astoundingly large number, few can name the number which precedes it. Most reckoners, ancient or modern, assume that numbers beyond frequently encountered measurements exist but blissfully ignore them. The

Table 1 Hieroglyphics for powers of 10.

10^0	10^1	10^2	10^3	10^4	10^5	10^6
❘	∩	૬	⚱	❈	❧	𓁨

---

[5]Popular histories of mathematics often reproduce weakly cited counter-evidence. According to one edition, Chapter 64 of the *Book of the Dead* tallies 4,601,200 gods [4, p. 164]. Other editions confirm only one million. A grapheme introduced in the Ptolemaic era may express a sense of totality, or cyclical completion [40, p. 12, n. 6], or it may stand for millions, thereby increasing the old grapheme of the god Ḥeḥ to tens of millions [21, p. 280]. Occasionally, decorative hieroglyphs combine the graphemes of large numbers into images which may be read as "a hundred thousand million years" or even "ten million hundred thousand million years" [5, p. 507, s.v. ḥeḥ], but these graphemes are more likely artistic hyperbole than mathematical quantity.

Table 2 Egyptian subitization patterns.	1	2	3	4	5	6	7	8	9
								...	...
					...	...	....	...	...
	.	..	...	....				...	...
					..	...	...	..	...

Greek number naming system did not originally extend beyond 10,000, but Classical Sanskrit introduced a new word for each power of ten up to 10^{12}. Like the modern scientific notation, the name for the unit of trillions refers to twelve—the power of ten which the unit denotes. Although no number larger than one trillion has been found in Classical Sanskrit, greater units could have been generated by the same rule. An early Buddhist composition, the *Lalitavistara Sūtra*, describes a number generation contest between the mathematician Arjuna and Buddha. Arjuna named the units up to 10^{53} but Buddha expanded the method geometrically to 10^{421} before sagely abandoning the contest in pursuit of wrestling and archery. Like linguistic communities, mathematical communities are irrepressible at communicating necessary information. Archimedes expanded the traditional Greek system beyond 10,000 in much the same way as Arjuna, but Archimedes bested both Arjuna and Buddha by naming numbers up to $10^{8 \times 10^{64}}$. Perhaps his wrestling partner and archery coach were busy. Archimedes trumped Buddha, but he recognized the limitations of human numeracy: he never employed numbers larger than 10^{63}.

Whereas Egypt, India, Greece, and Rome generated new words to describe the ever increasing units, Babylon employed a different strategy. Although the Babylonian system of numerals is often described as "base 60" or "sexagesimal," this characterization is not perfectly accurate. Cuneiform numerals did not contain a set of 59 unique digits. Rather, the Babylonian notation used vertical wedges for units and horizontal wedges for tens. Just as the Egyptian scribes organized their hieroglyphic graphemes for readability, the Babylonian scribes subitized both the vertical wedges of the units and the horizontal wedges of the tens. The positional notation, however, was sexagesimal. For numbers greater than 60, the vertical wedge was reemployed as 60. This system of notation lacks a radix point, so only context distinguishes 141 from 8,460 or 507,600, or some other summation of the corresponding powers of 60 (Tables 3, 4 and 5).

Introductions to the history of mathematics often dwell on this ambiguity, but polysemy is not the interesting point of floating-radix, sexagesimal notation. Languages permit ambiguities and clarify them by context and convention. Two other points deserve greater notice. First, the cuneiform notation is entirely removed from the verbal expression of either numeric grapheme, namely the two vertical wedges and the horizontal wedge. In fact, the trend toward a sexagesimal base is not reflected in the spoken languages of Mesopotamia. The words for seventy, eighty, and ninety derive from the words for seven, eight, and nine. The lexemes for numbers continued to invoke new units with the change of the powers of 10 (*meatum* for 100, *līmum* for 1000). Even though the grapheme for one hundred consists of one vertical wedge and four horizontal wedges, the lexeme marked a change based in

Table 3 Cuneiform powers of 10.

Table 4 Mesopotamian subitization patterns (rotated ≈ 30°counterclockwise for tens).

1	2	3	4	5	6	7	8	9

Table 5 Babylonian positional notation, where n depends on context.

60^n	60^{n+1}	60^{n+2}	60^{n+3}	...
[1-59]	[0-59]	[0-59]	[0-59]	...

the powers of ten and did not relate to the appearance of the grapheme. Nor did the graphemes represent a bijection with the lexemes. A single vertical wedge could be read (and perhaps vocalized) as *išten* (one) or *šūšum* (sixty, from *šeššum*, meaning six). Secondly, the inclusion of this numbering system into cuneiform compositions embedded a secondary semiotic system into the language. This embedding relates directly to mathematical ontology. These cuneiform signs were not shorthand for a verbalization. Their interpretation and manipulation relied on an independent set of rules. While these two changes do not constitute a definition of a mathematical entity, they may serve as a bellwether that mathematical expressions have separated from natural language.

4.2   Number and Zero

Like countability, linguists acknowledge number as a grammatical category which can potentially apply to any category of word, but nothing requires the application of this grammatical category number to any word. Languages vary in their identification of grammatical number. For the most part, English marks only singular or plural. Egyptian and Greek had fossilized "dual" constructions for paired objects, such as hands or eyes. Sanskrit and most Semitic languages never abandoned the category of dual. Linguists have even described languages which use trial and quadral as grammatical categories. Some languages include more than one type of plural category, such as the grammatical category "paucal" which refers to a small number of items. In contrast to most natural languages, the nouns of mathematics— that is, numbers—cannot be marked according to whether they are singular, plural, or some other variation. From a linguistic perspective, most mathematical expressions are "numberless." This numberless aspect of integers reflects an assumption about numbers as mathematical entities: because they are distinguished only by quantity, when that quantity is the same, the entities are interchangeable. Although identifying a language with no grammatical category of number invites teleological errors similar to searching for a language without verbs, nouns, or adjectives, a numberless language seems plausible because some languages, such as Chinese, essentially limit the category of number to pronouns. For example, the pronoun 他 (tā) has a plural form 他們 (tāmen), but 們 (men) cannot be appended to common nouns. This limitation does not preclude such languages from discussing numbers. Linguistic communities are irrepressible at communicating necessary information. However, the extreme truncation of grammatical number suggests that such a language could easily exist. Somewhat paradoxically, then, if mathematics is assumed to be a language, from a grammatical perspective, it constitutes a numberless language.

The grammatical state of numberlessness differs greatly from a mathematical state of numberlessness. Mathematical numberlessness has become another standard opening for surveys of the history of mathematics: the story of zero. The standard story of zero acknowledges the Babylonian need to introduce a place holder to separate the various powers of sixty in the floating-radix, sexagesimal notation but then retracts this "achievement" on the grounds that the Babylonians never fully realized the glory of their accomplishment. The standard story gives short shrift to the Egyptians, who, because they wrote with clumsy pictographs, were incapable of writing nothing. Greece is chastised *pro forma* for its shortsightedness, and India and the Maya receive snippets of Orientalizing adulation to confirm a "global perspective" which conforms to modern, Western political biases. All of which is important, of course, because zero defines "true" mathematics. Without recognition of the importance of zero, the fundamental operations of arithmetic cannot be stated because zero cannot be excluded from divisors without explicit acknowledgement of its existence. As a result, natural numbers could not be separated from the integers, set theory could never develop, and civilization would never advance. The story varies in the choice of historical anecdotes which it articulates, but the perspective rarely changes.

The frequent emphasis on the story of zero reveals the extent to which the history of mathematics has been pressed into a Whiggish view of mathematical development. By limiting the analogy between language and mathematics to modern mathematics, the comparison of nouns and numbers reveals several similarities. First, the integers correspond to nouns. The identification of zero and its exclusion from divisors forms a key grammatical rule which creates categories among the mathematical nouns. Natural numbers can be divisors; the integers are limited to dividends. The rational numbers, then, present an analogy to clauses. The divisor stands as the head of a noun phrase; the operation of division functions as the verb; and the dividend fills the role of an object. The analogy defines the natural numbers, integers, and rational numbers with a small set of grammatical rules. Mathematics may be rebuilt by considering mathematical objects as noun phrases of increasing complexity; set theory survives; and civilization may flourish.

However, as noted in the consideration of the mathematical verbs, neither Babylon nor Egypt recognized division as a fundamental operation. Babylonians considered multiplicative inverses to be a fundamental operation. The Babylonian tables of multiplicative inverses do not report the lack of a multiplicative inverse for zero, but they also usually omit numbers relatively prime to 60, such as 7 or 13. In this way, Babylonians classified commensurate and incommensurate inverses much like modern arithmeticians classify rational and irrational numbers.

On the other hand, because Egyptian mathematicians did not use sexagesimal notation, they were not equally concerned with commensurability. Egyptian mathematicians had two types of fractions. The first of these fractions derived from the inverse of a fundamental operation, namely doubling. This set of fractions are the erroneously named "Horus-eye" fractions and consist of the first six mediations, or, in more modern mathematical parlance, the first six powers of $\frac{1}{2}$ [30]. These fractions, and their graphemes, developed from a system of volumetric fractionalization. Perhaps, at one time, Egyptian mathematicians would have approximated all fractions by adding successive mediations. Just as a binary representation could approximate any multiplier, a summation of dyadic fractions can approximate any rational number with an arbitrary degree of precision. By the time of the earliest Egyptian texts on mathematics, though, this fractional system had been limited to six units and restricted to measures of grain. The second set of fractions invoked multiplicative inverses directly by setting the grapheme for a mouth over the digits of the number for which the inverse was sought. Most historical surveys of mathematics do not rigorously separate these two classes of fractions in their explanations of "Egyptian fraction." These modern surveys introduce "Egyptian fractions" as unitary fractions, or aliquot fractions, or fractions with a numerator of one, often for the purpose of introducing mathematical exercises. No such entity existed for the Egyptian reckoner.

For Egyptian mathematicians, multiplicative inverses did not constitute an artificially limited set of rational numbers. Instead, the multiplicative inverses constituted another set of mathematical entities, over which the fundamental operations also reigned. Instead of mixed numbers and improper fractions, Egyptian numbers had an integer portion and an inverse portion.[6] When these inverses were doubled by multiplication, the number of the multiplicative inverse was halved. When those numbers were odd, equivalent expressions for the multiplicative inverses of odd numbers had been tabulated. One example of such a table forms the beginning of the

---

[6]The fraction $\frac{2}{3}$ appears idiosyncratically as the only portion of an Egyptian number which is not easily classified as either an integer or an inverse. (The fraction $\frac{5}{6}$ is sometimes cited as another example but this fraction is actually a ligature of $\frac{2}{3}$ and $\frac{1}{6}$.) This idiosyncrasy may be resolved by morphology. The fraction $\frac{2}{3}$ was vocalized as *rwy*, a grammatical dual which literally means "the two parts." The grapheme is better interpreted as $\frac{1}{3} + \frac{1}{3}$ rather than a ratio of 2 to 3.

RMP. This table omits zero for two reasons. First, zero is even. Secondly, zero has no multiplicative inverse. The expectation that Egyptians might distinguish between natural numbers and integers imposes modern categories as much as the insistence that they used "fractions." In fact, an Egyptian word *nfr* has been found to indicate the starting point in technical drawings and as the final statement of closed accounts (pBoulacq 18) [25, pp. 113–115]. Through names such as Nefertiti, *nfr* (good) has become one of the few relatively widely known Egyptian words. However, in its uses related to zero, *nfr* relates to another usage for negation [3].

4.3 Negation

Languages employ a wide variety of techniques for negation, with the negation of the main clause seemingly universally permitted. Depending on the language, though, both nouns and verbs can be negated. Likewise, both nouns and verbs can have positive and negative polarities. The method of negation may vary with tense, mood, or aspect. Negations may rely on grammatical constructions, intonation, or contextual cues. In English, ambiguities relating to these shades of meaning often enable humor or trickery. ("Have you stopped beating your wife?") The grammar of negation depends on an axiomatic system, but this system reflects culture as much as mathematics. Some languages (such as Greek or French) permit doubly negated constructions as intensifiers. Other languages reject such constructions as contradictory. Still other systems permit such constructions for different purposes. Western constructivist mathematicians distinguished double negations from positive statements, but the earlier Buddhist Madhaymaka school developed the interpretation of a negation of a negation in the details of paraconsistent logic. For them, a doubly negated statement might undermine the categorizations and logic on which the statement and its negation were built. In logical operators, $\neg\neg A \neq A$. In a concrete example, A may be the statement, "Harry is a dog;" $\neg A$ would be "Harry is not a dog;" $\neg\neg A$ would be the double negative "Harry is not not a dog." Perhaps Harry is a wolf. Grammarians and schoolmarms to the contrary, contemporary English often employs exactly such constructions when modifying the scope of negation. Thus, if logicians insist that $\neg\neg A \neq A$ is nonsensical, their complaint echoes those who would dismiss the example as ungrammatical because of its double negative. Such constructions occur in English and are meaningful. In linguistic terms, even negations are potentially subject to contrastive focus reduplication. If it is allowed that Harry is not not a dog, it must also be permitted that Harry is not a dogdog. The denial of the potential value of such constructions divides the function of negation in language and mathematics. This functional division results from a deliberate limitation of the rhetorical range of mathematical expressions.

Whereas languages often develop negation to reflect the negation of contingencies and possibilities, modern mathematicians depend on negation according to formal logic (which has diverged from natural language). If mathematical negation is infrequent relative to linguistic negation (or at least, the potential for linguistic negation), this fact relates to a tacit division made between natural languages and mathematics. Linguists assume that language is communicative, but communication can be descriptive or executive. Mathematical texts often compose the definitions of terms in descriptive language but operate in executive language. By nature, definitions are positive. Likewise, useful executive expressions also tend to be positive because prohibitive imperatives produce few changes in mathematical expressions. Often, mathematical negation is categorical and depends on formal logic. For example, many proofs arrange ideas to generate a contradiction. At other times, though, mathematicians consider conditions which have a value of zero to express a special property. Like Egyptian scribes, modern mathematicians still find value in equating negation with zero. Taken in the other sense, though, as the identification of an additive inverse, negation parallels the third syntactic category, adjectives.

5   The Adjectives of Mathematics

In the case of adjectives, mathematics differs greatly from natural languages, but the adjectives of natural languages vary more than other syntactic categories. In some languages (like Chinese), adjectives resemble verbs in their negation or their inflections for tense. In other languages (like Latin), adjectives resemble nouns because they are inflected for gender, number, and case. In a small number of languages, adjectives have the properties of nouns and verbs. In another small group of languages (like English), adjectives are distinct from both nouns and verbs. Adjectives perform two tasks [12, vol. 1, p. 113]. First, they state a property. Secondly, they specify the referent of a noun. That is, in the phrase "the green leaves," the adjective specifies those leaves and not the brown ones. In some languages, adjectives perform other tasks, such as serving as a parameter in comparative constructions or functioning adverbially, but stating properties and specifying referents are fundamental for adjectives. Whereas a rough parallel could be forced into place for both verbs and nouns, the mathematical analog to an adjective eludes simple identification.

5.1   Properties of real numbers

Certainly, adjectives exist with mathematical definitions. Prime, negative, and square all have precise definitions which reflect the mathematical culture which became interested in these properties. These adjectives usually derive from the natural language in which mathematical texts are composed. Generally, mathematicians rely on mathematical definitions of adjectives from their natural language to perform the first task of adjectives, stating a property about numbers. The mathematical expression of these ideas may be standardized, but it develops with reference to the natural languages of mathematical communities. Despite the familiarity of the opening phrase of many proofs "Let p be a prime," this phrase constitutes an English statement. A German mathematician could compose "Es sei p eine Primzahl." Lest the term "prime" be considered a linguistic universal among mathematics, Japanese renders the same idea as 甲は素数だと言いましょう with no reference to Western terminology.[7] And, as in natural languages, these definitions evolve contemporaneously with their use. Pre-Greek mathematicians indicated no special interest in prime numbers, but Mesopotamian mathematicians excluded numbers non-commensurate with 60 from tables of reciprocals. The underlying ideas are similar, but the goals of the mathematicians are not. Even in modern mathematics, the definition of prime has evolved [6]. No notation analogous to an adjective class indicates that a number is prime, but the analogy of mathematics to language invokes more than semiotic similarity of notation to script. Some portion of modern mathematics is still rendered in natural language.

If mathematics constitutes a language in more than a semiotic sense, the linguistic elements need not be limited to notation, but some notations do communicate mathematical properties. For example, a preceding "minus sign" denotes negative numbers. Some users of this sign elevate it to distinguish it from the operator for subtraction. Like many elements of mathematical notation, this sign is overdetermined. A simple horizontal line indicates negativity, subtraction, or the complement of two sets. These uses are clearly related, but the composers of mathematical texts are free to introduce new notations with *ad hoc* definitions. For example, a bar over a variable may indicate the mean value of that variable, a finite sequence, algebraic closure, or a complex conjugate. Egyptian mathematical texts occasionally indicated subtractive quantities with red ink, but cuneiform texts did not have recourse to this method. In fact, cuneiform mathematics did not acknowledge a class of negative numbers. The lack of a given mathematical adjective does not indicate the lack of mathematics. Different languages do not exclusively contain adjectives which map bijectively to the adjectives of other languages. These differences in language reflect differences in culture.

Linguistic descriptions of adjectives often acknowledge differences in the semantic content of adjective classes. The core semantic types form a set of four adjective classes: dimension, age, value, and color. Not all languages have all four adjective classes, but in languages with few adjectives, the adjectives usually embrace these types [12, vol. 2, pp. 73–76]. Just as the mathematical analog to verbs lacks most

---

[7]The naked inclusion of non-Western languages too often provokes consternation. The phrase may be transliterated as "*kō ha sosū da to īmashō*" and translated as "Let's say that 甲 is a prime." The grapheme 素 derives from a pictogram of hands braiding raw silk and the semantic range of the sign includes plain, poor, foundation, and root. The use of 甲 *kō* as a variable is admittedly somewhat contrived for modern Japanese. Probably, a variable from the Roman alphabet would be used, most likely a in slight contrast to the Western preference for p. Mathematics has become an international endeavor.

aspects of natural language and the analog to nouns lacks the grammatical category of number, mathematical adjectives present an unusually narrow range of adjective classes. Ultimately, all mathematical adjectives relate to dimension.

The small set of dimensional adjectives of mathematics may be limited to positive and negative. The other frequently used dimensional adjectives—greater, lesser, and equal—have been described as "relations." The atemporality of mathematical verbs obviates the second class of adjectives and key difference between numbers and nouns nearly precludes mathematical adjectives of the third class. Because every number has a magnitude and an implicit ordering, any given number always has a single fixed relationship to every other number. Even when nothing can be stated about the relationship between two numbers because nothing is known about them, they exist in a common frame of reference. The same is not true of nouns. Broad categories of nouns may be classified together and marked by grammatical gender, but no relationship exists between the categories, let alone a global ordering. No implicit relationship exists between the nouns "crocodile" and "heresy," nor can any linguistic relationship be stated without reference to an external category such as gender. Mathematical comparisons, on the other hand, relate to dimension but allow no accommodation for the value judgments of the third adjective class. If a metaphorical interpretation of the fourth adjective is permitted, mathematics does preserve a limited set of adjectives related to abstract principles. These adjectives, such as figurate, prime, or perfect, rarely merit mathematical notation. Rather, these conditions are described abstractly in the portion of mathematics composed in natural languages. Mathematical objects may be identified with or excluded from the set of elements which share those descriptions. Although rarely abbreviated to mathematical notation, these adjectives also represent a restriction of natural language. Because a number is objectively prime or square or negative, comparatives and superlatives do not apply to these categorizations.

5.2   Referents of mathematical objects

Mathematicians have made use of the implicit ordering of numbers. As noted in the discussion of the aphorism, Schofield identified the subscript index of variables as a set of mathematical adjectives. Schofield was correct that this use of indices fulfilled the second task common to adjectives—they specify the referent of a noun. Insofar as each of these indexes invokes another instance of the same set of values, this use resembles the noun-like adjectives of natural languages. (Another explanation might equate such variables with an open category of pronouns or determiners, but these categories are not as universally used in natural languages as adjectives.) The meaning of subscript indices, however, can vary. In one case, the indices may refer to the values returned for an ordered progression of a function through the integers; in another case, they may merely represent the individual values associated with the elements of a group in no particular order. Either case specifies the referent of a mathematical object, namely the variable.

Despite their similarity to noun-like adjectives, these purely mathematical indexes do not state a property, the first task common to adjectives. Rather, these indices introduce new mathematical nouns which fulfill predetermined characteristics. The mathematical use of indices does not reveal the property which each element satisfies. Consider the variable b. Various values which satisfy b may be identified and a set B emerges. The elements b_1, b_2, $\cdots$ b_n, and so on may be identified as $b_1 = 4$ and $b_2 = 64$, but whether $b_n =$ any even number, any square number, or any power of 2 is not communicated by the mathematical adjective. The subscript adjective functions more like a determiner (such as an article, demonstrative, or quantifier) than an adjective because it serves only to separate the instances of the noun which fill the category.

A similar notation, superscript, often relates to exponentiation. In early European mathematical texts, repeated multiplications were laboriously repeated, but modern notation indicates number of multiplications by superscript numbers. This practice has liberated mathematical expressions from a connection to the dimensions of the physical world and ended the special privilege once extended to the category of square and cubic numbers. The notation considers all powers analogous, even though specialized adjectives may be drawn from natural language to discuss squares and cubes. Because this notation refers to a set of operations performed on the numbers, this practice resembles the verb-like adjectives of natural languages. Although exponentiation is acknowledged as an operation, the notation n^2 is commonly rendered into English as "n squared." The analogy of superscript notation of exponentiation with mathematical adjectives is limited, though, because exponentiation reveals a mathematical property of the range, not the domain. In effect, the verb-like mathematical adjective of exponentiation functions like a variable because mathematical nouns are distinguished only by magnitude.

5.3   Variables and Generalization

Thus, mathematical adjectives, whether noun-like or verb-like, do not state the property of a mathematical noun directly but, rather, function as referents which indicate membership in a set of mathematical objects which share a property. Mathematicians still must distinguish between the nouns which satisfy this property. Although mathematicians cannot discuss the category of squares or primes directly, they can generalize the properties of these categories through variables. Thus, a square number is s^2, with s being an element of the set of integers; a prime p is any integer for which the statement $m \times n = p$ implies that m or n (but not both) must be p. A strong tradition of shared notation for variables has arisen. Thus, n is often an integer and p is often a prime, despite the fact that variables are a relative late-comer to mathematical expressions. Ancient mathematical statements consider problems phrased in terms of known and unknown sides, areas, and quantities on which operations are performed. Francois Viète (1540–1603) formulated the idea of variables which occur in an analytical argument but to which no specific numerical value has been assigned. Despite the fact that he displayed a linguistic awareness by assigning variables to vowels and parameters to consonants, Viète considered mathematics an art (*ars*) [42, p. 7]. When mathematicians use variables to represent values which satisfy a category, these variables often reflect the mathematical adjectives derived from natural language. Thus, p is often prime; n is often an integer. Adjacent letters often introduce other mathematical objects of the same category. A proof which requires two primes usually invokes primes p and q; two integers are often m and n (with o omitted due to its typographical similarity to 0.) In this way, variables also resemble determiners because they enable mathematicians to refer to "this prime" and not "that prime." Historically, determiners have been classified among adjectives, but the linguistic consensus has changed. Mathematicians could achieve the same communicative goals with subscripts (such as p_1 and p_2 for p and q), but they rarely employ this rhetorical strategy.

Mathematicians have avoided developing adjectives analogous to natural languages for good reason. Mathematics is assumed to express a generalization [11, pp. 134–136]. Generalization may be a rhetorical goal of mathematicians, but it is not a grammatical feature of a language. This feature, along with formalization, underpins the universality towards which mathematicians aspire [11, pp. 137–140]. The development of mathematical adjectives to distinguish between instances of the same number risks undermining the goal of generalization. The only property relevant in numeric statements is magnitude, and this property is inherent in the lexeme, or "name," of the mathematical entity itself. The sign of the number, whether it is positive or negative, forms an extremely truncated class of adjective which relates directly to magnitude and may be equally well expressed by operators. Set theory enabled the formalization of many adjectives which had once been expressed in portions of mathematical texts which closely resembled natural language. The fact that this technique of description was not invoked by early mathematicians but was developed deliberately may be evidence of the linguistic status of mathematics.

6 Mathematical Language

The analogy of operations with verbs demonstrates that the verbs of mathematical systems have developed with reference to natural languages and vary by culture. The analogy of numbers with nouns shows that mathematical expressions are grammatically numberless. A search for mathematical adjectives reveals that while mathematicians use adjectives to describe relationships among numbers, they create these portions of mathematical compositions in something very near natural language because distinguishing between two instances of the same magnitude runs contrary to the presumed goals of mathematics. In each case, mathematics differs from natural languages not so much by what can be stated in a mathematical way but by which simplifications and truncations have been applied to a natural language. If a natural language produces an axiomatic system, then correspondences with the language in which the axioms originated ought to be found in that system. This proposition seems to be confirmed by the fact that throughout the development of this history of mathematics, mathematical texts are clarified by consideration of similarities between mathematical texts and non-mathematical texts in the same language.

A related proposition states that if a language corresponds to an axiomatic system, then that system, in some way, must have produced that language. Taken as a hypothesis, this proposition has buoyed the hopes of a universal grammar. Presumably, according to this perspective, at some far off point of abstraction, linguistic, mathematical, and cognitive systems converge. Concurrent with the pursuit of a universal grammar, though, linguists endeavored to mathematize their descriptions and analysis of languages. As early as 1914, Leonard Bloomfield proposed that each sentence contained two constituents and each of these constituents in turn contained two constituents [2, p. 110]. Roman Jakobson reworked phonological oppositions as another set of binary expressions [9]. In retrospect, these models better reflect the successes of binary expressions in the computing of Claude Shannon and the expansion of mathematical modeling in the sciences than an empirically derived structure of language. For example, the assumption of binary models does little to explain the Japanese trend toward a tripartite division, as expressed by the determiners この (kono, indicating something close to the speaker), その(sono, indicating something close to the listener), and あの(ano, indicating something removed from both the speaker and the listener) and the related pronouns これ (kore), それ (sore), and あれ (are). For some time, the mathematicization of language was a useful analogy, but analogies (and mathematical models) are descriptive creations, not prescriptive discoveries. These tools are not reifications of reality and may be abandoned when they are no longer useful.

The analogy between language and mathematics reveals some connections but an analysis of the metaphor reveals unstated presumptions. The preservation of an analogy between nouns and numbers necessitates the exclusion of the grammatical category of number in a way not found among natural languages. Despite the faults of the analogy, an analysis of the aphorism that mathematics is a language reveals that mathematics resembles a recently identified linguistic practice—controlled natural language [18, p. 4]. On the whole, mathematics permits fewer categories and developments than natural languages.[8] Operators lack tense, mood, and other aspects. Numbers lack grammatical number. Adjectives may be limited to sign. This systematic reduction of linguistic complexity fits the description of a controlled natural language. Some computer systems use controlled natural languages to weld the precision of machine languages with the intelligibility of natural languages.

---

[8]The obvious exception is Indian mathematics which has developed an overabundance of mathematical synonyms. Because Indian mathematical texts are almost exclusively poetic, these variations may be explained by *causa metris*. However, in many cases, this explanation is facile. To date, a systematic study of whether the variation in vocabulary of Indian mathematics remains a desideratum. Only careful comparison will reveal if these variations indicate regional variations, the temporal development of Sanskrit, or nuances among mathematical objects.

The resulting languages stand between constructed languages like Astraglossa and Lincos and natural languages. The role of controlled natural languages is not limited to numeric topics. Commentaries, citations, and scholastic compositions also conform to the strategies of controlled natural language. Just as the traditions of commentary and citation vary between cultures, so do differences mark mathematical systems. As a comparison of Egyptian and Babylonian mathematics shows, different cultures adopted different mathematical conventions, just as different branches of mathematics adopt different conventions to this day. In the case of antiquity, these differences reflected differences not only in mathematical culture but also in language.

In the case of mathematics, several cultures have developed a similar set of linguistic strategies for the study of quantities. Whether the decision to control natural language for the discussion of quantities stemmed from a common origin or whether multiple cultures adopted the same tool for the problem is not obvious. Mesopotamia and Egypt seem to share similar perspectives on the study of quantities. However, the mathematics of one culture may not be immediately understood by members of another culture despite modern claims of universality. The greatest difficulty in understanding mathematical expressions comes not from mathematical cultures geographically distant from modern mathematicians, but from cultures temporally removed from modern mathematics. The first reason for this trend is that modern mathematics has been an international affair for more than a century. Most mathematics have been done in a handful of languages relative closely related to each other. Even those recent works of modern mathematics undertaken in non-Western languages have been inspired by "Western" mathematical culture. The second reason for this trend is that mathematics is often learned through sharp value judgments of "correct" and "incorrect." Coupled with a belief in technological progress, the foreign mathematics of antiquity too easily looks "wrong," or "primitive." Among the variants of the aphorism, a limited version proposed by Yuri Manin is the most defensible: "[T]he basis of all human culture is language, and mathematics is a special kind of linguistic activity [26, p. 17]." The modern understanding of these conventions is a reflection of the understanding of mathematical culture and the process of controlling different base languages.

Acknowledgements Academics frequently praise books, but scholarly reflections seldom suit the tastes or developmental needs of children. Michel de Montaigne seems apt to inspire future Nietzsches when he declares that "obsession is the wellspring of genius and madness." In contrast, [10] by Philip Davis allows those who doubt their genius but prize their sanity to *pafnuty*, that is, "to pursue tangential matters with hobby-like zeal." Children as young as eleven have been known to incorporate pafnutying into their pedagogical formation. Through this slim volume, Davis tames madness with whimsy and humanizes genius by levity. Should an early exposure to the admixture of language and mathematics distract the student, that is, if the child becomes a recidivist pafnutier, youthful exuberance may be regulated by [11]. If the present argument resembles too much Montaigne's madness or Davis' skeptical classicist, it must be realized that despite several introductions and a shared institutional affiliation, this offering did not result from tutelage under Davis. Rather, it is offered as an homage to a past and present inspiration.

References

1. A. Adler. Mathematics and creativity. *The New Yorker*, 47(53):39–45, feb 1972.
2. L. Bloomfield. *An introduction to the study of language*. H. Holt, 1914.
3. M. Brose. Die mittelägyptische nfr-pw-negation. *Zeitschrift für Ägyptische Sprache und Altertumskunde*, 136(1):1–7, aug 2009.
4. E. A. T. W. Budge. *The Gods of the Egyptians*. Methuen, London, 1904.
5. E. A. T. W. Budge. *An egyptian hieroglyphic dictionary*, volume 1. John Murray, London, 1920.
6. C. K. Caldwell and Y. Xiong. What is the smallest prime? *Journal of Integer Sequences*, 15(9), 2012. Article 12.9.7 and arXiv:1209.2007.
7. A. B. Chace. *The Rhind mathematical papyrus : British Museum 10057 and 10058*. Mathematical Association of America, Oberlin, 1929.
8. J.-P. Changeux and A. Connes. *Conversations on mind, matter, and mathematics*. Princeton University Press, 1995.
9. E. C. Cherry, M. Halle, and R. Jakobson. Toward the logical description of languages in their phonemic aspect. *Language*, 29(1):34–46, 1953.
10. P. J. Davis. *The thread : a mathematical yarn*. Birkhäuser, Boston, first edition, 1983.
11. P. J. Davis and R. Hersh. *The mathematical experience*. Birkhäuser, Boston, 1981.
12. R. M. W. Dixon. *Basic Linguistic Theory*. Oxford University Press, Oxford, 2010–2012.
13. T. W. R. E. L. Kaufman, M. W. Lord and J. Volkmann. The discrimination of visual number. *The American Journal of Psychology*, 62(4):498–525, oct 1949.
14. J. Evelyn. *Sculptura, or, The history, and art of chalcography and engraving in copper*. G. Beedle and T. Collins, Oxford, 1662.
15. J.-B.-J. Fourier. *Théorie analytique de la chaleur*. F. Didot, 1822.
16. H. Freudenthal. *Lincos; design of a language for cosmic intercourse*. Studies in logic and the foundations of mathematics. North-Holland Publishing Company, Amsterdam, 1960.
17. J. Friberg. *Unexpected links between Egyptian and Babylonian mathematics*. World Scientific, 2005.
18. N. E. Fuchs, H. F. Hofmann, and R. Schwitter. Specifying logic programs in controlled natural language. In *Specifying logic programs in controlled natural language*, volume 94.17 of *Berichte des Instituts für Informatik der Universität Zürich; Institut für Informatik der Universität Zürich*, 1994.
19. G. Galilei. *Il saggiatore*. Giacomo Mascardi, Rome, 1623.
20. K. Gödel. *The consistency of the axiom of choice and of the generalized continuum-hypothesis with the axioms of set theory*. Annals of mathematics studies. Princeton University Press, Oxford, 1940.
21. B. Gunn. Notices of recent publications. *The Journal of Egyptian Archaeology*, 3(4):279–286, oct 1916.
22. L. Hogben. Astraglossa, or first steps in celestial syntax. *Journal of the British Interplanetary Society*, 11(6):258–274, nov 1952.
23. J. Høyrup. *Lengths, widths, surfaces : a portrait of old Babylonian algebra and its kin*. Sources and studies in the history of mathematics and physical sciences. Springer, New York, 2002.
24. T. Jech. *Set Theory: The Third Millennium Edition, Revised and Expanded*. Springer Monographs in Mathematics. Springer, Berlin Heidelberg New York, 2003.
25. B. Lumpkin. Africa in the mainstream of mathematics. In A. B. Powell and M. Frankenstein, editors, *Ethnomathematics : challenging eurocentrism in mathematics education*, SUNY series, reform in mathematics education., pages 101–118. State University of New York Press, Albany, 1997.
26. Y. I. Manin. Good proofs are proofs that make us wiser. In M. Aigne, editor, *Berlin intelligencer : International Congress of Mathematicians*, Mitteilungen der Deutschen Mathematiker-Vereinigung, page 16–19, Berlin, aug 1999. Springer.
27. O. Neugebauer. *Mathematische Keilschrift-Texte*, volume 1 of *Quellen und Studien zur Geschichte der Mathematik, Astronomie und Physik*. J. Springer, Berlin, 1935.

28. T. E. Peet. *The Rhind mathematical papyrus, British museum 10057 and 10058*. The University Press of Liverpool and Hodder & Stoughton, London, 1923.

29. C. Proust. Mesopotamian metrological lists and tables:forgotten sources. In F. Bretelle-Establet, editor, *Looking at It from Asia: The Processes that Shaped the Sources of History of Science*, volume 265 of *Boston Studies in the Philosophy of Science*, pages 245–276. Springer, 2010.

30. J. Ritter. Closing the eye of horus: The rise and fall of horus-eye fractions. In J. M. Steele and A. Imhausen, editors, *Under One Sky: Astronomy and Mathematics in the Ancient Near East*, volume 297 of *Alter Orient und Altes Testament*, pages 297–323. Ugarit-Verlag, Münster, 2002.

31. M. Rukeyser. *A turning wind*, chapter Gibbs. The Viking Press, 1939.

32. M. Rukeyser. *Willard Gibbs*. Doubleday, Doran & Company, 1942.

33. R. P. Runyon, A. Haber, and P. Reese. *Student workbook to accompany Fundamentals of behavioral statistics*. Addison-Wesley Publishing Company, fourth edition, 1980.

34. C. Sagan. *The cosmic connection: an extraterrestrial perspective*. Anchor Press, Garden City, NY, 1973.

35. C. Sagan. The recognition of extraterrestrial intelligence. *Proceedings of the Royal Society of London. Series B, Biological Sciences*, 189(1095):143–153, may 1975.

36. C. Sagan. *Cosmos*. Random House, New York, 1980.

37. C. Sagan. *Contact: a novel*. Simon and Schuster, New York, 1985.

38. R. Schofield. Sampling in historical research. In E. Wrigley, editor, *Nineteenth-century society: essays in the use of quantitative methods for the study of social data*, Cambridge Group for the History of Population and Social Structure, pages 146–190. Cambridge University Press, 1972.

39. R. L. E. Schwarzenberger. The language of geometry. *Mathematical Spectrum*, 4(2):63–68, 1972.

40. K. Sethe. *Von Zahlen und Zahlworten bei den alten Ägyptern und was für andere Völker und Sprachen daraus zu lernern ist*. K. J. Trübner, Strassburg, 1916.

41. J. Thomson. *Britain: being the fourth part of Liberty, a poem*. A. Millar, Oxford, 1736.

42. F. Viète. *In artem analyticem isagoge*. Mettayer, Tours, 1591.

43. N. West. *Miss Lonelyhearts, a novel*. New Directions Books, 1933.

44. L. Wheeler. *Josiah Willard Gibbs, the history of a great mind*. Yale University Press, New Haven, 1951.

45. E. T. Whittaker and G. Robinson. *The calculus of observations; a treatise on numerical mathematics,*. Blackie and Son Limited, London, first edition, 1924.

Mathematics as multimodal semiosis

Kay L. O'Halloran

Abstract In this chapter, mathematics is considered to be a multimodal semiotic process involving the use of language, images, and mathematical symbolism, each with their own systems of meaning which integrate to create mathematical knowledge. The specific relations between linguistic, visual, and symbolic systems have led to semantic expansions in mathematics, resulting in a hierarchical knowledge structure which extends beyond the reach of other forms of human communication.

1 Introduction

Mathematics is a subject that is one of the finest, most profound intellectual creations of humans, a subject full of splendid architectures of thought. It is a subject that is also full of surprises and paradoxes. Mathematics is said to be nothing more than organized common sense, but the actuality is more complex. As I see it, mathematics and its applications live between common sense and the irrelevance of common sense, between what is possible and what is impossible, between what is intuitive and what is counterintuitive, between the obvious and the esoteric. The tension that exists between these pairs of opposites, between the elements of mathematics that are stable and those that are in flux, is a source of creative strength [11, p. viii].

Davis [11, p. xix] describes mathematics as "one of the greatest human intellectual accomplishments," with inherent features which are not easy to characterize. Indeed, mathematics is a remarkable achievement and a major "source of creative strength" [11, p. viii] which has, if one considers technology and other scientific products derived from mathematical formulations, (literally) changed the nature of life on earth. But how is this possible? That is, what is the nature of mathematics that has resulted in new domains of knowledge, which are so usefully employed in the natural sciences? The aim of this chapter is to explore this question by conceptualizing mathematics as a multimodal semiotic process (i.e., involving sets of inter-related sign systems) which draws upon and integrates language, images, and symbolism to create meanings which extend beyond those possible with other

K.L. O'Halloran (✉)
Faculty of Humanities, School of Education, Curtin University, Level 4, Building 501,
Kent Street, Bentley, Perth, WA 6102, Australia
e-mail: Kay.Ohalloran@curtin.edu.au

© Springer International Publishing Switzerland 2015
E. Davis, P.J. Davis (eds.), *Mathematics, Substance and Surmise*,
DOI 10.1007/978-3-319-21473-3_14

forms of human communication. In this regard, mathematics is considered to be a specialized tool for thinking, specifically designed to move beyond our everyday experience of the world to an abstract semiotic realm for restructuring thought and reality. Natural language also functions to organize and structure human experience on an abstract semiotic plane (e.g., [17, 50]) but, unlike mathematics, it lacks the 'meaning potential' [23] to effectively model and predict events in the physical world. The purpose of this paper is to explore how mathematics achieves its unique functionality, as a designed system, which is the result of centuries of human effort.

In what follows, multimodal social semiotics, the theoretical approach to mathematics adopted in this paper, is presented, followed by a discussion of the functions of natural language, images, and symbolism, and their integration in mathematics texts. There are three key ideas. First, human sign systems are tools for structuring thought and reality. Second, by developing new integrated, written systems in mathematics which combined textual forms (linguistic, symbolic) with visual forms (graphs, diagrams, and figures), it was possible to construct new views of the physical world. Third, by its very nature, mathematics as a multimodal hierarchical knowledge system can be usefully employed to describe and predict events in the material world, but it has limitations in terms of modeling and predicting the human socio-cultural world. From this perspective, an important challenge today is to develop new semiotic tools for modeling the social world—a general "science of [human] meaning" [6]—to replicate the effectiveness of mathematics in the natural sciences.

2   Theoretical Foundations: Multimodal Social Semiosis

Semiotics is the study of sign systems and processes (e.g., [8, 39]), and 'social semiotics' is the branch of semiotics which studies human signifying processes as social practices [17, 30, 49]. Social semiotics is concerned with different sign systems (e.g., linguistic, visual, aural, and gestural), and their integration in texts and interactions, interpreted within the context of situation and culture. 'Multimodal social semiotics' is specifically concerned with the study of relations within and across semiotic resources, both as multimodal systems (i.e., the process) and as multimodal texts (i.e., the product) (e.g. [32]). That is, multimodal social semiotics is concerned with modeling semiotic resources as inter-related systems of meaning, and analyzing how choices from the different systems work together to create meaning in multimodal texts. In this case, the semiotic resources of interest are language, image, and mathematical symbolism, which are the fundamental resources through which mathematical knowledge is created. Other resources, such as three-dimensional objects, calculating devices and digital technology, are also briefly considered.

Following Halliday's [17] and Halliday and Matthiessen [28] social semiotic theory, the basic premise is that language and other semiotic resources, such as images, symbolism, gestures, actions, and sounds, are sign systems which have evolved to fulfill particular functions: (a) to record experience as happenings and events; (b) to logically connect those happenings and events; (c) to create a stance toward the happenings and events while enacting social relations; and (d) to organize the message itself. These functions, called the experiential, logical, interpersonal, and textual 'metafunctions,' respectively [22, 23], are the four strands of meaning which are communicated in any social interaction—that is, humans construct, reason about, and evaluate happenings and events while negotiating social relations. In reality, these four strands of meaning combine, but they can also be considered individually for theoretical and analytical purposes.

In order to (differentially) fulfill the four metafunctions, language and other semiotic resources have each evolved a unique infrastructure (or architecture), formulated as 'systems of meaning' in systemic functional theory [21]. For example, language has a multidimensional architecture [22], which includes lexical and grammatical systems which operate at different levels (i.e., sound, word, clause, sentence, and paragraph). In this case, the term 'grammar' does not refer to a set of rules that govern language use; rather, it is a description of the potential of language to create certain meanings, from which selections are made in texts. In this respect, Halliday [27] views "language as system" and "language as text" as "two aspects of one single phenomenon" (p. ii). That is, the linguistic system has a meaning potential, modeled as sets of inter-related networks of options, and the text is the process and product of selecting from that potential. Linguistic patterns are built up culturally over time so that any instance of language use is conditioned by previous choices. That is, linguistic choices are routinely deployed so they form recognizable cultural configurations (e.g., casual conversation, news bulletin, academic paper, and so forth). In multimodal semiotics, Halliday's formulations of metafunction, system and text are extended to other semiotic resources, which are also seen to have evolved their own infrastructure to fulfill particular functions, often in collaboration with language. In this regard, systemic functional theory forms a useful framework for viewing mathematics as a (multimodal) semiotic tool for restructuring thought and reality.

In what follows, language, image, and mathematical symbolism are viewed as semiotic resources which integrate to create mathematical knowledge. In this case, 'the whole is other than the sum of the parts,' as maintained by Gestalt psychologist Kurt Koffka [31], because the precise relationships between the three resources create a semantic space which extends beyond the potential of each individual resource. Indeed, the "creative tension" of mathematics [11] originates (at least in part) from the transitions between language, image, and symbolism, where 'meta–metaphorical phenomena,' referred to as semiotic metaphors (or inter-semiotic metaphors) are created (e.g., [40]). The semantic potential of semiotic metaphor extends beyond traditional conceptions of metaphor which are language-based; for example, conceptual metaphors are seen to map one domain of experience onto another (e.g., "life is a journey") (e.g. [34]). Derivatives of conceptual metaphor have been applied to multimodal texts, resulting in the formulation of 'visual metaphor' and 'multimodal metaphor' [15]. However, semiotic metaphors, arising from the transition from one semiotic resource to another, are fundamentally different because, rather than mapping one experiential domain onto another, these metaphors involve a grammatical reconfiguration which changes the essence of how experience is organized, as explained below.

Before investigating the integration of natural language, images, and symbolism in mathematics and the metaphorical transformations that result, the three resources are considered individually in order to examine their semiotic functions. As the discussion is concerned with how mathematics restructures human experience, the focus is directed toward experiential meaning (i.e., happenings in the form of entities, processes, and circumstances) and logical meaning (i.e., the connections between those happenings). The nature of interpersonal meaning (i.e., social relations and truth-value) and textual meaning (i.e., the organization of the message) in mathematics is also briefly considered.

3   Linguistic Formulations of Human Experience

As Halliday [24, p. xvii] explains, "the grammar of every language contains a theory of human experience; it categorizes the element of our experience into basic phenomenal types, construing these into configurations of various kinds, and these configurations in turn into logical sequences." Thus, by organizing human experience in such a manner, the words and grammar of everyday discourse are used to construe happenings in a variety of ways, using the available options in the language system. For example, William Golding [16] uses language to create the world views of two early human species in *The Inheritors*, a novel concerned with the extinction of one of the last Neanderthal tribes by homo sapiens. In investigating the linguistic patterns in the novel, Halliday [20] focuses on the grammar of transitivity (i.e., process types, participants and circumstance, and active and passive voice) to show how Golding uses these systems to characterize the relatively simple Neanderthal world—a world dominated by material actions which have little or no impact on the environment, unclassified objects which lack basic description, and an absence of reasoning about unfolding events. "Thus the picture is one in which people act, but they do not act on things; they move, but they move only themselves not other objects" [20, p. 108]. As Halliday observes, even verbs that typically involve an impact on some other object (e.g., "grab") invoke a sense of hopelessness and futility in the Neanderthal world (e.g., "he threw himself forward and grabbed at the branches with hands and feet"). As Lukin [37, pp. 362–363] explains: "The 'syntactic tension expresses this combination of activity and helplessness', a world view in which 'there is no cause and effect', where 'people do not bring about events in which anything other than themselves, or parts of their bodies, are implicated', where 'people do not act on the things around them; they act within the limitations imposed by the things' [20, pp. 109, 113, 114]."

In the final section of Golding's novel the world view shifts to the relatively more sophisticated world of early homo sapiens, where the horizons of early humans are broadened beyond the confines of the immediate context. "This transformation, as Halliday shows, [is] encoded lexicogrammatically; ... transitive structures predominate, agency belongs to humans not inanimate things, humans intentionally act on external objects, their actions are more varied, and they produce results, and things in the world are increasingly taxonomized" [37, p. 363]. In this way, readers are able to recognize the beginnings of human social life: for example, religion, war, and control of the physical environment.

Natural language is also used to create a scientific view of the world; a world that differs dramatically from the early human worlds described by Golding. For example, in a study of the emergence of scientific writing from the time of Chaucer's discussion of the astrolabe in the 14th century and the origins of modern science in the writings of Newton, Priestley, Dalton, Darwin and Maxwell in the 17–19th centuries. Halliday [26] demonstrates how a major semantic shift from a commonsense view of the world to an abstract scientific view took place. That is, everyday experience as a series of events or happenings was gradually replaced with a virtual world of cause and effect. Linguistically, the abstract scientific view was constructed by recoding processes (i.e., verbs) as metaphorical (virtual) entities (i.e., nouns), and logical connections (i.e., conjunctions) as causal processes (i.e., verbs), in order to relate the virtual entities to each other. For example, 'a interacts with b, thus transforming z' is recoded as 'the interaction of a and b results in the transformation of z,', so that: (a) the processes 'interacts' and 'transforms' become the virtual entities 'interaction' and 'transformation'; and (b) the conjunction 'thus' is recoded as the causal processes 'results in.'. As Halliday [26] explains, scientific writers from Newton onwards favored such metaphorical modes of expression "– one in which, instead of writing 'this happened, so that happened', they write 'this event caused that event'" [26, p. 174] because their aim was to construct a logical argument:

> Newton and his successors were creating a new variety of English for a new kind of knowledge: a kind in which experiments were carried out; general principles were derived by reasoning from these experiments, with the aid of mathematics; and these principles in turn tested by further experiments. The discourse had to proceed step by step ... [and] the most effective way of doing this, in English grammar, is to construct the whole step as a single clause, with the two parts turned into nouns, one at the beginning and one at the end, and a verb in between saying **how** the second follows from the first [original emphasis] [26, p. 174].

Halliday [25] uses the term 'grammatical metaphor' to describe the semantic shift whereby processes, logical conjunctions, and other grammatical elements (e.g., qualities, circumstances) are reconstrued metaphorically to transform the commonsense world of actions and events into an abstract semiotic world specifically designed for thinking and reasoning: "Grammatical metaphor creates virtual phenomena – virtual entities, virtual processes – which exist solely on the semiotic plane; this makes them extremely powerful abstract tools for thinking with. Thus what grammatical metaphor does is to increase the power that a language has for **theorizing**" [original emphasis] [24, p. xvii]. Over time, increasing amounts of information were encoded into noun group structures to condense and propel logical thinking even further; for example, the grammatical metaphors in the titles of recent articles in *Nature* (Volume 521, Number 7552, May 2015) include: "counteraction of antibiotic production and degradation," "the crystallography of correlated disorder," and "spin–motion entanglement and state diagnosis with squeezed oscillator wavepackets," which are typical of the extended noun structures found in scientific writing. Scientific language is the "product of conscious design," where "grammatical metaphor reconstrued the human environment, transforming the commonsense picture of the world into one that imposed regularities on experience and brought the environment more within our power to control. ... It is presented most clearly in the discourse of the natural sciences, which is where it evolved" [24, p. xvii]. While scientific language also introduced a large number of technical terms, the change was more profound than a new vocabulary; that is, the underlying grammatical change, whereby virtual entities are logically related to each other, transformed our view of the world.

Newton and his successors based their science on mathematics, however, the tool through which their results and findings were derived. Fundamentally, mathematical symbolism restructures the world in very different ways to scientific language, as discussed below.

4   The Symbolic Creation of a World View

> Diophantus, looking over the clutter of words and numbers in algebraic problems, saw that an abstraction could be a great simplification. And so, Diophantus took the first step toward introducing symbolism into algebra [29, p. XIII].

Histories of mathematics reveal that mathematical symbolism evolved from rhetorical forms which were linguistic in nature, to syncopated forms which contained linguistic elements and some symbols, to the modern symbolic system found today (e.g., [5]). Although it would seem logical to have introduced symbols for arithmetic operations after symbols were created for entities and numbers in early mathematical systems, this step was not always taken due to the existence of mechanical calculating devices (e.g., counting rods, pebbles, counting boards, and so forth) (e.g. [9]). While most calculating devices were symbolic in nature to some degree (e.g., the abacus), they operated across different modes, including three-dimensional space, action, and movement, making it difficult to develop a unified symbolic system for mathematics. Eventually the Hindu-Arabic number system was fully adopted in 16th century Europe (see discussion of algorists versus abacists in [44, 45]), leading to written forms of arithmetic algorithms and the later development of modern mathematical notation. Historians note that mathematical notation was the subject of much debate and controversy, and the result was often a matter of politics rather than the usefulness of the symbol itself [7].

Even though the symbolism evolved from linguistic forms, the grammatical systems which evolved in modern mathematical notation differ from those found in natural language. There were several reasons for the departure from the grammatical systems of language, and the forms of scientific writing which later evolved. First, language operates in both written and spoken modes, and so possesses graphological and phonological systems for encoding meaning via words and sounds. However, modern mathematical notation developed in written form, so new visual-based grammatical strategies could be developed (e.g., use of spatial notation, lines, brackets, textual layout, and so forth). The invention of the printing press in the 15th century permitted these new grammatical systems to be standardized and widely circulated. Indeed, the printing press was a major factor in the scientific revolution because, as Eisenstein [14] points out, Newton (for example) taught himself mathematics from books he bought or borrowed: "At least in my view the changes wrought by printing provide the most plausible point of departure for explaining how confidence shifted from divine revelation to mathematical reasoning and man-made maps" [14, p. 701]. The printing press, and the subsequent shift to standardized forms of symbolic notation, paved the way for the development of mathematical symbolic notation as a semiotic tool developed for written (and today digital) modes of communication.

Second, language and mathematical symbolism, as designed semiotic systems, evolved to fulfill different purposes. Scientific writing was designed to develop a logical argument about scientific findings, hence the creation of virtual entities which could be used to reason about a world of cause and effect. Mathematical symbolism, however, provided the results upon which those arguments rested, and so the functions and subsequent nature of this semiotic tool are necessarily different to language. In fact, it can be argued that scientific language evolved in the manner described by Halliday [25] due to the intellectual advances and the remodeling of the world made possible through mathematics. This claim is explored below by examining mathematical notation as a grammatical system for encoding meaning, which may be compared to the linguistic system from which it evolved.

In language, a limited number of key entities are configured around a single process (often with circumstantial information), which in turn is logically related to a similar configuration, so that happenings unfold in sequence, one after the other. As discussed earlier, scientific writing partially overcomes the limitations of the serial encoding of happenings by creating virtual entities and processes so logical thinking can be intensified. However, despite the recoding of reality achieved through grammatical metaphor, the restrictions of the serial encoding of happenings remains in linguistic descriptions of experience. While mathematical symbolic statements also unfold in a linear (logical) fashion, happenings are encoded in each statement as multiple configurations of generalized entities and processes which interact different ways that are not necessarily serial in nature. That is, mathematics organizes experience in terms of the relations between multiple (generalized) entities interacting with each other via multiple (generalized) processes at any given instance. The configurations of entity and process interactions are encoded in an economical and unambiguous fashion, using deep levels of embedding and other grammatical strategies (e.g., positional notation, spatial notation, specialized symbols, rule of order for operations, ellipsis, use of brackets, and so forth), so that the symbolic configurations are easily rearranged to derive results.

For example, the equation $x^3 + \frac{xy}{3} = \frac{y^2}{4} - x^2y + 2\left(y^2 + 3\right)$ involves configurations of entities (variables x and y and integers 2, 3, and 4) and processes (addition, subtraction, multiplication, and division), which are economically encoded using exponents, lines for division, brackets, and ellipsis of the multiplication sign. Using such grammatical resources, mathematical symbolism creates a dynamic representation of multiple relations in a format that is easy to comprehend and manipulate; that is, the symbolism is specifically designed for logical thinking. Scientific language is dense and static (given the amount of information encoded in the virtual entities and processes) compared to the fluidity and intricacy of mathematical symbolism, but unlike spoken language which Halliday [18] characterizes in similar terms, the encoding of multiple generalized entities and processes in the symbolism extends its meaning potential for capturing patterns and relations beyond the capabilities of language. As such, mathematics is designed to manage the complexity of the universe, as the ultimate tool for mapping and thinking about patterns and relations.

In summary, language imposes a certain way of experiencing reality, where key entities are identified and assigned roles within a semantic configuration which features a single process, so the world is constructed as a series of happenings which unfold in a linear sequence. Mathematical symbolism has no such restrictions with regard to ordering experience; rather, the world consists of many interacting entities and processes, forming relations which can be reconfigured in order to establish results. Moreover, mathematics is a 'hierarchical' (i.e., cumulative) knowledge structure [3, 4], where knowledge is built upon earlier results. In this world, the focus is the experiential and logical domains, aided by textual resources (e.g., spatial layout) that organize mathematical knowledge in very specific ways. In order to foreground the experiential and logical domains in mathematics, the interplay of social relations is curtailed by holding interpersonal meaning and truth-value (e.g., possibility, probability, obligation, and potentiality) within limited spheres of activity. That is, the interpersonal realm is subdued (or constrained) in mathematics, in order to focus on the experiential and logical domains. Indeed, unlike language, mathematics is not designed to map the human social world where interpersonal meaning is often the primary consideration, along with other domains of experience (e.g., sensing, feeling, behaving, and so forth).

Mathematical symbolism and language are designed for different purposes, and accordingly, they have different grammatical systems and strategies for managing complexity. One key aspect of mathematical symbolism is the direct link to mathematical images, which creates a semantic circuit between language, symbolism, and image—the defining feature of mathematics for reformulating reality. In what follows, mathematical images are considered, before examining mathematics as an integrated form of multimodal semiosis.

5   The Visual Creation of Reality

Seeing comes before words. The child looks and recognizes before it can speak. But there
is also another sense in which seeing comes before words. It is seeing which establishes
our place in the surrounding world; we explain that world with words, but words can never
undo the fact that we are surrounded by it. The relation between what we see and what we
know is never settled [2, front cover].

Images, along with natural language and other semiotic resources, function to
structure thought and reality. That is, in much the same way that writers are able
to use linguistic resources to create different world views, painters, photographers,
illustrators, and others use visual resources to structure the world in alternative
ways. For example, Berger [2] considers paintings of female nudes (also advertising
images) in western culture and shows how the images differ in terms of the choices
which the painters make with respect to facial expression, gaze, body posture, skin
texture, clothes, use of objects (e.g., mirrors), the immediate environment, and
so forth. Berger [2] demonstrates that, with the exception of a small number of
grand masters who depict the woman as herself, the paintings perpetuate cultural
perceptions about women as objects of male gaze and idealization. In doing so,
Berger [2, p. 54] explains the difference between nakedness and nudity: "To be
naked is to be oneself. To be nude is to be seen naked by others and yet not
recognized for oneself. A naked body has to be seen as an object in order to become
a nude"; that is, "nudity is a form of dress" [2, p. 54]. The difference between nudity
and nakedness then is how the painter (or photographer, artist or image creator)
chooses to portray the woman, selecting from a range of available options.

Images function to create the world around us, rather than reflect it, as Berger [2]
demonstrates in his study of paintings and other visual genres. In this regard images
are semiotic resources which are used to make experiential, logical, interpersonal,
and textual meanings (e.g., [1, 33, 43]). As such, images may also be conceptualized
as system and text, where the system is the meaning potential (e.g., participants,
objects, visual processes and circumstances, and so forth) and the text is the actual
image itself (e.g., the painting, with participants and their gaze, facial expression,
and so forth), constructed through choices from systems, which constitute the visual
grammar. As for language and other resources, visual systems can be organized
according to a ranked scale; for example, figure (e.g., a woman) and associated
components (her face, hands, and body), the episodes in the image (what she is
doing, and what others are doing), and the entire work (the picture itself) [43].

Images have a unique meaning potential, which is related to the way in which we
perceive world as a series of parts, all of which are related to the whole, following
Gestalt theory [31]. In particular, mathematical images (e.g., graphs, diagrams,
and other visualizations) are powerful resources for reasoning because they permit
mathematical entities, processes, and circumstances to be viewed in relation to each
other, offering a powerful lens to understand patterns and relations. Despite the
potential of the visual image for understanding mathematical relations, this semiotic
resource has traditionally occupied a secondary place compared to the symbolism

which has been privileged as the most significant resource for deriving mathematical results [10]. In what follows, the pivotal contributions of mathematical images are considered, before turning to mathematics as a multimodal construction.

The history of the semiotics of mathematics discourse (e.g., [40]) reveals how the human participants and the context of situation and culture (e.g., everyday activities, machines, human figures) were gradually removed from mathematical images, so the focus became the lines, the curves, and the mathematical objects themselves. Descartes' [13] analytic geometry (see Figure 1), where algebra and geometry became united, was a major turning point for the shift from the everyday world to the abstract world of mathematics so clearly observable in Newton's writings. For the first time, algebraic equations could be viewed using a visual coordinate system and geometrical shapes could be described using algebra, opening up the potential of the visual for viewing patterns which were encoded symbolically and vice versa, as depicted in Figure 1 where the focus shifts back and forth between the diagram and symbolism constantly. Images play an important role in mathematics, because the parts are seen in relation to the entire mathematical construct, opening up new avenues of reasoning. Significantly, mathematical images also relate to our lived-in sensory experience, providing a bridge from perceptual understanding of the world to the abstract semiotic realm of mathematics, which in turn lead to further abstractions, given the close links between the images and the mathematical notation.

Given that the focus of the visual representations is mathematical symbolic entities, processes, and relations, the mathematical image is also a purpose-built semiotic tool for encoding meanings that are unambiguous and precise. As there are various ways of framing mathematical abstractions (e.g., scale, color, perspective, and so forth), the creative tension which Davis [11] mentions is also generated through visual representations: for example, Tufte [47, 48] shows how images are used to visualize information and quantitative data and to provide explanations using visual systems of reasoning. Today, dynamic and three-dimensional visualizations of patterns and relations across space and time offer new opportunities for exploring mathematical patterns and relations. However, another major source of creative tension in mathematics arises from shifts between textual (linguistic and symbolic) and visual forms of semiosis [35, 36, 40–42], as discussed in the following section.

6 Multimodal Semiotics: Language, Images, and Symbolism

As Hawking [29, p. XI] explains, revolutions in the way humans have perceived the world have always gone hand in hand with revolutions in mathematical thought: for example, "Isaac Newton could never have formulated his laws without the analytic geometry of Rene Descartes and Newton's own invention of calculus." Descartes [13] and Newton [38] derive their results in a running text format, constantly moving back and forth between language, geometrical diagrams, and symbolism as the need arises (e.g., see Figure 1). This format is the precursor to modern mathematics,

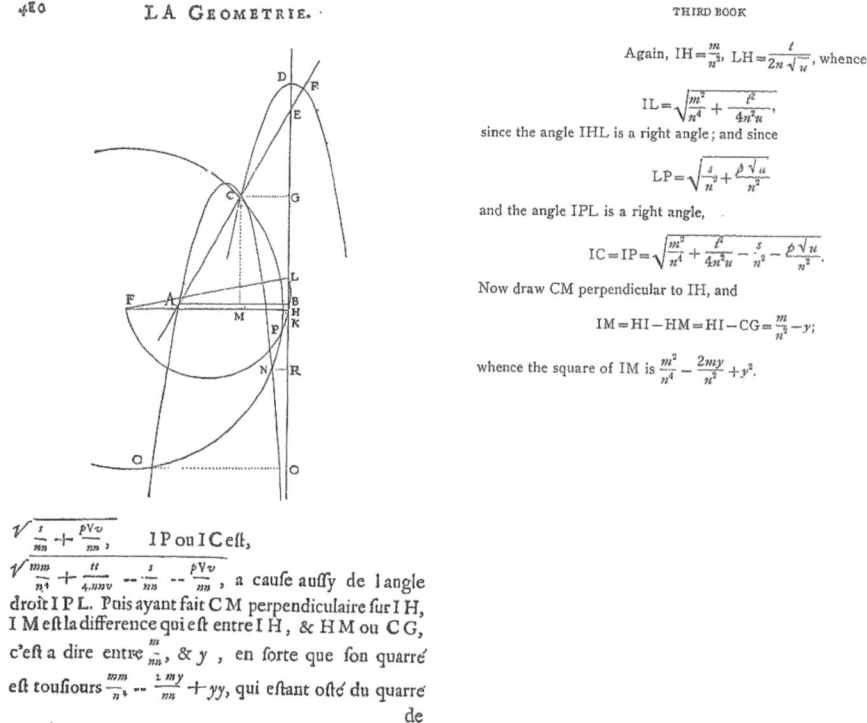

Fig. 1 Descartes [13, pp. 234–235], with Facsimile of First Edition (left)

where symbolic expressions and equations are set apart from the remainder of the text. Regardless of format, the intent is clear—to reason logically in the most effective way possible, drawing upon linguistic, visual, and symbolic resources as required.

In order to investigate how and why such transitions between the three resources take place, the semantic expansions which occur as a result are explored, based on the key concepts of meaning potential and semiotic metaphor. As explained earlier, meaning potential is the ability of a resource to create meaning arising from its architecture; that is, the grammatical systems from which selections are made. Semiotic metaphor is defined as a shift in the functional status of a choice (e.g., entity, process, and entity/process configuration) when that element is resemiotized into another semiotic form; for instance, when linguistic choices are visualized, or visual choices are symbolized (e.g., [41]). Semiotic metaphor also includes the introduction of new metaphorical entities and processes which occur in the transition from one semiotic resource to another. Semiotic metaphor is based on Halliday's [25] concept of grammatical metaphor in language (see above), but in this case, the semantic realignment takes place across different semiotic resources. In examining

In general, consider any two points P and Q with coordinates (x_1, y_1) and (x_2, y_2) respectively. By completing the right-angled triangle PQR, we have the coordinates of R as (x_2, y_1).

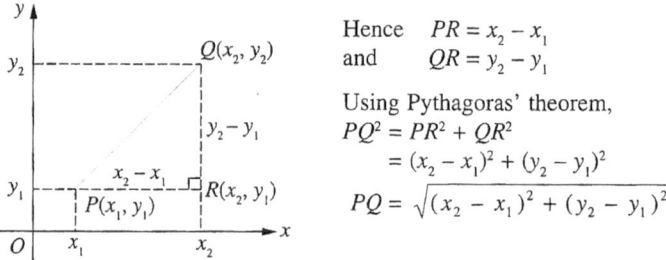

Hence $PR = x_2 - x_1$
and $QR = y_2 - y_1$

Using Pythagoras' theorem,
$$PQ^2 = PR^2 + QR^2$$
$$= (x_2 - x_1)^2 + (y_2 - y_1)^2$$
$$PQ = \sqrt{(x_2 - x_1)^2 + (y_2 - y_1)^2}$$

Fig. 2 Distance in Cartesian Coordinate Plane [46, p. 91]

such metaphorical constructions, it is apparent that inter-semiotic transitions give rise to semantic expansions that extend beyond those possible with language or any single semiotic resource.

For example, the general formula for the distance between any two given points $P(x_1, y_1)$ and $Q(x_2, y_2)$ is given by $PQ = \sqrt{(x_2 - x_1)^2 + (y_2 - y_1)^2}$ in analytic geometry. The semantic expansions of meaning that take place across language, image, and symbolism in the derivation of this formula (e.g., Figure 2) are discussed below.

Initially, the first steps for deriving the distance formula are described linguistically, given the potential of language to organize everyday experience in terms of different process types; in this case, mental processes ("consider") and material actions ("completing") and possession ("have"). These linguistic instructions (with symbolic participants) result in a transition to the mathematical image, where "distance" is resemiotized as the line segment connecting $P(x_1, y_1)$ and $Q(x_2, y_2)$ and the (virtual) right-angled triangle PQR with coordinates of R (x_2, y_1) is drawn. In addition to accessing the meaning potential of the image, where the parts of the right-angled triangle are perceived in relation to the whole, the vertices and sides of the triangle are exactly describable using the xy coordinate system. That is, the problem is reconstrued visually as parts (i.e., vertices, sides) related to the whole (i.e., the right triangle) within a symbolic coordinate system. The length of the line segments PR and QR are resemiotized as configurations of mathematical entities and processes $(x_2 - x_1)$ and $(y_2 - y_1)$. From there, there is a shift to the mathematical symbolism, where the formula $PQ = \sqrt{(x_2 - x_1)^2 + (y_2 - y_1)^2}$ is derived using the Pythagorean theorem.

It is apparent that the shift from language to image to symbolism permits the meaning potentials of the three resources to be accessed, with a move from everyday reality to the abstract semiotic realm of mathematics. Beyond this, the transitions mean that linguistic elements (e.g., "distance") are reconstrued visually (i.e., line segment in visual image). This in turn permits the introduction of a new entity (the right triangle), where the length of the sides and the relations between the length of the sides can be perceived and expressed using mathematical symbolism. The semiotic metaphors which occur include (a) the introduction of new visual entities, i.e., triangle *PQR* and its component parts (vertices and sides); and (b) the lengths *PR* and *QR* as configurations of mathematical entities and processes, i.e., $(x_2 - x_1)$ and $(y_2 - y_1)$; and (c) the relations between the lengths of the sides as a configuration of mathematical entities and processes, i.e., $PQ^2 = (x_2 - x_1)^2 + (y_2 - y_1)^2$. That is, the problem is resolved via the semantic expansions involving the introduction of new visual entities (the triangle and its associated parts) and the reconstrual of the visual elements (the sides of the triangle and their relations) as mathematical symbolic configurations. Indeed, mathematics is the result of shifts across linguistic, visual, and symbolic resources where experience is restructured by reconfiguring semiotic phenomena in new ways and introducing new semiotic phenomena that did not previously exist.

The study of the semiotic transitions between language, image, and symbolism in mathematics also sheds light on the functions and roles of the three resources and their resulting grammatical strategies for encoding meaning. For example, information is packed into the noun group structure in "the general formula for the distance between any two given points $P(x_1, y_1)$ and $Q(x_2, y_2)$." This may be compared to the symbolic formula "$PQ = \sqrt{(x_2 - x_1)^2 + (y_2 - y_1)^2}$" which is a configuration of mathematical entities and processes. That is, scientific language has extended noun group structures, but the symbolism retains a dynamic formulation of happenings in order to derive results, as demonstrated in this example (also Figure 1). In other words, language provides a bridge from everyday experience to the abstract semiotic plane, but the image and symbolism permit thinking to extend beyond linguistic semiotic boundaries, basically to rewrite human experience of the universe.

7 Conclusions

"Can there be knowledge without words, without symbols [and images]?" [12, p. 44]

As Davis et al. [12, p. 44] point out, mathematics has been a human activity for thousands of years, and most cultures developed some form of mathematical activity for investigating quantity, space, and patterns. Hawking [29, p. XI] reflects how mathematics has been responsible for great insights into nature, "such as the realization that the earth is round, that the same force that causes an apple to fall here on earth is also responsible for motion of heavenly bodies, that space is finite and not eternal, that time and space are intertwined and warped by matter and energy, and that the future can only be determined probabilistically." In order to make these advances, mathematics developed specific 'tools of the trade' [12], giving rise to semiotic metaphor as the ultimate form of multimodal metaphor for meaning expansion.

But mathematics is more than a tool and language for understanding the universe. As Hawking [29, pp. XII–XIII] and others point out, mathematics has affected our world view in its own right, in terms of introducing concepts such as infinity, continuous functions, logic as a system of processes, the power and limits of digital computing, and unproven laws and logical consistency of systems. Beyond reformulating the world and introducing new mathematical concepts, the types of metaphorical construal developed in scientific language through grammatical metaphor is found in most discourses of modern life, from politics and economics to social life. Indeed today, our everyday view of the world is largely a metaphorical construct, originating from the need to organize the material universe. Lastly, computers are having a major social impact on how we live and communicate: "From the point of view of social impact, the major mathematical breakthrough since the end of World War II is the digital computer in all its ramifications. This breakthrough, involved a combination of mathematics and electronic technology" [11, p. xxiii].

Despite the widespread impact of mathematics and science on human life today, a division still exists between sciences and the humanities, where scientists generally understand more about the humanities than vice versa. As Davis et al. [12, p. 54] explain, "part of the reason for this lies in the fact that the locus of the humanities is to be found in sound, vision, and common language. The language of science with its substantial sublanguage of mathematics poses a formidable barrier to the humanist." However, the major challenges in our globally connected world of today are social, political, and economic in nature (e.g., poverty, healthcare, social, and cultural unrest). In order to address these issues, it would seem that the sciences, mathematics, and the humanities must necessarily unite to develop new semiotic tools which permit the human realm to be understood. This could be achieved by harnessing the potential of digital technology, the derivative of mathematics and classical science, combined with the insights from humanities and the wealth of knowledge about human life which exists today, as the product of digital technology (e.g., Wikipedia and other structured knowledge systems about human life). Indeed, it is difficult to imagine how big problems in the world today can be resolved, unless the focus on the experiential and logical domain is expanded to include the interpersonal domain, and this necessarily involves bringing together science, mathematics, and the humanities, aided by digital technology. From there, it may be possible to understand human and physical worlds as an integrated system.

As Halliday [19, p. 128] claims, "[i]f the human mind can achieve this remarkable combination of incisive penetration and positive indeterminacy, then we can hardly deny these same properties to human language, since language is the very system by which they are developed, stored and powered." However, if we add multimodal mathematics and science to human language, and explore the potential of new digital semiotic tools, then we may be a step closer to understanding and preserving human life and the earth itself.

References

1. Bateman, J. (2014). *Text and Image: A Critical Introduction to the Visual/Verbal Divide.* London & New York: Routledge.
2. Berger, J. (1972). *Ways of Seeing.* London: BBC and Penguin Books.
3. Bernstein, B. (1999). Vertical and Horizontal Discourse: An Essay. *British Journal of Sociology of Education, 20*(2), 157-173.
4. Bernstein, B. (2000). *Pedagogy, Symbolic Control and Identity: Theory, Research and Critique* (revised ed.). Lanham: Rowan & Littlefield.
5. Boyer, C. B., & Merzbach, U. C. (2011). *A History of Mathematics* (3rd ed.). New Jersey: John Wiley & Sons.
6. Butt, D. (2015). The 'History of Ideas' and Halliday's Natural Science of Meaning. In J. Webster (Ed.), *The Bloomsbury Companion to M. A. K. Halliday* (pp. 17-61). London: Bloomsbury.
7. Cajori, F. (1993). *A History of Mathematical Notations (Two Volumes).* New York: Dover Publications [first published 1928 and 1929].
8. Cobley, P. (Ed.). (2010). *The Routledge Companion to Semiotics.* London & New York: Routledge.
9. Cooke, R. (2005). *The History of Mathematics: A Brief Course.* New Jersey: John Wiley & Sons.
10. Davis, P. J. (1974). Visual Geometry, Computer Graphics and Theorems of Perceived Type. *Proceedings of Symposia in Applied Mathematics, 20,* 113-127.
11. Davis, P. J. (2006). *Mathematics and Commonsense: A Case of Creative Tension.* Wellesley MA: A. K. Peters Ltd.
12. Davis, P. J., Hersh, R., & Marchisotto, E. A. (2012). *The Mathematical Experience, Study Edition.* Boston: Birkhäuser Boston
13. Descartes, R. (1954 [1637]). *The Geometry of Rene Descartes* (D. E. Smith & M. L. Latham, Trans. 1637 ed.). New York: Dover.
14. Eisenstein, E. L. (1979). *The Printing Press as an Agent of Change.* Cambridge: Cambridge University Press.
15. Forceville, C. J., & Urios-Aparisi, E. (Eds.). (2009). *Multimodal Metaphor.* Berlin & New York: Mouton de Gruyter.
16. Golding, W. (1955). *The Inheritors.* London: Faber and Faber.
17. Halliday, M. A. K. (1978). *Language as Social Semiotic: The Social Interpretation of Language and Meaning.* London: Edward Arnold.
18. Halliday, M. A. K. (1985). *Spoken and Written Language.* Geelong, Victoria: Deakin University Press [Republished by Oxford University Press 1989].
19. Halliday, M. A. K. (1993). Language and the Order of Nature. In M. A. K. Halliday & J. R. Martin (Eds.), *Writing Science: Literacy and Discursive Power* (pp. 106-123). London: The Falmer Press.
20. Halliday, M. A. K. (2002 [1971]). Linguistic Function and Literary Style: An Inquiry into the Language of William Golding's The Inheritors. In J. Webster (Ed.), *Linguistic Studies of Text and Discourse: Collected Works of M. A. K. Halliday (Volume 2)* (pp. 88-125). London: Bloomsbury.
21. Halliday, M. A. K. (2003a). Introduction: On the "architecture" of Human Language *On Language and Linguistics.: Volume 3 in the Collected Works of M.A.K. Halliday.* (pp. 1-29). London and New York: Continuum.
22. Halliday, M. A. K. (2003b). On the 'Architecture' of Human Language. In J. Webster (Ed.), *On Language and Linguistics. The Collected Works of M.A.K. Halliday (Volume 3)* (pp. 1-31). London and New York: Continuum.

23. Halliday, M. A. K. (2003 [1985]). Systemic Background. In J. Webster (Ed.), *On Language and Linguistics: The Collected Works of M. A. K. Halliday Volume 3* (pp. 185-198). London and New York: Continuum.

24. Halliday, M. A. K. (2004). Introduction: How Big is a Language? On the Power of Language. In J. Webster (Ed.), *The Language of Science: Collected Works of M. A. K. Halliday (Volume 5)* (pp. xi-xxiv).

25. Halliday, M. A. K. (2006). *The Language of Science: Collected Works of M. A. K. Halliday (Volume 5)*. London & New York: Continuum.

26. Halliday, M. A. K. (2006 [1993]). Some Grammatical Problems in Scientific English. In M. A. K. Halliday & J. R. Martin (Eds.), *Writing science: Literacy and Discursive Power* (pp. 159-180). London: Falmer Press.

27. Halliday, M. A. K. (2008). *Complementarities in Language*. Beijing: Commercial Press.

28. Halliday, M. A. K., & Matthiessen, C. M. I. M. (2014). *Halliday's Introduction to Functional Grammar*. London & New York: Routledge.

29. Hawking, S. (Ed.). (2005). *God Created the Integers: The Mathematical Breakthroughs that Changed History*. London: Running Press.

30. Hodge, R., & Kress, G. (1988). *Social Semiotics*. Cambridge: Polity Press.

31. Hothersall, D. (2004). *History of Psychology*. New York: McGraw-Hill.

32. Jewitt, C. (Ed.). (2014). *The Routledge Handbook of Multimodal Analysis* (2nd ed.). London: Routledge.

33. Kress, G., & van Leeuwen, T. (2006). *Reading Images: The Grammar of Visual Design* (2nd ed.). London: Routledge.

34. Lakoff, G., & Johnson, M. (1980). *Metaphors We Live By*. Chicago: The University of Chicago Press.

35. Lemke, J. L. (1998). Multiplying Meaning: Visual and Verbal Semiotics in Scientific Text. In J. R. Martin & R. Veel (Eds.), *Reading Science: Critical and Functional Perspectives on Discourses of Science* (pp. 87-113). London: Routledge.

36. Lemke, J. L. (2003). Mathematics in the Middle: Measure, Picture, Gesture, Sign, and Word. In M. Anderson, A. Sáenz-Ludlow, S. Zellweger & V. V. Cifarelli (Eds.), *Educational Perspectives on Mathematics as Semiosis: From Thinking to Interpreting to Knowing.* (pp. 215-234). Ottawa: Legas.

37. Lukin, A. (2015). A Linguistics of Style: Halliday on Literature. In J. Webster (Ed.), *The Bloomsbury Companion to M. A. K. Halliday*. London: Bloomsbury.

38. Newton, I. (1729). *The Mathematical Principles of Natural Philosophy* (A. Motte, Trans.). London: Printed for Benjamin Motte.

39. Nöth, W. (1990). *Handbook of Semiotics*. Indianapolis: Indiana University Press.

40. O'Halloran, K. L. (2005). *Mathematical Discourse: Language, Symbolism and Visual Images*. London and New York: Continuum.

41. O'Halloran, K. L. (2008). Inter-Semiotic Expansion of Experiential Meaning: Hierarchical Scales and Metaphor in Mathematics Discourse. In C. Jones & E. Ventola (Eds.), *New Developments in the Study of Ideational Meaning: From Language to Multimodality* (pp. 231-254). London: Equinox.

42. O'Halloran, K. L. (2011). The Semantic Hyperspace: Accumulating Mathematical Knowledge across Semiotic Resources and Modes. In F. Christie & K. Maton (Eds.), *Disciplinarity: Functional Linguistic and Sociological Perspectives* (pp. 217-236). London & New York: Continuum.

43. O'Toole, M. (2011). *The Language of Displayed Art* (2nd ed.). London & New York: Routledge.

44. Stone, W. E. (1972). Abacists Versus Algorists. *Journal of Accounting Research, 10*(2), 345-350.

45. Swetz, F. J. (1987). *Capitalism and Arithmetic*. La Salle, Illinois: Open Court.
46. Teh, K. S., & Looi, C. K. (2001). *New Syllabus Mathematics 3*. Singapore: Shinglee Publishers Pte Ltd.
47. Tufte, E. R. (1997). *Visual Explanations: Images and Quantities, Evidence and Narrative*. Cheshire, Connecticut: Graphics Press.
48. Tufte, E. R. (2001). *The Visual Display of Quantitative Information*. Cheshire, Connecticut: Graphics Press.
49. van Leeuwen, T. (2005). *Introducing Social Semiotics*. London: Routledge.
50. Whorf, B. L. (1956). *Language, Thought and Reality: Selected Writings of Benjamin Lee Whorf*. MA: MIT Press.

Problems in philosophy of mathematics: A view from cognitive science

Steven T. Piantadosi

Abstract I argue that many of the problems in mathematical ontology result from our use of natural language, a communication system that is imprecise. I discuss evidence from cognitive science illustrating the pervasive fuzziness of human conceptual and linguistic systems. When we use these systems to discuss the nature of reality, we often succumb to an illusion of conceptual precision that makes it very hard to realize that our mental tools are too imprecise to make useful progress on may questions that sound deep.

The success of mathematical thinking relies in great part on the creation of a language for expressing precise ideas. This language contains in the simplest cases notions like disjunction, implication, and the concepts of calculus, and in most complex cases the vocabulary of group theory or the entire *Principia Mathematica*. Mathematics must create its own language because natural language is not up to the task—natural language is loaded with ambiguity, vagueness, and conceptual representations that decades of effort in cognitive science have yet to adequately formalize. Even when mathematicians borrow natural language terms, they must refine the meanings in order to leave no uncertainty about what was intended. Mathematical language explicitly distinguishes, for instance, "if" vs. "iff," "or" vs. "xor," and a meaning of "some" that ignores the awkwardness in natural language of using it when "all" is also true.

The primary argument of this paper is that at least a few problems in philosophy—including some in the ontology of mathematics—arise when the precision of natural language is overestimated and we mistake the fuzziness of natural linguistic and conceptual systems for the precision needed for mathematical ones. Attempting to answer natural language questions about the reality of mathematical objects is as futile as trying to decide whether it is really true that "Bill or John played the accordion" when they both did. This question is not about the nature of the universe; it is at best a question about what "or" means to people or how "or" is most naturally used. The problem becomes even worse for the topics

S.T. Piantadosi (✉)
The Computation and Language Lab, University of Rochester, Rochester, NY, USA
e-mail: spiantado@gmail.com

© Springer International Publishing Switzerland 2015   305
E. Davis, P.J. Davis (eds.), *Mathematics, Substance and Surmise*,
DOI 10.1007/978-3-319-21473-3_15

in mathematical ontology—questions like existence, sameness, and truth—which sound reasonable in natural language, but which are imprecise at their core. This critical view follows a rich tradition in philosophy, touching on analytic philosophy, ordinary language philosophy, and Logical Positivism in general, as well as the works of Bertrand Russell, Rudolf Carnap, and Ludwig Wittgenstein in particular.

My article will attempt to provide a cognitive perspective on these issues, arguing that there is now substantial evidence that natural human concepts are imprecise, and in fact imprecision is likely useful to their normal usage and in the context in which we evolved. At the same time, we appear to have poor metacognition about the precision of our concepts. These two facts conspire create what I will call an *illusion of conceptual precision*. The strength of this illusion makes it very easy for us to ask questions that seem answerable—perhaps even deep—but have no substance beyond our own fuzzy mental constructs.

1   Human concepts are fuzzy, probably for a good reason

Natural language provides convenient labels for concepts and 50 years of cognitive science has yet to make substantial progress on formalizing what type of thing concepts might be. One important debate concerns whether concepts are represented with logical rules, stored examples (called exemplars or prototypes), or something else entirely. Proponents of the rule view call on our ability to represent abstract, logical concepts like *even numbers*, which appear to be most naturally formalized in a logic-like language, not unlike that found in mathematics. Such a *Language of Thought* [12] would explain humans' systematic patterns of thought, as well as the structured, compositional form that many of our beliefs take [13]. For instance, we are able to think about our own thoughts ("I doubt I believed yesterday that Sam would eat a mouse.") and construct richly interconnected representations in natural language ("My doctor's sister's husband never forgets to take a pair of glasses with him when he visits a town with at least one movie theater."). It appears difficult to explain our ability to think such thoughts without hypothesizing a structured representation system for mental language, perhaps one like those used in logic [26]. If mental representations were logical structures, that might give hope that mental concepts could be firmly grounded with the same precision and quality of meaning as mathematical systems.

Unfortunately, cognitive systems appear to be much more complex, particularly below the level of sentences. Although a few words are amenable to rule-like definitions, there are probably not many [12]. Philosopher Jerry Fodor even argues that even a term like the transitive verb "paint" cannot be captured with a definition even if it is taken for granted that the noun "paint" is known [14]. It is tempting to think that "X paints Y" if X covers Y with paint. But, as Fodor points out, this won't work because if a paint factory explodes, it can cover something with paint without painting it. It seems that X has to be an agent. But then, an agent could accidentally cover their shoes with paint, in which case it wouldn't be quite right to say "X painted their shoes." There seems to be a necessary notion of intention—"X paints Y" if X is an agent who covers Y with paint and X intends to do so. But then Fodor considers Michaelangelo dipping his brush in paint. He doesn't seem to have painted *the brush* even though he intended to cover it with paint. Fodor remarks, "I don't know where we go from here."

Even for concepts that seem to have natural definitions, the definitions still do not capture much of humans' knowledge. A classic example is that of a "bachelor," which could naturally be defined as an unmarried male. Critics of the definitional view point out that even rule-like concepts appear to have "fuzzy" boundaries. Most readers would agree that although a widower may technically be a bachelor, he certainly isn't a good one. The pope, another unmarried male, is another bad example. My baby nephew isn't really a bachelor yet, either, and—even stranger— it's not clear at what age he might become one. These examples demonstrate that we know a lot more than "unmarried male"—we know about typical and atypical features, as well as different senses, connotations, and analogies. Strikingly, the fuzziness of our concepts applies even to *mathematical* ones: subjects report, for instance that 2 is a "better" even number than 178 [1][1]. Even experts in mathematics would probably agree that $f(x) = x^2$ is a "better" function than the Dirichlet function.

Motivated by these effects, many theories of conceptual representation do not rely on systems of precise logical rules. Instead, categorization and conceptualiza- tion may work based on the similarity to stored examples, as in *prototype theory* (using a single stored example) or *exemplar theory* (using many) (for an overview see [24]). A new example could then be categorized by measuring its similarity to the stored example(s). In this case, we might not have a precisely defined notion of bachelor, but could still decide whether someone was a bachelor by comparing them to previous bachelors we have seen and remembered. In this case, there may be no fact of the matter—only varying degrees of bachelorhood.

Of course, prototype and exemplar theories also have challenges (see [24]). One is in particular with explaining the aspects of thinking which look structured and compositional (e.g., how might prototypes be combined to represent meanings like "short giraffe"?). Another is in explaining how our rich system of concepts interrelate (e.g., are the terms "parent" and "child" defined in terms of each other?). A third concerns how one might explain concepts that really do seem definitional, and whose features (surface similarity) are not diagnostic: for instance, an old woman with gray hair may be very similar to any previously seen grandmothers, yet not be one.

The correct theory of conceptual representation must make peace with all sides of human thinking—its rule-like nature, as well as its fuzzy, approximate, and graded

---

[1]Interestingly, these results were used to argue that typicality effects were poor evidence *against* definitions since graded judgments can appear for these clearly definitional concepts.

nature.[2] A fair generalization from this debate is that many of the concepts we use everyday seem to be imprecise, although the degree and nature of the imprecision varies considerably. Critically, if an imprecise system forms the foundation for cognition, there is little hope for formalizing concepts and determining fundamental truth behind many questions that can only be phrased in natural language. This is not to disparage them—human concepts and language have given rise to remarkable cognitive, cultural, and social phenomena.

The prevalence of conceptual fuzziness is why one could write a dissertation on the definition of a "kill," "treason," or "blame." This graded nature of concepts is reflected in our linguistic constructs, including phrases like "he's a pretty good chef" and "he's mostly trustworthy," and "she's probably your mother." The problem is particularly worrisome for the terminology that forms the foundation of science—we don't really have a precise notion of what makes something a cause[3], a computation, a representation, or a physical entity, despite the centrality of these notions for scientific theories. Difficulties with aligning our intuitive concepts with those that best describe the objective world can also be seen in debates about the definition of races, species, life, and intelligence. Imprecision is why it is unclear when a fetus becomes a human, and why there is debate about whether viruses are alive. Even the biological definition of "male" and "female" is surprising—it is not based on the readily apparent physical differences that our intuitive concepts might use. Instead, the biological definition is based on the size of gametes produced: females in a recently discovered species of insect use a "penis" (apparently another vague concept) to extract sperm from the male [48].

There is no accepted account of why natural language concepts lack precision. It could be that such undefined concepts are actually more useful than precise ones, in the same way that ambiguity permits efficient communication by allowing us to leave out redundant information [34]. Vague terms certainly permit communicative analogies and metaphors (e.g., "All politicians are criminals."). Alternatively, fuzziness might result from the necessities of language learning: the average child will acquire about 10 words per day on average from age 1 to 16. Perhaps there is not enough time or data to acquire specific, detailed meanings. Indeed, experts often come to have more refined word meanings in their domain of expertise. For instance, "osmosis" and "diffusion" are used interchangeably by many, except experts in biology or chemistry. This is consistent with the fact that meanings are always changing, showing that cultural transmission of meanings is noisy and imperfect. Despite the efforts of linguistic prescriptivists—the climate-change deniers of the language world—the word "literally" has come to mean "figuratively" just as "you" has come to mean "thou." Such linguistic change is gradual, inevitable, and self-organized, meaning that children can't be acquiring precise replicas of their parents' conceptual and linguistic systems.

---

[2]Some exciting recent work unifies stochastic conceptual representations with structure, Turing-completeness, and inference [17]

[3]Though see [31] for mathematical work formalizing this notion.

Alternatively, human language and concepts may be imprecise due to abstraction. We are able to have thoughts about people, events, and objects at multiple levels: I can think of my family as a family, a collection of individuals who obey certain relations, a collection of undetached-person-parts (Quine), a curious collection of symbiotic cells (Margulis), as DNA-bound information trying to replicate itself (Dawkins), or as extraordinarily complex interactions of atoms. Our thinking admits such varying levels of abstraction because they are necessary for understanding the causal relations in the world. Hillary Putnam's example is of the inability to put a square peg in a round hole: even though this fact about the world results from "just" the interaction of atoms, understanding of *why* this is true comes about from more abstract—and nonreductionist—notions of geometry [37]. Steven Pinker's example is that of World War I which—though it is "just" a complex trajectory of subatomic particles—is only satisfactorily explainable in terms of high-level concepts like nations and treaties.

Once the leap has been made to allow more abstraction, concepts quite naturally come to lose some of their precision. An abstract concept deals with generalities and not particulars. As a result, it seems very natural for systems of abstract representation to break down in unusual particular instances since the particulars are not tracked in detail. Moreover, the abstraction must be discovered through inference and generalization. This means that abstractions may additionally be fuzzy since they represent parts of reality that are not directly experienced.

2 The illusion of conceptual precision

Even as human cognition is a remarkable force of nature, human *meta*cognition often leaves much to be desired. Metacognition refers to our ability to understand the abilities and limitations of our own cognitive skills. In many domains, we perform poorly on metacognitive tasks, often overestimating our own capacity. 93% of US drivers believe themselves to be above average, for instance [45]. Patricia Cross [7] found that 94% of faculty members rate themselves as above-average teachers, noting that therefore "the outlook for much improvement in teaching seems less than promising." People often believe that they understand even commonplace systems like locks, helicopters, and zippers much better than they do, a phenomenon called the "illusion of explanatory depth" [39]. An "illusion of understanding" in political policies has been linked to political extremism [11].

Such metacognitive limitations likely extend to the core properties of our own concepts. This is why the marginal cases of "bachelor" are hard to think of—much easier to think of are the more typical bachelors. In many cases, our explicit knowledge of concepts appears to be much more like a knowledge of the average example rather than the decision boundary. This observation is a fairly nontrivial one in cognitive science: many classification techniques (e.g., support vector machines) represent only the boundaries between classes. Human cognition probably does not use such discriminative concepts, as we both don't know where the boundaries are very well. This is exactly what would be predicted under a prototype, exemplar, or generative view—we tend to know category centers or typical cases. This view is supported by the *typicality effect* in word processing, where typical examples tend to receive privileged processing, including faster response times, stronger priming, and most critically a higher likelihood of being generated in naming tasks [38]. This property may make sense if most of our uses of concepts are in labeling or referring. Perhaps it is not important to know exactly where the boundaries are when producing labels (words), since by the time you get to the boundaries of bachelors (e.g., the pope) there is often a better term available (e.g., "the pope") that we can use instead.

One unfortunate consequence of this illusion is that we often get lost arguing what a term *really* means, instead of realizing that any definition we pick will be arbitrary in some sense. This can be seen in debates about the definition of vulgarity or profanity, for instance. An inability to formalize our intuitive notions was once celebrated in the Supreme Court: in a famous case on the definition of pornography, Justice Potter Stewart famously declared "I shall not today attempt further to define the kinds of material ... and perhaps I could never succeed in intelligibly doing so. But I know it when I see it, and the motion picture involved in this case is not that." The "I'll know it when I see it" doctrine now has its own Wikipedia page and has been heralded as an exemplar of judicial commonsense and realism—even though it fails completely to solve the problem.

One remarkable failure of conceptual metacognition was the recent public uproar over changing the definition of "planet" in a way that excluded Pluto. The decision was simply one about the *scientific* definition of a term, and has no bearing on the public use of the term.[4] However, the public had very strong views that *of course* Pluto was a planet and that should not change. The problem arose because people believed that they had a notion of planethood which was correct (perhaps even objectively) and therefore there was something wrong with the changed definition. People's cognitive representation of a planet was not precise—I'll know a planet when I see one!—and this fact clashed with their strong beliefs about Pluto. The precise (scientific) definition[5] *happened* to demote Pluto from "planet" to "dwarf planet." The only way you could think the definition was "wrong" would be to beg the question, or to believe—almost certainly erroneously—that your cognitive representation was a better scientific definition than the astronomers'.

Curious conceptual metacognition also appears in children's development. Many humans—children included—are *essentialists*, meaning that we believe there is

---

[4]Under common usage, some people—most recently some viral-internet QVC hosts—believe that the moon is a planet. It can be forgiven since the moon doesn't seem like a "satellite" much anymore.

[5]The International Astronomical Union holds that a planet must (i) orbit the Sun, (ii) be nearly round, and (ii) have "cleared the neighborhood" around its orbit.

some unobservable essence to our psychological categories (see [15]). For instance, recent buyers were willing to pay over 1 million dollars for JFK's golf clubs, even though there is nothing physically different about the clubs that ties them to JFK. Instead, they seem to have some nonphysical essence linking them to JFK simply because he once possessed them. A case more typical to development is the difference between men and women: children may believe there to be an inherent difference, even though they do not know what it is. In development, such essences may be "placeholders" for things children do not yet understand [25], suggesting that our concepts may not even at their core be about properties of objects [15]. A concept can apparently be formed even in the absence of any real notion of what the concept might be.

Our sense of having precise and justifiable concepts when we really do not should be familiar in both mathematics and philosophy, where we often learn with great pains that we don't fully understand a concept. The concept of a *real number* is a nice example, where students will often get some conception of this term from high school. But a full, deep understanding of its properties requires more thought and study—and it's amazing to learn the number of things we *don't* know about real numbers once we first have the concept. Such non-obvious properties include, for instance, knowledge that decimal expansions of real numbers are not unique $(0.9999\ldots = 1)$, knowledge that the cardinality of real numbers is a different infinity than the reals, knowledge that *almost every* real number is uncomputable and incompressible, a randomly sampled real number (from, e.g., a uniform distribution on $[0, 1]$) is certain to have probability 0 of being chosen, and knowledge that any Dedekind-complete ordered field is essentially the same as real numbers. Deep understanding of concepts requires much more than our easily achieved conceptions about the world. But, we should not take for granted that it is possible at all.

3   The relevance of the illusion to philosophy

The toxic mix of fuzzy concepts and illusory precision is relevant to areas of philosophy that state their problems in natural language. This problem has been recognized by philosophers. In his *Tractatus Logico-Philosophicus*, Wittegenstein writes

4.002 Man possesses the capacity of constructing languages, in which every sense can be expressed, without having an idea how and what each word means—just as one speaks without knowing how the single sounds are produced. ...
Language disguises the thought; so that from the external form of the clothes one cannot infer the form of the thought they clothe, because the external form of the clothes is constructed with quite another object than to let the form of the body be recognized. ...
4.003 Most propositions and questions, that have been written about philosophical matters, are not false, but senseless. We cannot, therefore, answer questions of this kind at all, but only state their senselessness. Most questions and propositions of the philosophers result from the fact that we do not understand the logic of our language. ...

And so it is not to be wondered at that the deepest problems are really no problems.
4.0031 All philosophy is "'Critique of language'" ...

Rudolf Carnap's *Pseudoproblems in Philosophy* explains away philosophical problems like realism as simply insufficiently grounded in a scientific epistemology to even have meaning [5]. He writes of such philosophy, "A (pseudo) statement which cannot in principle be supported by an experience, and which therefore does not have any factual content would not express any conceivable state of affairs and therefore would not be a statement, but only a conglomeration of meaningless marks or noises."

Daniel Dennett describes the term "deepity" for phrases which sound profound but are in actuality meaningless [9]. A nice prototype is the sentiment "Love is only a word," which is true on one reading and seems deep on another, until you think about it. Deepak Chopra is a purveyor of household deepities, including "The idea of a world outside of awareness exists in awareness alone," "Existence is not an idea. Only its descriptions are," and "Only that which was never born is deathless." The ability to form such meaningless thoughts that sound profound is one consequence of the illusion of conceptual precision. The cultural celebration of such sentiments makes those inclined towards precision sympathetic with Hunter S. Thompson, whose response to "What is the sound of one hand clapping?" was to slap the questioner in the face [46].

The effect of imperfect meanings in natural language can also be seen in paradoxes of language and logic. One is Bertrand Russell's Barber paradox [40], concerning the barber who cuts a person's hair if they do not cut their own. Does the barber cut his own? In set theory, the paradox is to consider set S that contains all sets that do not contain themselves ($S = \{s \; s.t. \; s \notin s\}$). Does S contain itself? This problem provided a challenge to phrasing set theory in terms of natural language (*naive set theory*) as well as other formal systems. Such logical catastrophes demonstrate that natural language is not precise enough for sound reasoning: it is too easy to construct sentences whose meaning *seems* precise, but which cannot be. Use of underspecified natural language concepts yields other puzzles, including the proof that there is no uninteresting integer. For if there were, there would have to be a smallest uninteresting integer and sure that number would be interesting (since it is the smallest uninteresting integer).[6] There is also the smallest number that cannot be described in less than sixteen words (yet I just described it). Relatedly, there are even sentences—called Escher sentences (see [36])—which give the illusion of having a meaning, when really they don't: "More people have eaten tomatoes than I have." This is a semantic illusion in the same way that a Necker cube is a visual illusion and Shepard tones are an auditory illusion.

---

[6]In 2009, Nathaniel Johnston found that 11630 was the smallest uninteresting number, with interestingness determined by membership in the Online Encyclopedia of Integer Sequences (OEIS) [41]. Johnston used this to create a sequence of the smallest uninteresting numbers, a great candidate sequence for membership in OEIS.

Not only are many of the terms of debate imprecisely grounded, but the illusion creates a social or cognitive pull that makes humans particularly willing to engage in meaningless discussions: we do not easily recognize our own imprecision. Here, I'll consider three examples in detail: the existence, sameness, and underlying nature of mathematical concepts. In each case, I'll argue that because the underlying language is natural language, core questions in these domains have yet to be usefully posed. These questions are as useful to argue about as whether Russell's barber *can* or *cannot* cut his own hair.

3.1   Existence

One of the core questions of mathematical ontology concerns the question of whether mathematical objects like numbers exist. The range of philosophical positions that have been argued for range from the Platonic form of mathematical realism to those who believe that pure mathematics is nothing more than a social construct that bears no relationship to reality.[7]

Adopting a broad cognitive perspective, however, one question is why we should focus this debate on mathematics. There are lots of other cognitive phenomena whose existence is not easy to decide with our fuzzy conceptualization of existence. Examples of this include: mothers, phonemes, objects, algorithms, centers of mass, emotions, colors, beliefs, symbols, reminders, jokes, accordions, control conditions, outrage, charity, etc. In each case—including common and abstract nouns—these terms certainly are used to describe components of reality. Yet it is not clear what sense *if any* these terms might have an objective existence that is independent of our cognitive systems. There are certainly arrangements of physical matter that everyone would call an accordion, but that doesn't make accordions (or objects, or planets) a *thing* in a metaphysical sense. It's not even clear that the physical world should be described as having independent objects or parts in any objective sense.[8]

If our representation of "exists" can't decide existence for these ordinary nouns, what is it good for? It could be that the meaning of "exists" is set to be useful for ordinary human communication about present physical objects, not for deciding metaphysical questions. The following list shows typical instances of "exists" used in the spoken language section of the American National Corpus [22]:

- … But uh, they lived in a little country town outside of Augusta called Evans, and Evans still **exists**, and that's where a lot of my relatives still, still are too …
- … and all the auxiliary personnel the school bureaucracy that **exists** mainly to perpetuate itself…

---

[7]Even more extreme, some have argued that pure mathematics is inherently sexist in its form and assumptions [23]. For discussion, see [42].

[8]This "insight" is behind a lot of new-age mumbo jumbo.

- ...the technology **exists** to check it if they want...
- ... do you know if that **exists** today in a cross country kind of collective...
- ... I personally feel like as long as the Soviet Union **exists** it's going to be a threat to the United States and...
- ... died down on that I thought it was interesting that recently here the Warsaw Pact no longer **exists** as a military force but it's merely an economic now ...
- ... yeah it's the best one that **exists** I guess so ...
- ... uh the opportunity **exists** for someone to get killed ...
- ... um it's kind of amazing the disparity that **exists** between ...
- ... but I'm I'm about uh thirty miles west of there so you have uh actually green trees and such that you don't notice that other part of New Jersey **exists** actually ...

As these examples illustrate, "exists" is typically used to talk about objects in concrete terms relevant to some shared assumptions by all speakers and listeners. In many cases, existence is not even asserted but *assumed* (presupposed) ("...it's kind of amazing the disparity that **exists** between..."). It is no wonder that whatever our conception of "exists" is, it is ill-suited to metaphysical discussions. The concept was acquired through usage like these—or more likely, even more simplified versions that occur in speech to children—without any sense of existence as is debated by philosophers.

Clearly what is needed to debate the existence of, say, natural numbers is a refined notion of (metaphysical) existence, one that is not confused by our ordinary commonsense notions. However, it may also be the case that our commonsense "exists" cannot be refined enough—perhaps the metaphysical "exists" is too far from our intuitive notion for us to analyze such questions.

3.2 Sameness

Other puzzles in mathematical ontology concern questions of sameness. For instance, is $f(x) = x + 4$ and $g(x) = \frac{x+4}{1}$ the same, or different? When can—or should—two mathematical objects be considered to be the same?

The problem here is another version of the problem encountered with existence: such questions ignore the fact that sameness is a fuzzy cognitive concept. In fact, one classic puzzle in philosophy deals with our notions of sameness for non-mathematical concepts—the problem of Theseus' ship. As the story goes, the ship has its planks removed one at a time and replaced. At what point does it stop being the same ship? A variant is the story of George Washington's axe which had its handle replaced twice and its head replaced once. The ship puzzle is actually one of practical import for the USS Constitution, which is the oldest naval vessel still working. It was commissioned by George Washington in the 1790s, but most of its parts have been replaced. At what point will it stop being the same ship?

Part of what makes sameness interesting semantically is that our conception of it is flexible. We can discuss whether or not all action movies have "the same plot" or "all my friends have the same haircut" while ignoring the obvious trivial senses in which those statements are false. This debate even occurs somewhat in cognitive science with the debate about whether learners could ever get something fundamentally "new"—is it the case that everything learners create must be a combination of what they've had before (nothing is new, everything is the same) [12], or can learners create something fundamentally unlike their early representations [4].[9]

One common response to Theseus' ship is to consider the semantics of sameness or identity ("is"): indeed, other languages distinguish notions of sameness. German, for instance, has separate words for exact identity ("das selbe") and similar but different copies ("das gleiche"). Programming languages have similar constructs (e.g., Python's "is" versus "equals"). Scheme includes at least three equivalence operators (*eq?*, *eqv?*, and *equal?*) for testing different senses of equality, in addition to numerical equality (=). For the mathematical examples f and g above, we might consider variants of sameness including sameness of intension versus extension. Given that the semantics of English are ambiguous and we could in principle distinguish the relevant meanings, why think that there's any objective truth to the matter about sameness for Theseus or mathematical objects? In all cases, the puzzle arises from using a word whose imprecision is easy to miss.

3.3   The nature of mathematics

Some of the most basic questions about mathematics concern what mathematics is and what mathematical objects are. At the very least, mathematical terms are good descriptions of neural representations in mathematicians' heads. When a mathematician says that they discovered a proof by noticing a 1-1 correspondence between integers and rational numbers, that statement surely describes some cognitive process that relates mental representations. Work in cognitive science has sought to discover the cognitive origins of many ordinary mathematical ideas, like numbers [4] and geometry [8, 43]. Work in neuroscience has identified neural signatures of approximate number (e.g., estimation) [27–30], its hallmarks in other species [2, 3, 10], and derived some core characteristics of human numerical estimation as an optimal solution to a well-formalized problem of representation [32, 35, 44]. In all of these senses, mathematical concepts are good scientific descriptions of certain types of cognitive systems, a view close to psychologism in the philosophy of mathematics.

---

[9]One way to make progress in the debate is to formalize precisely what might be meant by learning, novelty, and sameness, showing, for instance, how learners could create particular representations in learning that are not the same as what they start with, even though the capacity to do so must be "built in" from the start (for work along these lines in numerical cognition, see [33]).

But of course, much of the debate about mathematics is whether there is *more* than what is in mathematicians heads. Psychologism was notably argued against by Edmund Husserl [21], who emphasized the unconditional, necessary (a priori), and non-empirical nature of logical truths, in contrast to the conditional, contingent, and empirical nature of psychological truths (see also [18]). Indeed, there is a sense in which mathematical ideas do seem fundamentally non-psychological, even though they are realized primarily in computational systems like brains and computers.

Perhaps the mystery of what mathematical concepts essentially are reduces yet again to imprecision—in this case, imprecision about the concept of mathematics itself. For many, mathematics means arithmetic (math majors are constantly asked to calculate tips on restaurant bills), even though some of the best mathematicians I know were terrible at adding and subtracting. It takes experience and broad training to realize the extent of mathematical thinking—that it is not even about numbers, for instance, but about concepts and their relationships. These concepts are sometimes numbers, but only in the beginning. Advanced mathematics moves on to study the properties and relationships for creatures much more abstract than integers, including functions, spaces, operators, measures, functionals, etc. One of my favorite examples of the breadth of mathematical thinking is Douglas Hofstadter's *MU* system [20], a Post canonical system with four rewrite rules: (i) a *U* can be added to any string ending in *I*, (ii) any string after an *M* can be doubled, (iii) Any *III* can be changed to *U*, and (iv) any *UU* can be removed. Hofstadter's question is: can *MI* be changed to *MU* using these rules? The solution—a simple feat of mathematical thinking—is to notice the invariance that (ii) can double the number of *I*s, and (iii) can reduce that number by three. Therefore, the rules leave the property of whether the number of *I*s is divisible by 3 unchanged, so they cannot change a string *MI* with one *I* (not divisible by 3) to one with zero *I*s (divisible by 3), *MU*.

Given the range of activities that are described as mathematics, it is not so surprising that there are a range of ideas about what mathematics is at its core. Indeed, many of the positions on the nature of mathematics seem quite reasonable. There is a sense in which mathematical concepts exist independent of minds (Platonism), a sense in which mathematical truths are empirically discovered (Empiricism), a sense in which mathematics is "just" about formal systems (Formalism), a sense in which it is about our own concepts (Psychologism), a sense in which it is a useful lie or approximation (Fictionalism), and a sense in which they are socially—or cognitively—constructed in that we could have chosen other formal systems to work with. Proponents of any view would likely argue that their version of mathematics is fundamentally incompatible with others, and that might be true.

It needn't be the case that only one of these views is correct. Like the terms discussed throughout this paper, the cognitive representation of mathematics and mathematical objects might themselves be imprecise. What is considered to be a core feature of mathematics might depend largely on the context of the comparison. When compared to physics, chemistry, or biology, a defining feature of mathematical thinking might be its focus on a-priori truth. When mathematics is compared to systems of violable rules like music, the relevant feature of mathematics might be on its stricter adherence to formalism. When compared to natural sciences, the key feature of mathematics might be that it takes place in the minds of mathematicians. The psychological underpinnings that support basic mathematical notions like integers, points, and lines might have resulted historically from our particular cognitive systems; at the same time, the insights required for mathematical progress might be most similar to art.

Such a fluidity with essential features is well known in psychology. Amos Tversky and Itamar Gati argue for a *diagnosticity hypothesis* [47] where people may choose features to compute similarity over that are most diagnostic in clustering a set of objects. In their simple example, England and Israel are judged to be more similar when presented in the context of Syria and Iran than when presented in the context of France and Iran. The intuition behind this is that people may judge similarity based on a diagnostic feature like religion (Muslim-vs-Not) when Syria and Iran are present, and a different feature (European-vs-Not) when France is in the mix. Cognitive similarity also obeys other kinds of interesting, nontrivial properties, including being asymmetrical: North Korea is more similar to China than China is to North Korea. These studies show that people's conception of the relevant features to a concept depend considerably on context, so it may be a waste of time to argue about what that feature "really is." In this light, it is not so surprising that philosophers come to different conclusions about the core properties of mathematics and mathematical objects. The mistake is in thinking that any one of them is right.

4 Conclusion

The perspective I have argued for straddles a line between major mathematical positions and does not commit to strong statements about mathematical ontology, realism, or language. The reason for this is that strong positions are not justified. Strong views are only sensible when a precise question has been asked. It is useful to consider and evaluate the strong positions for whether $2 + 2 = 4$ or Goldbach's conjecture is true. It is not useful to debate the strong positions for whether a baby can be a bachelor. Natural linguistic and conceptual systems are simply not up to the task: such questions are clearly questions of definitions and we pick definitions that are useful. Even worse in cases like bachelor, we may not pick them at all, instead relying on fuzzier or stochastic notions that are so far hard for anything but human-raised-humans to acquire.

Natural language terms like "planet," "bachelor," and "mathematics" can of course be interrogated as part of cognitive science. There is a superbly interesting question about how people might represent words and what words might mean for the cognitive system. But, natural language terms cannot be interrogated usefully as part of metaphysics. The reason is that the meanings of natural language are only about the forces that shaped our cognitive science. We evolved moving far from the speed of light, on a timescale of decades, and a distance scale of a few meters. Our language and cognitive systems often intuitively capture phenomena on this level of the universe, but are hopelessly inadequate outside of this domain. For instance, we have a difficult time conceptualizing the very small or very large, or relativistic effects like time dilation that are seen closer to the speed of light. And even for things inside the domain of our evolutionary heritage, our cognitive and linguistic systems did not grow to *accurately* represent the world, but rather to *usefully* do so. Survival utility explains why we have percepts of colors in a multidimensional space (rather than a linear wavelength, like the physical space), why we like BBQ and sex, and why we feel pain and joy even though there is no pain and joy outside of our own brains. Our natural concepts were not made for truth, and cannot be trusted for truth. This insight lies behind the success of the scientific revolution.

In large part, the illusion may explain why the sciences have made remarkable progress on understanding the world but philosophical problems linger for centuries. Could it be the case that determining whether natural numbers "exist" is a harder problem that of determining the origin of species or the descent of man? It seems, instead, that such problems are simply ill-posed or incoherent at their core, and this prevents useful progress.[10] The problem may in part be that natural language does not have the tools to tell us that its questions are impossible to answer. This contrasts with mathematics, where a nice example is provided by proofs showing that Cantor's *Continuum Hypothesis* is independent of the axioms of Zermelo-Fraenkel set theory [6, 16]. It is remarkable that mathematicians were able to resolve this deep question by showing that the assumptions did not determine an answer. Similarly strong answers cannot even be aspired to for philosophical questions that are stated without axioms, like those under consideration in mathematical ontology and areas of philosophy more generally. It is not easy to see what method might tell us that our questions are unanswerable, other than stepping back and realizing that the questions are (critically) asked in natural language, and natural language has misled us. Progress may only be made by realizing that the hypothesis under consideration is too incoherent to even be evaluated—or, in the words of Wolfgang Pauli, "not even wrong."

It may be temping to try to resolve questions of mathematical ontology or philosophy more generally by refining natural language terms. This has been the program of several approaches to philosophy of mathematics and philosophy of science. However, it is hard to see how dedication to terminological precision could ultimately end up with anything other than a formal mathematical system. If we provided notions of existence or sameness that were grounded in a logical system, it feels as though we would be no closer to discovering the truth about the fundamental questions we thought we asked. Analogously, resolving the question

---

[10]Philosophical problems do sometimes get solved, although often by the natural sciences. A recent example from cognitive science is work on Molyneux's Question of whether a person who had never seen could, upon activation of their vision, visually recognize objects they had only touched. Is there an innate amodal representation of shape/space? Studies of curable blindness have provided a negative answer [19].

of whether prime numbers exist in the context of an axiomatization of number theory is not going to tell us whether prime numbers exist according to our intuitive conceptualization of "real" (metaphysical) existence. If the psychological notions of existence are inherently imprecise, it is likely that there is no clearly definable metaphysical question lurking behind our psychological intuitions, just as there is no such question for "or" and "bachelor." Only psychological questions can be asked about these systems, not questions about the nature of reality. In this case, the appearance of a profound and usefully debatable question would only be an illusion.

Acknowledgements I am grateful to Celeste Kidd, Ernest Davis, Julian Jara-Ettinger, Mark Frank, and Lance Rips for providing extremely helpful comments or discussion relevant to this work.

References

1. Sharon Lee Armstrong, Lila R Gleitman, and Henry Gleitman, *What some concepts might not be*, Cognition **13** (1983), no. 3, 263–308.
2. E.M. Brannon and H.S. Terrace, *Representation of the numerosities 1–9 by rhesus macaques*, Journal of Experimental Psychology: Animal Behavior Processes **26** (2000), no. 1, 31.
3. J.F. Cantlon and E.M. Brannon, *Basic math in monkeys and college students*, PLOS Biology **5** (2007), no. 12, e328.
4. Susan Carey, *The Origin of Concepts*, Oxford University Press, Oxford, (2009).
5. Rudolf Carnap, *Pseudoproblems in Philosophy* (1928).
6. Paul J Cohen, *The independence of the continuum hypothesis*, Proceedings of the National Academy of Sciences of the United States of America **50** (1963), no. 6, 1143.
7. K Patricia Cross, *Not can, but will college teaching be improved?* New Directions for Higher Education **1977** (1977), no. 17, 1–15.
8. Stanislas Dehaene, Véronique Izard, Pierre Pica, and Elizabeth Spelke, *Core knowledge of geometry in an Amazonian indigene group*, Science **311** (2006), no. 5759, 381–384.
9. Daniel C Dennett, *Intuition pumps and other tools for thinking*, WW Norton & Company, 2013.
10. J. Emmerton, *Birds' judgments of number and quantity*, Avian visual cognition (2001).
11. Philip M Fernbach, Todd Rogers, Craig R Fox, and Steven A Sloman, *Political extremism is supported by an illusion of understanding*, Psychological science (2013).
12. J.A. Fodor, *The language of thought*, Harvard University Press, Cambridge, MA, 1975.
13. J.A. Fodor and Z.W. Pylyshyn, *Connectionism and cognitive architecture: a critical analysis*, Connections and symbols A Cognition Special Issue, S. Pinker and J. Mehler (eds.) (1988), 3–71.
14. Jerry Fodor, *The present status of the innateness controversy* (1981).
15. Susan A Gelman, *Psychological essentialism in children*, Trends in Cognitive Sciences **8** (2004), no. 9, 404–409.
16. K Gödel, *The consistency of the continuum hypothesis*, Princeton University Press, 1940.
17. Noah D Goodman and Daniel Lassiter, *Probabilistic semantics and pragmatics: Uncertainty in language and thought*, Handbook of Contemporary Semantic Theory, Wiley-Blackwell (2014).
18. Robert Hanna, *Husserl's arguments against logical psychologism*, Prolegomena **17** (2008), 61.
19. Richard Held, Yuri Ostrovsky, Beatrice de Gelder, Tapan Gandhi, Suma Ganesh, Umang Mathur, and Pawan Sinha, *The newly sighted fail to match seen with felt*, Nature neuroscience **14** (2011), no. 5, 551–553.
20. Douglas Hofstadter, *Gödel, Escher, Bach: An eternal golden braid*, Basic Books, New York, 1979.

21. Edmund Husserl, *Logical investigations vol. 1.*, Routledge & Kegan Paul, 1900.
22. Nancy Ide and Catherine Macleod, *The American National Corpus: A standardized resource of American English*, Proceedings of corpus linguistics (2001).
23. Luce Irigaray, *Is the subject of science sexed?* Cultural Critique **1** (1985), 73–88.
24. Eric Margolis and Stephen Laurence, *Concepts: core readings*, The MIT Press, (1999).
25. Douglas L Medin and Andrew Ortony, *Psychological essentialism*, Similarity and analogical reasoning (1989), 179–195.
26. R. Montague, *English as a formal language*, In: Bruno Visentini (ed.): Linguaggi nella societá e nella tecnica. Mailand (1970), 189–223.
27. A. Nieder and S. Dehaene, *Representation of number in the brain*, Annual Review of Neuroscience **32** (2009), 185–208.
28. A. Nieder, D.J. Freedman, and E.K. Miller, *Representation of the quantity of visual items in the primate prefrontal cortex*, Science **297** (2002), no. 5587, 1708–1711.
29. A. Nieder and K. Merten, *A labeled-line code for small and large numerosities in the monkey prefrontal cortex*, The Journal of Neuroscience **27** (2007), no. 22, 5986–5993.
30. A. Nieder and E.K. Miller, *Analog numerical representations in rhesus monkeys: Evidence for parallel processing*, Journal of Cognitive Neuroscience **16** (2004), no. 5, 889–901.
31. Judea Pearl, *Causality: models, reasoning and inference*, Vol. 29, Cambridge Univ Press, (2000).
32. Steven T. Piantadosi, *A rational analysis of the approximate number system*, Psychonomic Bulletin and Review. (in press).
33. Steven T. Piantadosi, J.B. Tenenbaum, and N.D Goodman, *Bootstrapping in a language of thought: a formal model of numerical concept learning*, Cognition **123** (2012), 199–217.
34. Steven T. Piantadosi, H. Tily, and E. Gibson, *The communicative function of ambiguity in language*, Cognition **122** (2011), 280–291.
35. RD Portugal and BF Svaiter, *Weber-Fechner Law and the Optimality of the Logarithmic Scale*, Minds and Machines **21** (2011), no. 1, 73–81.
36. Geoffrey Pullum, *Plausible Angloid gibberishs*, Language Log 2004. http://itre.cis.upenn.edu/~myl/languagelog/archives/000860.html
37. Hilary Putnam, *Reductionism and the nature of psychology*, Cognition **2** (1973), no. 1, 131–146.
38. Eleanor Rosch, *Cognitive representations of semantic categories*, Journal of Experimental Psychology: General **104** (1975), no. 3, 192.
39. Leonid Rozenblit and Frank Keil, *The misunderstood limits of folk science: An illusion of explanatory depth*, Cognitive Science **26** (2002), no. 5, 521–562.
40. Bertrand Russell, *The philosophy of logical atomism*, The Monist (1919), 345–380.
41. Neil James Alexander Sloane et al., *On-line encyclopedia of integer sequences*, AT&T Labs, 1999.
42. Alan Sokal and Jean Bricmont, *Fashionable nonsense: Postmodern intellectuals' abuse of science*, Macmillan, 1999.
43. Elizabeth Spelke, Sang Ah Lee, and Véronique Izard, *Beyond core knowledge: Natural geometry*, Cognitive Science **34** (2010), no. 5, 863–884.
44. J.Z. Sun and V.K. Goyal, *Scalar quantization for relative error*, Data compression conference (2011), pp. 293–302.
45. Ola Svenson, *Are we all less risky and more skillful than our fellow drivers?* Acta Psychologica **47** (1981), no. 2, 143–148.
46. Hunter S Thompson, *Kingdom of fear: loathsome secrets of a star-crossed child in the final days of the American century*, Simon and Schuster, (2003).
47. Amos Tversky and Itamar Gati, *Studies of similarity*, Cognition and categorization **1** (1978), no. 1978, 79–98.
48. Kazunori Yoshizawa, Rodrigo L Ferreira, Yoshitaka Kamimura, and Charles Lienhard, *Female penis, male vagina, and their correlated evolution in a cave insect*, Current Biology **24** (2014), no. 9, 1006–1010.

Beliefs about the nature of numbers

Lance J. Rips

Abstract Nearly all psychologists think that cardinality is the basis of number knowledge. When they test infants' sensitivity to number, they look for evidence that the infants grasp the cardinality of groups of physical objects. And when they test older children's understanding of the meaning of number words, they look for evidence that the children can, for example, "Give [the experimenter] three pencils" or can "Point to the picture of four balloons." But when people think about the positive integers, do they single them out by means of the numbers' cardinality, by means of the ordinal relations that hold among them, or in some other way? This chapter reviews recent research in cognitive psychology that compares people's judgments about the integers' cardinal and ordinal properties. It also presents new experimental evidence suggesting that, at least for adults, the integers' cardinality is less central than their number-theoretic and arithmetic properties.

1 The Structural Perspective and the Cardinal Perspective on Numbers

Numbers don't lend themselves to psychologists' usual way of explaining how we know about things. The usual explanation is perception. We gain knowledge of many physical objects—such as squirrels and bedroom slippers—by seeing them, and we gain knowledge of many other things by reasoning from perceptual evidence. But numbers and other mathematical objects leave no perceptible traces. Although we might consider mathematical objects as abstractions from things that we can perceive, this explanation faces difficult problems. How could such a psychological process be sufficiently powerful to give us knowledge of *all* the natural numbers (not to mention the numbers from other systems)? We can't possibly abstract them one-by-one. Not all cognitive scientists have given up the abstraction story (see, e.g., [25]), but it's safe to say that it is no longer the dominant view.

L.J. Rips (✉)
Psychology Department, Northwestern University, 2029 Sheridan Road, Swift 314, Evanston, IL 60208, USA
e-mail: rips@northwestern.edu

© Springer International Publishing Switzerland 2015   321
E. Davis, P.J. Davis (eds.), *Mathematics, Substance and Surmise*,
DOI 10.1007/978-3-319-21473-3_16

Most psychologists who study the development of number knowledge no longer think that we grasp numbers like ninety-five by generalizing from encounters with groups of ninety-five squirrels, ninety-five slippers, and other groups of that size. But these groupings still play a crucial role in current theories. According to these theories, children's ability to count objects is the pivotal step in their acquisition of true number concepts. Learning to count groups of objects—for example, saying "one, two, three, four, five: five squirrels," while pointing to the squirrels one by one—is supposed to transform children's primitive sense of quantities into adult-like concepts of the small positive integers [3]. Moreover, developmental psychologists' standard method for assessing children's knowledge of the meaning of number words like "five" is to ask the children to "Give me five balloons [or other small objects]" from a pile of many, or to ask the children to "Point to the card with five balloons" in the presence of one card with a picture of five balloons and a second card with a picture of four or six [46].

This emphasis on counting (and on picking out the right number of objects) is due to psychologists' belief that the fundamental meanings of number words are cardinalities, in line with earlier theories by Frege [11] and Russell [41]. The meaning of "five," for example, is the set of all sets of five objects (or some similar construction). What children learn when they learn how to count objects is a rule for computing the cardinality of collections for the number words they know. This rule—Gelman and Gallistel's "Cardinal Principle" [13]—is that the meaning of the final word in the count sequence is the cardinality of the collection. So the meaning of "five" is the set size you get when "five" is the final term in a correct application of the counting procedure. Asking a child to give you five balloons from a larger pile is a test of whether the child can use this counting procedure to arrive at the right total.

However, we should consider other possible paths to knowledge of number. Many contemporary philosophers argue that the meaning of a number is given by the position of the number in an appropriate number system [28, 30, 43]. Stewart Shapiro claims, for example, that "there is no more to being the natural number 2 than being the successor of the successor of 0, the predecessor of 3, the first prime, and so on . . . " [43, pp. 5–6], and Michael Resnik puts the point this way:

> The underlying philosophical idea here is that in mathematics the primary subject-matter is not the individual mathematical objects but rather the structures in which they are arranged. The objects of mathematics, that is, the entities which our mathematical constants and quantifiers denote, are themselves atoms, structureless points, or positions in structures. And as such they have no identity or distinguishing features outside a structure. [30, p. 201]

This structural perspective suggests that people's understanding of particular numbers depends on their knowledge of the relations that govern these numbers. In the case of the natural numbers, the key relations would include the facts that every number has a unique immediate successor, that every number except the first has a unique immediate predecessor, and so on. From this point of view, children wouldn't be able to grasp five merely by connecting the word "five" to the set of all five-membered sets, since this connection wouldn't succeed in establishing the critical relations that five bears to the other naturals (e.g., being the successor of four, the successor of the successor of three, and so on).

The distinction between the cardinal perspective and the structural perspective doesn't mean that no connection exists between them. *Frege's Theorem* establishes that the Dedekind-Peano axioms that define the structure of the natural numbers (no number precedes zero, every number has a unique immediate successor, etc.) are derivable in second-order logic from definitions of zero, the ancestral relation, and natural number itself, together with a central fact about cardinality called *Hume's Principle*. This is the idea that the number of things of one kind is the same as the number of things of another kind if and only if there is a one-one relation between these things. For example, the number of squirrels in your backyard is equal to the number of bedroom slippers in your closet if and only if there is a one-one relation between the squirrels and the slippers. (See [17, 47] for expositions of the proof of Frege's Theorem.) You could take this result to mean that cardinality (in the form of Hume's Principle) provides the basis for the structure of the naturals (in the form of the Dedekind-Peano axioms). However, Frege's Theorem by itself does not settle the issue of whether the cardinal perspective or the structural perspective is conceptually prior in people's understanding of the naturals. A version of Hume's Principle is provable from the Dedekind-Peano axioms and the same definitions [16]. So the formal results seem to give us no reason to favor the cardinal perspective over the structural perspective as the conceptually fundamental one (see [23] for a discussion of this issue).

The structural perspective has recently been defended by Øystein Linnebo [21] as a thesis about "our actual arithmetic practice." Linnebo points out that the cardinal perspective seems to predict incorrectly that zero should be easy for children to grasp. Even very young children understand expressions like "allgone cookies" or "no more cookies"; so they understand the cardinality associated with zero cookies [14]. But they take until the end of preschool to understand zero as a number, alongside other integers [45]. Why the age gap if zero is a cardinality [38]? Similarly, older grade-school children understand that the number of natural numbers is infinite [15], but they probably don't gain the concept of the number $\aleph_0$ for this cardinality until much later (if they ever do).

Linnebo has a second argument in favor of the structural perspective, one that's closer to the theme of the present chapter. He maintains [21, p. 228] that reference to a number in terms of its cardinality—for example, to five as the number of squirrels in the yard—doesn't seem "particularly direct or explicit":

> Rather, the only perfectly direct and explicit way of specifying a number seems to be by means of some standard numeral in a system of numerals with which we are familiar. Since the numerals are classified in accordance with their ordinal properties, this suggests that the ordinal conception of the natural numbers is more basic than the cardinal one.

As Linnebo notes, the appeal to directness in thinking about numbers relies on intuition, but he conjectures that results from cognitive experiments might back the claim for the immediacy of the ordinal conception (which we have been calling the "structural perspective").

This chapter asks whether any psychological evidence favors the structural perspective or the cardinal perspective.[1] We can begin by looking at theories and data on how children learn the meaning of the first few positive integers to see whether the evidence supports developmentalists' emphasis on cardinality. I then turn to recent experiments that have compared adults' judgments of cardinality to their judgments of order for further insight on whether our understanding of numbers depends more tightly on one or the other of these two types of information. Finally, the chapter describes some new studies that directly probe the properties that people take as central to number knowledge.

2 The Origins of Number Knowledge

As an example of the role that the cardinal perspective plays in theories of number knowledge, let's consider Susan Carey's influential and detailed proposal about how children learn the meanings of their first few number words [3]. Figure 1 provides a summary of the steps in this process, which Carey calls *Quinian bootstrapping*. In trying to understand the children's progress through this learning regime, let's start by figuring out what the process is supposed to achieve.

At two or three years old, kids are able to recite the numerals in order from "one" to some number like "ten" or "twenty," but they don't yet know how to produce numerals for arbitrary integers in the way you do. They have just a short, finite list, for example, "one, two, three, four, five, six, seven, eight, nine, ten." At this age, they don't understand the cardinal meanings of the words on this list; so if you ask them to give you two balloons from a pile, they can't do it. Then, over an extended period of time—as long as a year or so—they first work out the meaning for the word "one," then for the word "two," then "three," and sometimes "four," again in the sense of being able to give you one, two, three, or four objects in response to a command. At that point, something clicks, and suddenly they're able to give you five things, six things, and so on, up to ten things (or to whatever the last numeral is on their count list). The Quinian bootstrapping theory is supposed to explain this last step, when things finally click: It extends kids' ability to enumerate objects in response to verbal requests from three or four to ten. Post-bootstrap, kids still don't know many of the important properties of the positive integers, but at least they can count out ten things, more or less correctly. In other words, what they've learned is how to count out objects to determine the right cardinality.

---

[1]This issue might be put by asking whether people think of the first few naturals as cardinal or ordinal *numbers*. However, "cardinal number" and "ordinal number" have special meanings in set theory, and these meanings don't provide the intended contrast. In their usual development (e.g., [8]), the ordinals and cardinals do not differ in the finite range, with which we will be concerned in this chapter. (Both ordinal and cardinal numbers include transfinite numbers—for which they *do* differ—and so extend beyond the natural numbers.)

Step 1 (pre-linguistic representations):

Representation a:
Representations of individual objects: object$_1$, object$_2$, object$_3$, ...

Representation b:
Representations of sets: {object$_1$}, {object$_1$, object$_2$,...}

Step 2 (initial language learning):

Representation c:
Count List: <"one," "two," "three," ... "ten">

Representation d:
Singular/plural: "a" (singular)   "some" (plural)

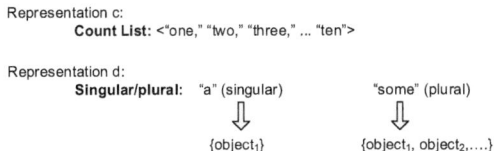

Step 3 (one-knower stage):

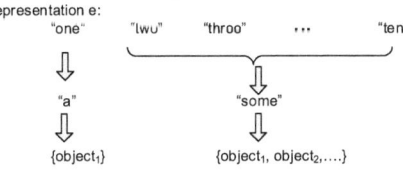

Step 4 (two-knower stage):

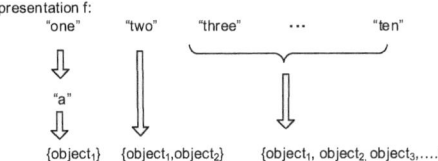

Steps 5 and 6 (three and four-knower stages): similar to the representation in Step 4

Step 7 (pre-bootstrap stage):

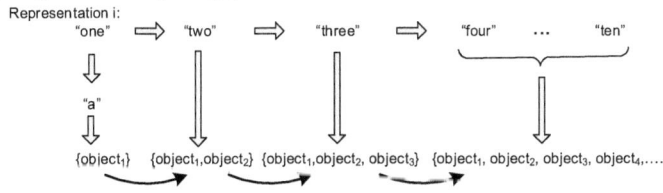

Step 8 (post bootstrap, Cardinal-Principle-knower stage):

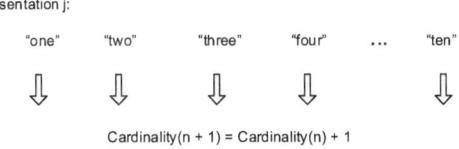

Cardinality(n + 1) = Cardinality(n) + 1

Fig. 1 Steps in the acquisition of the meaning of number words for the first few positive integers, according to Carey's [3] Quinian bootstrapping theory (figure adapted from [32]).

What are the steps that children are supposed to go through in graduating from their pre-bootstrap state of knowledge of number to their post-bootstrap knowledge? Here's a quick tour:

Step 1: At the beginning of this process, children have two relevant innate mental representations. Representations of type (a) in Figure 1 are representations of individual objects—for example, representations of each of three balloons. Representations of type (b) are representations of sets. I'll use $object_i$ to denote the mental representation of a single individual i, $\{object_1, object_2\}$ for the representation of a set containing exactly two individuals, and $\{object_1, object_2, \ldots\}$ to indicate the representation of a set containing more than one individual, but whose total size is unknown. Children at this stage also have representations for approximate cardinality, but they play no role in Carey's bootstrapping story. We'll discuss this approximate number system in Section 3 of this chapter.

Step 2: Language learning at around age two puts two more representations in play. The representation of type (c) in Figure 1 is the memorized list of number words, in order from "one" to some upper limit, which we'll fix for concreteness at "ten." We'll see that this list of numerals doesn't become important until quite late in the process, but it's in place early. The second new representation of type (d) is a mapping between the indefinite determiner "a," as in "a balloon" (or other singular marker in natural language), and the symbol for a singleton $\{object_1\}$.[2] Similarly, there's a mapping between the word "some" (and other plural markers in language) and the symbol for a set of unknown size, $\{object_1, object_2, \ldots\}$. This is the way the children learn the singular/plural distinction in number-marking languages like English, French, or Russian—the difference between "book" and "books."

Steps 3–6: In the next few stages, children learn the meanings of the words "one" through "three" or "four" by connecting them with mental representations of sets containing the appropriate number of things. First, the children think that the word "one" means what "a" means (i.e., $\{object_1\}$) and that the rest of the number words mean what "some" means (i.e., $\{object_1, object_2, \ldots\}$). At this stage, they can give you one balloon if you ask them to, but they're unable to give you two, three, four, or a larger number of balloons. They think all these latter number words mean the same thing, namely, some. But over a period of about a year, they learn to differentiate this second representation. "Two" comes to be connected with $\{object_1, object_2\}$. However, "three," "four," and so on, are still associated with an arbitrary set of more than two elements. Then they learn "three" and occasionally "four" in the same way.

Step 7: Finally, kids are able to notice that a relation exists between the sequence of numerals in the count list and the sequence they can form from their set-based

---

[2]You might wonder about the use of sets in this construction: Do young children have a notion of set that's comparable to mathematicians' sets? Attributing to children at this age a concept of set in the full-blooded sense would be fatal to Carey's claim that Quinian bootstrapping produces new primitive concepts (e.g., the concept FIVE) that can't in principle be expressed in terms of the child's pre-bootstrap vocabulary. We know how to express FIVE in terms of sets (see, e.g., [8]). So the notion of set implicit in representations like $\{object_1, object_2\}$ is presumably more restrictive than ordinary sets.

representations. They understand this connection because their parents and teachers have leaned on them to count small groups of objects. When children count to "two" while pointing to two squirrels in a picture book, their representation of one thing, $\{object_1\}$, is active, followed by their representation of two things, $\{object_1, object_2\}$. So kids begin to see a relation between the order of the numerals in their count list and the cardinality of the sets that these numerals denote.

Step 8: At long last, then, the children can figure out that advancing by one numeral in the count list is coordinated with adding one object to a set. In other words, what they have learned is how counting, by ticking off the objects with the count list, manages to represent a total. They now have a rule that directly gives them the appropriate number of objects for any of the numerals on their count list, and they no longer need to keep track of the set representations for each numeral. They've learned that:

$$cardinality\,(numeral(n)) = cardinality\,(numeral(n - 1)) + 1, \qquad (1)$$

where $numeral(n)$ is the nth numeral in the sequence $<$ "one," "two," ..., "ten" $>$ and $cardinality(m)$ is the cardinality associated with numeral m.

Recent research puts some kinks in the bootstrap of Figure 1. Many children who seem to have passed through Steps 1–8 by the usual criterion of being able to give the experimenter up to six or eight objects, nevertheless can't say whether a closed box containing ten objects has more items than a box containing six ("The orange box has ten fish in it. The purple box has six fish in it. Which box has more fish?") [20]. Similarly, children in the same position are often unable to say whether a single object added to a box of five results in a box with six objects or with seven objects [6]. The children who fail these tasks know how to recite the count sequence to at least "ten," and they know, for example, that the numeral following "five" is "six" rather than "seven." In terms of (1), they know that $numeral(n - 1)$ immediately precedes $numeral(n)$ for the numerals on their count list. So whatever the children have learned about the positive integers at this point doesn't seem equivalent to Rule (1). You might therefore wonder whether a child's ability to give the experimenter six or eight objects in response to a request is enough to show that the child really knows the cardinal meanings of the number words "one" through "ten." Assuming that the child complies with the request to give eight by counting out the objects ("one, two, ... , eight: eight balloons"), does the child realize that this procedure yields the total number of items? (See [12, 16] for worries about children's initial counting.)

A second pertinent finding from recent research is that children appear to understand the gist of Hume's Principle (the number of F's equals the number of G's iff there's a one-one relation between the F's and the G's) only after they are able to "Give me six" [18, 42]. Before this point, for example, children are able to affirm that a lion puppet with five cupcakes has just as many as a frog puppet with five cupcakes. But then if explicitly told that lion has "five cupcakes," they are often unable to say whether frog has five or six. (The experimenters discouraged the children from counting; so the children had to base their answers on the just-as-many relation.) These results confirm an earlier observation by Gelman and Gallistel [13]: If children have to determine the number of objects in each of two groups of the same cardinality, they sometimes ignore an obvious one-one relation between the groups of objects and instead count both groups separately. This finding would be unexpected if Hume's Principle provided the basis of cardinality for children, in the way it does in the context of Frege's Theorem (see Section 1). But if children's grasp of cardinality at the point when they're able to "Give me six" comes from counting, the results in the preceding paragraph imply that it's only in an anemic sense (at least initially) that has no implications about the relative cardinality of the numerals on the child's count list.

But let's not be too snarky about counting. Let's suppose that counting is one route children could take to learn (eventually) how number words connect to cardinalities. The issue here is: Why assume that children's understanding of the positive integers derives from cardinality (no matter whether they comprehend cardinality through counting, one-one relations, or some other way)? Today, you might dial Gerry on your cell, check the price of a six-pack of Gumball Head at the 7-Eleven, tune your FM to the public radio frequency, check the street address of Pizzeria Due, adjust the toaster to setting 4, note the speed limit on East 55th Street, calculate some averages for a presentation, check the temperature before going out, remind yourself of the date of your trip to Omaha, look online for a shirt with 32" sleeves, note the page of *Fahrenheit 451* that you've managed to reach. None of these encounters with numbers involves cardinalities. So why privilege cardinality? Of course, situations exist where cardinality is crucial. Conducting an inventory, a survey, or an election may be examples. But it is not clear that these contexts provide a reason for thinking that cardinality is the central feature of integer knowledge.[3]

From a structural perspective, what's striking about Steps 1–8 is that they don't teach children much about the positive integers that they didn't already know. At Step 2, children already know the count-list sequence of the number words < "one," "two," ..., "ten">. They learn in Steps 3–6 how to assign the first few words in this sequence to a representation of a set of appropriate size: $\{object_1\}$, $\{object_1, object_2\}$, $\{object_1, object_2, object_3\}$. Then they finally learn in Steps 7–8 the rule in (1) that pairs the rest of the words in the sequence with a set size. But that's it. They've learned to correlate one 10-item sequence (of the first few number

---

[3]Many quantitative contexts, including some just mentioned, involve measurement of continuous dimensions (e.g., temperature, time, and length) rather than counting. Current research in developmental psychology suggests that infants are about equally sensitive to the number of objects in a collection (for $n > 3$) and to the continuous extent (e.g., area) of a single object when it is presented alone (see [9] for an overview). This is presumably because the same kind of psychophysical mechanism handles both types of information (see Section 3.1). However, infants' accuracy for the number of objects in a collection (again, for $n > 3$) is better than that for the continuous extent of the objects in the same collection [5]. Of course, children's knowledge of continuous extent, like their knowledge of number, has to undergo further changes before it can support adult uses of measurement (see, e.g., [24]). Some intricate issues exist about children's understanding of continuous quantity that are the topic of current research [39], but because this chapter focuses on knowledge of natural numbers, this brief summary will have to do.

words) with a second, structurally identical 10-item sequence (successive set sizes or cardinalities). Although Rule (1) is perfectly general in applying to any natural number n, the domain of n is bounded at "one" through "ten," since those are the only numerals that the children know at this point. As a result, the structure of the rest of the positive integers is undetermined [31, 36]. For example, the children don't know on the basis of Steps 1–8 that the integers don't stop at 12 or 73 or that they don't branch at 13 into two independent sequences or that they don't proceed to 100 and then circle back to 20. They can't rule out any of these possibilities because, according to this theory, they don't know the structure of the integers beyond 10.

Carey is clear that Rule (1) is not the only principle that children have to master before they can understand the positive integers [3]. Just after Step 8, they clearly don't yet know the cardinal meanings of "seventy-eight," "seventy-nine," and so on, since these terms are not yet part of their vocabulary. And they still have additional work to do before they understand general facts about the positive integers, such as the fact that every positive integer has a unique successor. But should we even credit them with understanding the meaning of the terms in their count list, "one," "two," ..., "ten," on the basis of the knowledge they gain in Steps 1–8? From the structural perspective, the answer might well be "no" (see [38] for arguments along these lines). From this perspective, children at Step 2 know at least one finite structure of 10 elements. Although Carey treats this count list as a "placeholder structure" and as "numerically meaningless," it is the same structure that they arrive at through Quinian bootstrapping. The only difference is that it is correlated with mental representations of cardinalities. But if you are inclined to say that children's initial $<$ "one," "two," ..., "ten" $>$ is numerically meaningless, shouldn't you say that the same sequence in association with the isomorphic $< \{object_1\}, \{object_1, object_2\}, ..., \{object_1, object_2, ..., object_{10}\} >$ is also meaningless? Likewise, for Rule (1), when restricted to the numerals on the children's count list. Of course, this doesn't mean that children have learned *nothing* of importance in proceeding through these steps. They've learned how to calculate the correct number of items in response to the numerals they know. But do they have a better grip on the numbers five or six or ... ten than they had at the start?

As you might expect, Quinian bootstrapping isn't the only proposal on the table about how children learn the cardinal meanings of the small positive integers (see [29] for an alternative and [35] for a critique). However, I won't pursue comparisons to these alternative proposals, since my purpose is to contrast the structural perspective with the cardinal perspective, and the latter perspective is shared by nearly all developmental theories of number learning.[4]

---

[4]The same is true in education theory, as Sinclair points out in her chapter in this volume. See that chapter for an alternative that is more in line with the structural perspective.

3   Judgments of Order versus Numerosity

Perhaps we can get a better purchase on the cardinal and structural perspectives by looking at the way people make direct judgments about cardinal and structural relations. For example, if people find it easier or more natural to determine the cardinality of sets of five than to determine the relation between five and six (e.g., $5 < 6$), then the cardinal perspective may provide a better fit than the structural perspective to people's apprehension of numbers. But although several studies exist that are relevant to this comparison, some inherent difficulties muddy the implications these studies have for the issue at hand.

3.1   Distance Effects

To appreciate the difficulties in untangling these perspectives, consider a well-known and well-replicated finding from the earliest days of cognitive psychology [1, 2, 27]. On each trial in this type of experiment, adult participants see two single-digit numerals (e.g., "3" and "8"), one on each side of a screen, and their task is to press a button on the side of the larger number as quickly as possible. (The larger number is randomly positioned at the right or left across trials; so participants can't anticipate the correct position.) One finding from this experiment is that the greater the absolute difference between the numbers, the faster participants' responses. For example, participants are reliably faster to respond that 8 is larger than 3 than that 5 is larger than 3. This may seem surprising, given adults' familiarity with the small positive integers.

The standard explanation for this distance effect is that people automatically map each of the numerals to a mental representation that varies continuously with the size of the number. They then compare the two representations to determine which is larger. In comparing "3" and "5," for example, they mentally represent 3 as an internal quantity of a particular amount, represent 5 as an internal quantity of a larger amount on the same dimension, and then compare the two quantities to determine their relative size. Imagine, for example, that 3 is represented as some degree of neural activation in a particular brain region, and 5 as a larger degree of activation in an adjacent region. Then people find the correct answer to the problem by comparing these degrees of activation. On this account, the distance effect is due to the fact that people find it easier to compare quantities that are farther apart on the underlying dimension. Just as it's easier to determine the heavier of a 3 kg and an 8 kg weight than the heavier of a 3 kg and a 5 kg weight by hefting them, it's easier to determine the larger of 3 and 8 than the larger of 3 and 5. The mental system responsible for this type of comparison goes by a number of aliases (e.g., "mental number line," "analog magnitude system," and "number sense"), but these days most researchers call it the "approximate number system." So will I.

Much evidence suggests that the approximate number system is present in human infants and in a variety of non-human animals (see [7] for a review). Experiments along these lines present two groups of dots, tones, or other non-symbolic items to determine the creatures' sensitivity to differences in the cardinality of these groups. Distance effects appear with these non-symbolic items that echo those found with adults and numerals: The larger the difference in the cardinality between the two groups of objects, the easier they are to discriminate. So you could reasonably suppose that the approximate number system is an innate device specialized for detecting cardinalities (e.g., the number of edible objects in a region), that this device persists in human adults, and that the symbolic distance effect for numerals, described a moment ago, is the result of the same system. What's important in the present context is that what appears to be a primitive system for dealing with cardinality—the approximate number system—may underlie adults' judgments of ordinal relations (e.g., $3 < 8$). If so, the cardinal perspective may be more fundamental than the structural one in human cognition.

However, these results do not necessarily support the cardinal perspective. Nearly all quantitative physical dimensions—acoustic pressure, luminosity, mass, spatial area, and others—produce distance effects of similar sorts. The perceptual system translates a physical value along these dimensions (e.g., mass) into a perceived value (e.g., felt weight) that can be compared to others of the same type, and comparisons are easier, the greater the absolute value of the difference. The same is true of symbolic stimuli [26]. If participants are asked to decide, for example, which of two animal names (e.g., "horse" or "dog") denotes the larger-sized animal, times are faster the bigger the difference in the animals' physical size. But this effect provides no reason to think that we encode animal sizes as cardinalities. So it is not at all clear that distance effects for numerals depend specifically on representing them in terms of cardinality. Instead, the effects may be due to very general psychophysical properties of the perceptual and cognitive systems.

3.2   Judgments of Order

To make some headway on the cardinal and structural perspectives, we need a more direct comparison of people's abilities to judge cardinal and structural properties. Is it easier for people to assess the size of a set associated with a positive integer than to assess the integer's relation to others? Linnebo's [21] second argument, mentioned in Section 1, seems to predict a negative answer to this question. Several recent studies have attempted a comparison of this sort, but the implications for our purposes are difficult to interpret because of some inherent features of the procedures.

Here's an example: Lyons and Beilock [22] compared a "cardinal task," similar to the standard "Which is larger?" method, described in the preceding subsection, to a novel "ordinal task." Table 1 summarizes the main conditions in the study. In the ordinal task, participants decided as quickly as possible whether triples of single-

Table 1 Summary of Main Conditions from Lyons and Beilock's Comparison of Ordinal and Cardinal Judgments (Adapted from [22, Figure 1])

Stimulus Items	Task Type				
	Ordinal Task (Are the items in either ascending or descending order?)			Cardinal Task (Which item is larger?)	
Numerals					
Close	2	3	4	2	3
Far	2	4	6	2	4
Dots					
Close	(dot triple)			(dot pair)	
Far	(dot triple)			(dot pair)	

Entries Provide an Example in which Participants Should Respond "True" in the Ordinal Task and Push the Right-hand Button in the Cardinal Task

digit numerals were correctly ordered (in either ascending or descending sequence) or incorrectly ordered. For instance, participants were to answer "yes" to <2, 3, 4> or <4, 3, 2> but "no" to <3, 4, 2>. The elements of the triples could be separated by an absolute difference of one (the "close" triples in Table 1, e.g., <2, 3, 4>) or two (the "far" triples, e.g., <2, 4, 6>). Lyons and Beilock also included a task in which participants made analogous judgments for triples of dots. For example, they decided whether a triple consisting of two dots followed by three dots followed by four dots was correctly ordered. For the cardinal task, participants decided which of two numerals (e.g., 2 or 3) was larger or which of two groups of dots was larger (e.g., a group of two or a group of three dots). As in the ordinal task, the items within a pair could differ by one or by two.

Lyons and Beilock found the usual distance effect in the cardinal task, both for numerals and dots. That is, participants were quicker to respond to the far pairs (e.g., 2 vs. 4) than to the close pairs (e.g., 2 vs. 3). For the ordinal task, however, the results were different for numerals than for the dots. Although the dots showed a distance effect, numerals showed the reverse: The close triples were faster than the far triples. (For related findings, see [10, 44].) The investigators conclude from these findings that the link between mental representations of numerals and cardinalities is less direct than what one might gather from the typical distance effects. Judgments of order for numerals rely on a process distinct from the one governing judgments of cardinality (presumably, the approximate number system). "At the broadest level, the meaning of 6 may thus be determined by both its relation to other symbolic numbers and the computational context in which it rests. This is in keeping with the view that the meaning of symbolic numbers is fundamentally tied to their relations with other symbolic numbers . . . " [22, p. 17059].

The emphasis on "relations with other symbolic numbers" goes along with what we have been calling the structural perspective. However, the relations in question raise an issue about the type of structure responsible for the data. Lyons and Beilock plausibly suggest that the reversal they observe for ordinal judgments of numerals—faster times for close triples in Table 1—is the result of familiarity with the list of count words. Because college-student participants have rehearsed the count list ("one, two, three, . . . ") on many thousands of occasions during their lives, they find it easier to recognize numerals as correctly ordered when they appear in adjacent positions on the list (e.g., <2, 3, 4>) than when they are not in adjacent positions (e.g., <2, 4, 6>). fMRI imaging evidence from the same experiment suggests that "one interpretation of these results is thus that ordinality in symbolic numbers is processed via controlled retrieval of sequential visuomotor associations . . . " [22, p. 17056]. In the case of ordinal judgments for dots, however, participants can't directly access the count list, but have to compare successively the cardinality of the three groups (e.g., two dots is less than three dots is less than four). These comparisons are similar to those performed in the cardinal task and give rise to similar distance effects.

This interpretation of the ordinal judgments for numerals returns us to the concerns raised at the end of Section 2. According to Lyons and Beilock [22], the structure responsible for the ordinal task with numerals is the rote connections that we form in reciting the count list ("sequential visuomotor associations"). This is the same "placeholder structure" that Carey [3] finds "numerically meaningless." We can be more charitable than Carey in granting this structure some mathematical significance (and, of course, adults can recite more of the count list than children can). But clearly, this structure isn't the same as the structure of the positive integers. We can't individually store the connections between each successive pair of integers since there are infinitely many of them. So the structure responsible for the ordinal judgments of numerals in such tasks must comprise just a short finite segment of the integers if Lyons and Beilock are right in their interpretation. Do people have a deeper structural understanding of the integers?

4 Two Studies of Number Knowledge

The experiments described in the preceding section are timed tasks that call on people's immediate impression of number, and they sometimes reveal unobvious aspects of those impressions. But if what controls the responses in those experiments is automated cognition, such as rote recitation of the count words, the results may mask deeper features of people's thinking. We might be able to find out more by quizzing people's number knowledge directly.

4.1 What Types of Properties Do People Connect to Numbers?

As one way to find out whether people understand numbers in terms of cardinal or structural features, I asked 20 undergraduates from an introductory psychology class to list properties for each of seven numbers. The numbers were: "zero," "minus eight hundred forty-nine," "square root of 2," "three," "seventy-one hundred ninety-three," "twenty-nine billion and ninety-one," and "eighty-three septendecillion and seventy-six." These were spelled out, as in the preceding list. I picked the last four of these items to coincide with those from an earlier experiment [33]; the rest were chosen to contrast with the positive integers. Results from the three biggest numbers did not differ greatly in this experiment, and of these, I'll report only those for 7193 (with a few exceptions, noted later).

Participants saw these numbers one-at-a-time on a computer screen, in a new random order for each participant. The participants were told that the experiment concerned their knowledge of number properties. They were asked to think of ten properties for each number and to type them into spaces provided on the screen. The instructions cautioned them that they should "try not to just free associate—for example, if a number happens to remind you of your father, do NOT write down 'father.'" After the participants had finished listing properties for all the numbers, a new set of instructions asked them to rate the importance of each of these properties. A participant saw a series of screens, each containing one of the original numbers and a property that the participant had listed earlier for that number. For example, if the participant had listed "is an odd number" for three in the first part of the experiment, then he or she saw in the second part a request to rate the importance of "is an odd number" for three. A 0–9 rating scale appeared on the same screen, with "0" labeled "not at all important" and "9" labeled "extremely important."

We can get some idea of the nature of the properties that the participants produced by classifying each property token into one of the following categories. Examples of actual properties from the participants' lists appear in parentheses after the category name:

> **Cardinality** (e.g., "is nothing" for zero)
> **Magnitude** (overall size, e.g., "is a big number," "small," "very large")
> **Number line** (e.g., "three integers away from zero on the number line," "exact middle of the number line," "is between 1 and 2 on the number line")
> **Number system membership** (e.g., "integer," "is a rational number," "is not a rational number")
> **Arithmetic comparison** (e.g., "smaller than 10," "is bigger than one million" "between 2 and 4")
> **Arithmetic operations** (e.g., "multiplied by itself will give 2," "any number added to zero is the same number," "zero divi[d]ed by 1 is zero")
> **Number-theoretic properties** (in a loose sense in which, e.g., "is an odd number," "is negative," "is not a prime," "has an imaginary square root," "is a factor of 21" are number-theoretic)
> **Numeral properties** (properties of the written shape or spelling of the number, e.g., "contains one digit," "has a comma in it," "7 in the thousands place," "spelled with four letters")

Non-numeric properties (e.g., "is significant," "is important to mathematics," "common")

Other (e.g., "no idea," "I would like to have this much candy," "I don't like square roots")

Figure 2 (solid lines and circles) shows the frequency of tokens in these ten categories for each number. The first thing to notice is the very small number of tokens for cardinality. Zero produced a few such mentions—variations on "nothing"—but no one listed "describes 7193 objects" (or anything of the kind) for 7193. In fact, none of the property categories that might be linked to the approximate number system—the cardinality, magnitude, and number line categories—received more than a few mentions, as you can see from the first three points of each of the Figure 2 curves. Of course, no one would expect participants to mention cardinality for -849 or $\sqrt{2}$, but it's notable that 0, 3, and 7193 also received small frequencies.

By contrast, mention of number systems (e.g., "is an integer," "is not rational") was relatively frequent, especially for $\sqrt{2}$, the sole non-integer and non-rational in this group. Similarly, number-theoretic properties (e.g., "is prime," "is divisible by 3") were popular responses for the integers.

Although participants did not often produce properties having to do with arithmetic comparison or operations, zero did yield a fairly large number of operations properties, no doubt because of its special role in addition and multiplication. For example, one participant mentioned, "the product of zero with any other number is zero itself," and another wrote, "anything plus zero is itself." Properties of numerals (either for number words or symbols) were not especially common, except for 7193 ("four digit number," "no repeated digits") and, to a lesser extent, -849 ("has one 8," "looks like this '-849'").

You might argue that the very low frequency for cardinality is due to the fact that each natural number has just one cardinality, whereas it has many properties of other types (e.g., many number-theoretic properties). So perhaps the frequency of mention just reflects the actual number of available tokens per type. But although a natural number has just one cardinality, it nevertheless denotes the size of an infinite number of sets. Participants could have said that three is the number of bears in the fairy tale, the number of stooges in the film comedies, the number of instrumentalists or vocalists in a trio, the number of vertices or angles or sides in a triangle, the number of people that's a crowd, the number of deities in the trinity, the number of events in a triathlon, the number of rings in a circus, the number of faces in the triple crown, the number of children in a set of triplets, the number of outs in a turn at bat, the number of isotopes of carbon, the number of novels in a trilogy, the number of wheels on a tricycle, the number of leaders in a triumvirate, the number of panels in a triptych. But no one mentioned any of these or any other three-membered groups in response to three. Moreover, for the numbers 3, 7193, 29,000,000,091, and $(8.3 \times 10^{55}) + 76$, none of the participants mentioned cardinality even once. Of course, we asked participants to list number "properties," and perhaps this way of phrasing the question militated against their writing down items related to cardinality. For example, the number of things in a

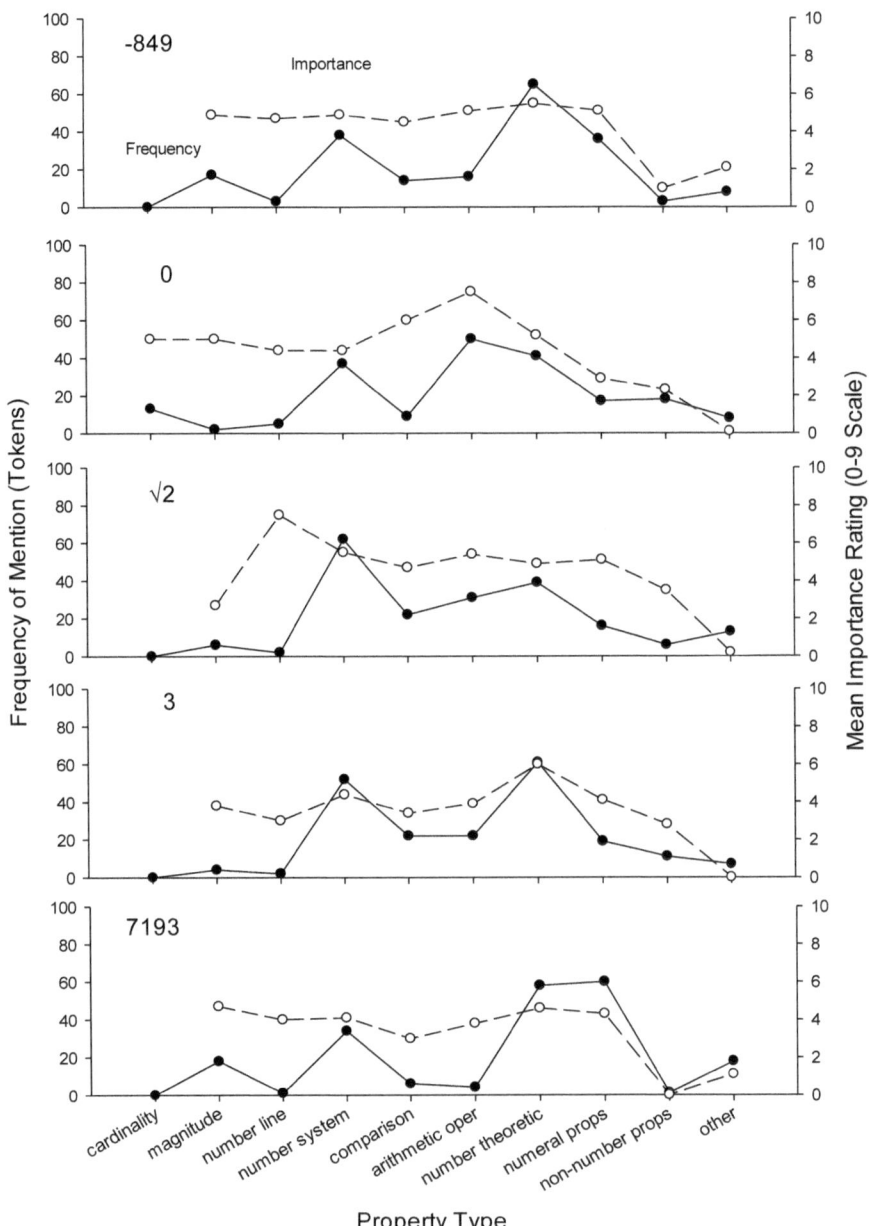

Fig. 2 Frequency of mention of properties of different types (solid lines and symbols) and mean rated importance of the same properties (dashed lines and open symbols) from Study 1. Points on the importance curves are missing if participants listed no properties of that type.

triple or of vertices in a triangle might have seemed too extrinsic to qualify as a property of three. But the question at issue in this study is, in fact, what participants believe to be intrinsic to their number concepts. What kind of information about a number is central to people's beliefs about the number's nature? If participants find "being an integer" and "being divisible by 3" to be number properties but not "being the number of items in a triple," then this suggests that cardinality may not be what organizes their conception of numbers.

Mean importance ratings for the same properties also appear in Figure 2 (dashed lines and open circles). The means for the cardinality and number-line properties are based on only a small number of data points, as the frequency curves show. Omitting these latter categories, a statistical analysis indicates that, across all the numbers, participants rated the number-theory, arithmetic-operations, and number-system properties as more important than the non-numeric properties. They also rated "other" properties lowest in importance, as you would hope. No further reliable differences in importance appeared among the property categories (adjusting for the number of comparisons). However, arithmetic operations are especially important for zero, probably for the reasons mentioned earlier, and number-theory properties (e.g., "are prime") are important for three. (7193 is also prime, but participants probably didn't recognize it as such.) These peaks contribute to a statistically reliable difference in the shape of the importance curves in the figure. Keep in mind, though, that the properties contributing to these importance ratings are the ones that the same participants produced in the first part of the experiment. It might be useful to look at an independent measure of the importance of number properties. The study in the next section provides a measure of this sort.

4.2   The Centrality of Number Properties

In a second study of number properties, a new group of participants decided whether a given property of an integer "was responsible for" a second property of the same integer. The properties included:

Cardinality (phrased as "being able to represent a certain number of objects")
Number system membership ("being an integer")
Arithmetic ("being equal to the immediately preceding integer plus one")
Position in the integer sequence ("being between the immediately preceding and the immediately following integers"), and
Numeric symbol ("being represented by a particular written symbol").

On one trial, for example, participants were asked to consider whether "an integer's ability to represent a certain number of objects is responsible for its being equal to the immediately preceding integer plus one." The instructions told participants, "By 'responsible' we mean that the first property is the basis or reason for the second property." The participants answered each of the responsibility questions by clicking a "yes" or a "no" button on the screen.

In a preliminary version of this experiment, I asked participants about the properties of specific integers from Study 1 (0, 3, 7193, and 29,000,000,091). For example, participants had to decide whether three's property of representing three objects was responsible for its being equal to $2 + 1$. The results from this pilot study, however, suggested that participants were interpreting these questions in a way that depended on the specificity of the properties. Asked whether being an integer was responsible for three's being able to represent three objects, for example, many participants answered "no," apparently because merely being an integer was consistent with being able to represent any number of objects. Some participants explicitly mentioned (in their written comments after the experiment) a sufficiency criterion of this sort: "If the second part had to be true because of the first part, then I selected that the first property was responsible for the second." To avoid this problem, the study reported here rephrased the properties to be about integers in general. For example, participants were asked "whether being an integer was responsible for the integer's ability to represent a certain number of objects."

Participants saw all possible pairs of the five properties listed above (in both orders). For example, the participants decided both whether "an integer's ability to represent a certain number of objects is responsible for its being equal to the immediately preceding integer plus one" and whether "an integer's being equal to the immediately preceding integer plus one is responsible for its ability to represent a certain number of objects." Thus, there were 20 key questions about integer properties. In addition, the experiment included five catch trials with questions that were intended to be obviously true or false (e.g., "Is an integer's ability to represent a certain number of items responsible for its color?"). The presentation order of the full set of 25 questions was random. Fifty participants were recruited from Amazon Mechanical Turk for this study, but three were excluded for making errors on three or more of the catch trials.

We can get some idea of the centrality of a numeric property by looking at participants' willingness to say that the property was responsible for the others. For example, if cardinality is a critical property, then participants should be willing to say that "the ability to represent a certain number of objects" is responsible for other properties. Of course, a participant's overall likelihood of endorsing a property in this way will depend on the other properties on our list and on the particular phrasing of these properties. Still, these scores provide a hint of the relative importance of the property types.

Figure 3 illustrates the results in a way that may help bring out the comparison of the properties' importance. The circles at the pentagon's vertices represent the properties listed earlier (cardinality, arithmetic, and so on). The arrows between the circles correspond to the participants' judgments of whether the property p_1 at the arrow's tail is responsible for the property p_2 at the arrow's head. The numbers on these arrows are the proportion of participants who agreed that p_1 was responsible for p_2. For example, the arrow from arithmetic to cardinality is labeled .62 and indicates that 62 % of the participants thought that the arithmetic property ("being equal to the immediately preceding integer plus one") was responsible for cardinality ("being able to represent a certain number of objects"). One measure

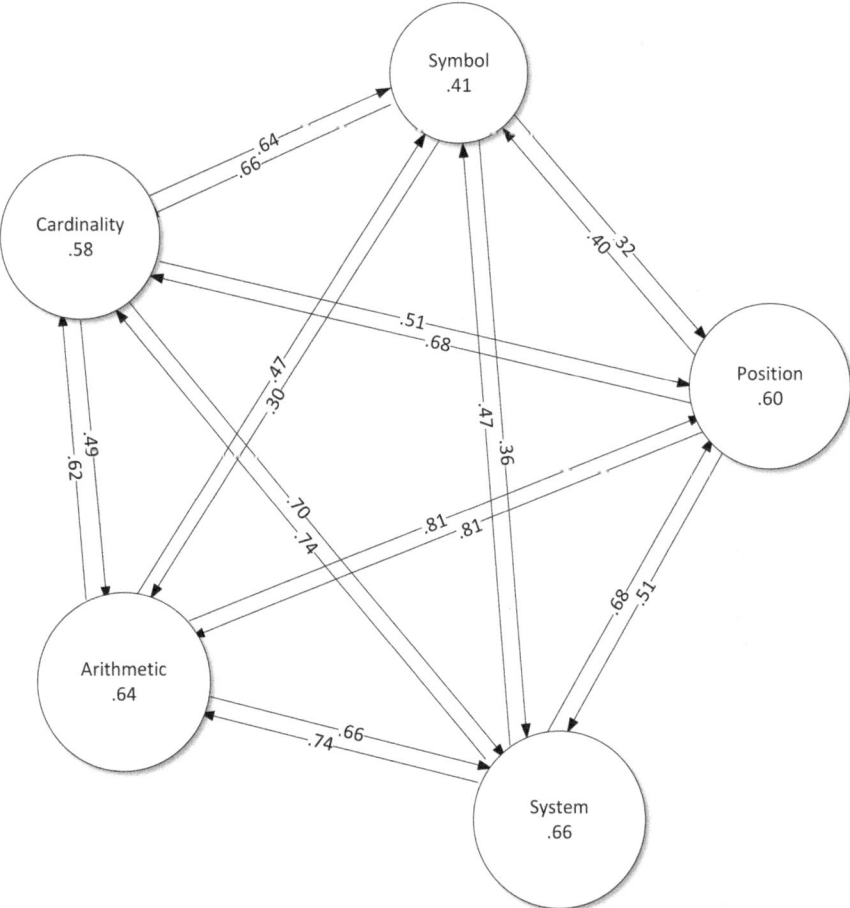

Fig. 3 Judgments of "responsibility" for number properties. Each arrow in the diagram represents participants' decisions as to whether the property at the base of the arrow is responsible for the property at the head of the arrow. Numbers on the arrows are the proportion of participants who said "yes." Numbers in the circles and their areas represent the means of the outgoing arrows.

of a property's centrality is its average responsibility for the other properties— the average of the proportions on the outgoing arrows from that property. These averages appear within the circles, and the areas of the circles are approximately proportional to these averages. The bigger the circle, the more central the property.

The mean responsibility score for the system property ("being an integer") is the largest of these items. This may be due in part to the fact that two of the other properties—the arithmetic and position properties—mentioned the word "integer." For example, participants were asked whether "being an integer" was responsible for "being between the immediately preceding and immediately following integers." However, most participants (74 %) also thought that being an integer was responsible for cardinality ("being able to represent a certain number of objects"), whose phrasing did not include the word "integer." This suggests that participants saw membership in the number system itself as an important factor.

The arithmetic and position properties also had relatively large responsibility scores in these data. The arithmetic property was just the +1 relation (i.e., "being equal to the immediately preceding integer plus one"), and the importance that participants attached to it may suggest that they view a successor function as central to the integers' identity. The same may be true of the positional property, which was specified in terms of being between the preceding and the following integer.

The numeric symbol property (e.g., "being represented by a particular written symbol") received lowest responsibility scores. Many participants probably answered "no" to questions about the symbols' responsibility because of the arbitrary nature of the symbol. In the pilot study mentioned earlier, one participant wrote in his or her debriefing comments, "I did not think that the symbol for any number could be responsible for anything, since the designated symbols for numbers are arbitrary." According to another, "Does a number's numerical symbolization really have to do with anything? After all, it could simply be an arbitrary symbol." Section 5 of this chapter discusses the implication of this result for the structural perspective.

Responsibility scores for cardinality ("being able to represent a certain number of objects") were about midway between those for the system property and symbol property. Cardinality does somewhat better here than we might have predicted from its very infrequent mention in the preceding study, where the participants had to produce their own properties for the numbers (see Figure 2, panels 2, 4, and 5). The results from Study 1 could be put down to the obviousness of the connection between an integer and its cardinality. Maybe it literally goes without saying that "three" denotes three entities. But the same is probably true of several of the other properties on our list. Another more likely possibility is that because the cardinality property fixes its referent in a nonarbitrary way, some participants in the present study may regard cardinality as sufficient for the other properties. For example, if a number has cardinality 3, then it *is* 3, and hence, is the successor of 2, is between 2 and 4, is an integer, and is symbolized by "3." Other participants, however, may regard cardinality as too "incidental" to an integer to allow it to be responsible for its purely mathematical properties, and this same feeling may explain why cardinality went unmentioned in Study 1. The final section tries to sketch what "incidental" could amount to in the mathematical domain.

5 Conclusions

What implications do the two studies of Section 4 have for the cardinal and structural perspectives? On the one hand, Study 1 suggests that cardinality isn't something that often occurs to people when they are asked to consider the properties of an integer. They don't immediately take a cardinality perspective under these circumstances.

Instead, the properties of the integer that come readily to mind are ones that might be called "number internal," such as being odd or positive or prime. In the case of special numbers, such as 0, these number-internal properties also include arithmetic or algebraic features, such as being an additive identity. On the other hand, Study 2 shows that some people acknowledge the importance of cardinality in relation to an integer's other properties. Once we know that an integer has a particular cardinality, say 7193, we can predict all its other properties, for example, that it is the successor of 7192 and the predecessor of 7194. Cardinality can, in this way, single out an integer uniquely, yet people may not typically use it in thinking about the integer as such.

Perhaps something of the same is true of numbers in general. In applied math, numbers denote specific times, spatial coordinates, velocities, temperatures, money, IQ, and what have you. But we don't think of a number as intimately connected with these instances (e.g., a particular position, time, or value). A shift in coordinates will assign a different number to the same spatial point. A change in currency will associate a different number with the same value for a good. Numbers provide models for time, space, utility, and other dimensions, as measurement theory makes clear (see, e.g., [19, 40]), but only up to certain well-defined transformations. The numbers themselves have an identity that's independent of their role in the model. In the same way, you could view the use of integers to denote cardinalities as another application of a number-based model to the size of sets.

Proponents of the cardinality perspective still have some cards to play in defending the idea that cardinality is conceptually fundamental. For example, they could take the view that cardinality provides children's first entry point to the natural numbers, even if children outgrow this perspective as they gain further knowledge of number properties in school. Or they could claim that cardinality is central to adults' thinking about the naturals, even though adults don't often mention it when explicitly asked about number properties. But although defenses of this kind may turn out to be right, not much evidence exists to back them. We noticed in Section 2 that it's hard to see how cardinality can by itself advance children's knowledge of the naturals. Mere apprehension of numerosity through the approximate number system can't provide the right properties for the naturals (e.g., the perceived difference in numerosity between 14 items and 15 items is smaller than the perceived difference between 4 and 5 in this system). Perhaps the more sophisticated use of cardinality in children's initial counting creates the link to the naturals, but we found in Section 2 that it's difficult to make this case without begging the question against the structural perspective (see [31, 37] for more on this theme). Similarly, the results we reviewed in Section 3 show that, although numerosity may influence adults' judgments about the integers in tasks like numeral comparison, there's a catch: The mental process most likely responsible for such effects is again the approximate number system, which delivers distorted information about cardinality. Maybe true cardinality underlies adult intuitions about the naturals, but what reasons support such an assumption?

The structural perspective may be better able to cope with the results. But we should enter a couple of qualifications about the support that these studies lend to the structural view. First, we've seen that participants didn't often mention properties connected to numerals (except for the larger numbers) in Study 1, and they didn't judge them as especially important in producing a number's other properties in Study 2. This seems reasonable in view of the arbitrary nature of the symbols. But if "the only perfectly direct and explicit way of specifying a number seems to be by means of some standard numeral in a system of numerals" [21], then you might have expected numerals to play a more important role in the results. Notice, though, that in these experiments no variations occurred in the manner in which the numbers were specified. In Study 1, a numeral was given to participants explicitly (as an English phrase), and they responded based on this numeral. In Study 2, no numerals appeared. So the results are still compatible with the possibility that numerals within a standard system provide an easier way to denote numbers than other possibilities, such as the set of sets containing that number of elements.

Second, in Study 1, the frequency of positional properties was fairly low. We classified these properties as arithmetic comparisons in that study (see Figure 2), and they included items such as "is greater than -1" (in the case of 0), "is smaller than four" (in the case of 3), and "is less than 10000" (in the case of 7193). Properties of this sort comprised only 7.3 % of tokens across the five numbers in Figure 2 and produced a mean importance rating of 4.5, which is the midpoint of the 0–9 scale. Why did participants fail to produce these positional properties and fail to rate them as important? The reason may be similar to the one that makes cardinality properties unpopular. People may feel that relations like being between 7192 and 7194 in the integer sequence are external to the number 7193 itself.

What come to mind more readily for the numbers in Study 1 are properties like being even or positive—properties we classified as "number-theoretic" in the loose sense of the first study. These properties were either the most often mentioned (for −849 and 3) or the second most often mentioned (for 0, $\sqrt{2}$, and 7193), as Figure 2 indicates. We can unpack properties like these in relational terms—being positive is being greater than 0, and being even is being divisible by 2—but participants may see them (at least at first thought) as something intrinsic to the numbers themselves. My guess is that much the same is true for most other individual concepts (e.g., UNCLE FRED, WALDEN COLLEGE): The properties of these concepts that are easiest to access are ones that we represent as monadic.

It may be an important fact about individual concepts that we mentally organize them around central properties of this sort and take other, more peripheral properties to be the products of the central ones. Uncle Fred may have traits like being genteel or pig-headed that we see as non-relational and intrinsic, but that are responsible for many other aspects of his personality and behavior. In planning our interactions with him, or in predicting what he will do at the party or believe about big government or want for his birthday, we consult and extrapolate from these central properties. The suggestion here is that this type of thinking may carry over to individuals in abstract domains, such as numbers. We don't personify numbers in the way Lewis Carroll did ("'Look out now, Five! Don't go splashing paint over me like that!' 'I couldn't help it,' said Five, in a sulky tone; 'Seven jogged my elbow.'" [4, chap. 8]). But we might think about five as centrally positive, odd, and prime because these properties are handy in making inferences about this number.

This isn't quite the structural perspective that we get in the philosophy of mathematics, but it has its own structural aspects. Structuralists in philosophy see numbers as atoms with no internal structure, as the quotations from Resnik [30] and Shapiro [43] in Section 1 make clear. Their content is exhausted by the position they have in a relevant number system, as given by an appropriate set of axioms. In one way, this seems right for psychological concepts of numbers, as well. Our mental representations of a natural number, for example, had better conform to the usual Dedekind-Peano axioms, since otherwise it's difficult to see in what sense they could represent that number [38]. In another way, however, mental representations about a natural number include a richer domain of facts that we use in dealing with typical mathematical tasks, including calculation and proof (see [34] for the distinction between "representations of" and "representations about" objects and categories). What the studies reported here suggest is that information of the latter sort, at least among college students, may organize itself around, not cardinality, but instead properties like primality that may be more helpful in mathematical contexts.

Acknowledgement Thanks to Ernest Davis, Jacob Dink, and Nicolas Leonard for comments on an earlier version of this chapter and to John Glines, Jane Ko, and Gabrielle McCarthy for their help with the experiments described in Section 4.

References

1. W.P. Banks, M. Fujii and F. Kayra-Stuart, "Semantic congruity effects in comparative judgments of magnitudes of digits," *Journal of Experimental Psychology: Human Perception and Performance*, vol. 2 (1976) 435–447.
2. P.B. Buckley and C.B. Gillman, "Comparisons of digits and dot patterns," *Journal of Experimental Psychology*, vol. 103 (1974) 1131–1136.
3. S. Carey, *The origin of concepts*, Oxford University Press, New York, NY, 2009.
4. L. Carroll, *Alice's adventures in wonderland*, Macmillan and Co., London, 1866.
5. S. Cordes and E.M. Brannon, "Attending to one of many: When infants are surprisingly poor at discriminating an item's size," *Frontiers in Psychology*, vol. 2 (2011) doi: 10.3389/fpsyg.2011.00065.
6. K. Davidson, K. Eng and D. Barner, "Does learning to count involve a semantic induction?," *Cognition*, vol. 123 (2012) 162–173.
7. S. Dehaene, *The number sense: how mathematical knowledge is embedded in our brains*, Oxford University Press, New York, 1997.
8. K. Devlin, *The joy of sets*, 2nd ed., Springer, Berlin, 1993.
9. L. Feigenson, "The equality of quantity," *Trends in Cognitive Sciences*, vol. 11 (2007) 185–187.
10. M.S. Franklin, J. Jonides and E.E. Smith, "Processing of order information for numbers and months," *Memory & Cognition*, vol. 37 (2009) 644–654.
11. G. Frege, *The foundations of arithmetic: A logico-mathematical enquiry into the concept of number*, 2nd rev. ed., Northwestern University Press, Evanston, IL, 1980.
12. K.C. Fuson, *Children's counting and concepts of number*, Springer-Verlag, New York, 1988.

13. R. Gelman and C.R. Gallistel, *The child's understanding of number*, Harvard University Press, Cambridge, Mass., 1978.
14. C. Hanlon, "The emergence of set-relational quantifiers in early childhood", in: *The development of language and language researchers: Essays in honor of Roger Brown*, F.S. Kessel, ed., Lawrence Erlbaum Associates Inc, Hillsdale, NJ, 1988, pp. 65–78.
15. P.M. Hartnett, "The development of mathematical insight: From one, two, three to infinity," *Dissertation Abstracts International*, vol. 52 (1992) 3921.
16. R.G. Heck Jr, "Cardinality, counting, and equinumerosity," *Notre Dame Journal of Formal Logic*, vol. 41 (2000) 187–209.
17. R.G. Heck Jr, "Frege's theorem: An overview", in: *Frege's theorem*, Oxford University Press, Oxford, UK, 2011, pp. 1–39.
18. V. Izard, A. Streri and E.S. Spelke, "Toward exact number: Young children use one-to-one correspondence to measure set identity but not numerical equality," *Cognitive Psychology*, vol. 72 (2014) 27–53.
19. D.H. Krantz, R.D. Luce, P. Suppes and A. Tversky, *Foundations of measurement*, Academic Press, New York, 1971.
20. M. Le Corre, "Children acquire the later-greater principle after the cardinal principle," *British Journal of Developmental Psychology*, vol. 32 (2014) 163–177.
21. Ø. Linnebo, "The individuation of the natural numbers", in: *New waves in philosophy of mathematics*, O. Bueno, Ø. Linnebo, eds., Palgrave, Houndmills, UK, 2009, pp. 220–238.
22. I.M. Lyons and S.L. Beilock, "Ordinality and the nature of symbolic numbers," *The Journal of Neuroscience*, vol. 33 (2013) 17052–17061.
23. J. MacFarlane, "Double vision: two questions about the neo-Fregean program," *Synthese*, vol. 170 (2009) 443–456.
24. K.F. Miller, "Measurement as a tool for thought: The role of measuring procedures in children's understanding of quantitative invariance," *Developmental Psychology*, vol. 25 (1989) 589–600.
25. K.S. Mix, J. Huttenlocher and S.C. Levine, *Quantitative development in infancy and early childhood*, Oxford University Press, Oxford, UK, 2002.
26. R.S. Moyer and S.T. Dumais, "Mental comparison," *Psychology of Learning and Motivation*, vol. 12 (1978) 117–155.
27. R.S. Moyer and T.K. Landauer, "Time required for judgements of numerical inequality," *Nature*, vol. 215 (1967) 1519–1520.
28. C. Parsons, *Mathematical thought and its objects*, Cambridge University Press, New York, 2008.
29. S.T. Piantadosi, J.B. Tenenbaum and N.D. Goodman, "Bootstrapping in a language of thought: A formal model of numerical concept learning," *Cognition*, vol. 123 (2012) 199–217.
30. M.D. Resnik, *Mathematics as a science of patterns*, Oxford University Press, Oxford, UK, 1997.
31. G. Rey, "Innate and learned: Carey, Mad Dog nativism, and the poverty of stimuli and analogies (yet again)," *Mind & Language*, vol. 29 (2014) 109–132.
32. L.J. Rips, "Bootstrapping: How not to learn", in: *Encyclopedia of the sciences of learning*, N.M. Seel, ed., Springer, Berlin, 2012, pp. 473–477.
33. L.J. Rips, "How many is a zillion? Sources of number distortion," *Journal of Experimental Psychology: Learning, Memory, and Cognition*, vol. 39 (2013) 1257–1264.
34. L.J. Rips, *Lines of thought: Central concepts in cognitive psychology*, Oxford University Press, Oxford, UK, 2011.
35. L.J. Rips, J. Asmuth and A. Bloomfield, "Can statistical learning bootstrap the integers?" *Cognition*, vol. 128 (2013) 320–330.
36. L.J. Rips, J. Asmuth and A. Bloomfield, "Do children learn the integers by induction?" *Cognition*, vol. 106 (2008) 940–951.
37. L.J. Rips, J. Asmuth and A. Bloomfield, "Giving the boot to the bootstrap: How not to learn the natural numbers," *Cognition*, vol. 101 (2006) B51-B60.

38. L.J. Rips, A. Bloomfield and J. Asmuth, "From numerical concepts to concepts of number," *Behavioral and Brain Sciences*, vol. 31 (2008) 623–642.
39. L.J. Rips and S.J. Hespos, "Divisions of the physical world: Concepts of objects and substances," *Psychological Bulletin*, vol. 141 (2015).
40. F.S. Roberts, *Measurement theory with applications to decisionmaking, utility, and the social sciences*, Addison-Wesley, Reading, Mass., 1979.
41. B. Russell, *Introduction to mathematical philosophy*, Allen & Unwin, London, 1919.
42. B.W. Sarnecka and C.E. Wright, "The idea of an exact number: Children's understanding of cardinality and equinumerosity," *Cognitive Science*, vol. 37 (2013) 1493–1506.
43. S. Shapiro, *Philosophy of mathematics: Structure and ontology*, Oxford University Press, New York, 1997.
44. E. Turconi, J.I.D. Campbell and X. Seron, "Numerical order and quantity processing in number comparison," *Cognition*, vol. 98 (2006) 273–285.
45. H.M. Wellman and K.F. Miller, "Thinking about nothing: Development of concepts of zero," *British Journal of Developmental Psychology*, vol. 4 (1986) 31–42.
46. K. Wynn, "Children's acquisition of the number words and the counting system," *Cognitive Psychology*, vol. 24 (1992) 220–251.
47. E.N. Zalta, "Frege's theorem and foundations for arithmetic", in: *Stanford encyclopedia of philosophy*, E.N. Zalta, ed., 2014.

What kind of thing might number become?

Nathalic Sinclair

Abstract Inspired by the work of Abraham Seidenberg, on the ritual origins of counting, as well as contemporary philosophical work on the embodied nature of mathematical concepts, this chapter proposes an alternate way of approaching number that places much more emphasis on ordinality, rather than cardinality, and that highlights number's temporality. This conception of number is then shown to be at play in the way young children engage with *TouchCounts*, an App designed to support early number sense.

This chapter will explore the possibility of number as an event rather than an object; that is, seeing number as temporal. The issue of time is a very delicate one in mathematics, which tends to excel at encapsulating anything that may be temporal (counting, plotting, rotating, etc.) into a timeless, seemingly immutable object (such as number). Indeed, it can be hard to think of mathematical objects such as number, set and point as anything but static, self-contained *things*. The title of this chapter suggests that we might indeed imagine number otherwise. It plays on the threefold assumptions of the question 'what kind of thing is number?', a question often posed by philosophers of mathematics[1]: one, number is a thing; two, things come in kinds; three, things and kinds of things are timeless (hence the apparent a temporality of 'is').

Another assumption of the oft-posed question 'what kind of thing is number?' is that number is a *singular* entity: an alternative phrasing of the title could have been 'what kinds of things are numbers?' Indeed, there have been several views, described by historians, philosophers, mathematicians and educators alike, on what a number is (or what numbers are). While some have tried to encapsulate various

---

[1] Benacerraf [2] plays off this question as well in his paper entitled "What numbers could not be", which explores the difficulties involved in thinking of numbers as sets of sets.

N. Sinclair (✉)
Faculty of Education, Simon Fraser University, 8888 University Drive,
Burnaby, BC, Canada V5A 1S6
e-mail: nathsinc@sfu.ca

© Springer International Publishing Switzerland 2015
E. Davis, P.J. Davis (eds.), *Mathematics, Substance and Surmise*,
DOI 10.1007/978-3-319-21473-3_17

347

types of numbers (whole, rational, real, complex, infinitesimal, hyperreal, etc.) into one overarching definition, others have focused on particular numbers.

In this chapter, I will be particularly interested in whole numbers, both for historical reasons and for pedagogical ones. Unlike current everyday uses of the word 'number', the early Greek term widely used for number, *arithmos*, was always a whole number, "and tied up with the actual procedure of counting" [4, p. 265]. Besides the question of whether to opt for the singular or the plural form, one might also question the common use of 'number' as a noun. Fowler [10] points out that number can be adverbial as well (as in, she fell twice, where 'twice' is a particular adverbial form of the number two).

In addition to being nouns and adverbs, numbers can also function as a verb in the sense that to number is to call into existence or to evoke. In the context of her work with the Mi'kmaw community, whose language (like most indigenous languages in Canada) is verb-based, Lisa Lunney Borden [3] has proposed a re-verbification of mathematics in which nominalisations such as 'number' could be infused with process, action and flux, thus moving away from the notion of fixed and rigid separate objects. This chapter can be seen as an initial step in this direction.

To set the stage for my arguments, I briefly outline the thesis proposed by the mathematician and historian of mathematics Abraham Seidenberg [19], on the ritual origins of counting. His thesis will set the stage for an alternate way of approaching the historical aspect of how number becomes, and will serve to highlight the temporal (and perhaps even rhythmic) nature of number. I then consider my chapter title question from the perspective of young children engaged with a multi-touch audio-visuo-tangible digital environment for counting, adding and subtracting called *TouchCounts*. My goal is not to argue for its pedagogical effectiveness but, rather, to show how the children's experiences offer new possibilities for what number can be and, as a consequence, how it might be learned, while also to echo strongly with Seidenberg's sense of the ritual origin of counting. Finally, I draw on ideas in contemporary philosophy that provide new ways of thinking about what number is and might be, in terms of both its logical and its epistemological nature. The main advantage of these ideas is that they provide some explanatory tools for understanding how the physical (which is usually recognised as being temporal) and the mathematical (usually not) can be related, and how mathematical concepts can emerge from human actions in the world.

1 "But how can one ask 'How many?' until one knows how to count?"

Primary school mathematics in many countries around the world, including Canada and the United States, take the question "How many?" as the driving force of the curriculum related to number. Children are routinely asked to say how many blocks or cookies or people there are in a given collection, and they usually do so by

Fig. 1 The circles on the left are easier to subitise than those on the right

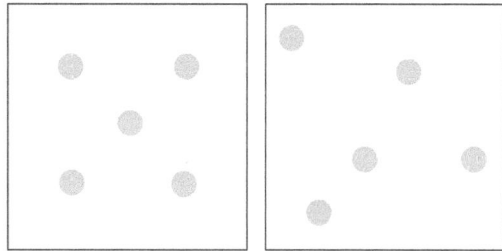

engaging the ritual of sequential counting, which involves pointing at or touching the objects while also saying a sequence of number words in a language. Children are also asked to compare numbers, but using the cardinal approach of having to decide which is greater or which is smaller, rather than which comes after which or which comes first in the spoken word sequence.

For small sets, young children are increasingly encouraged to use subitising to answer the question. Subitising (from the Latin word *subitus*, which means sudden), involves enumerating without counting, usually through visual perception (subitising 5 when the objects are configured as on a die can be much easier than for random configurations—see Figure 1). Subitising is valued in schools because it involves a cardinal conception of number, which in turn enables children to perform addition tasks like $5 + 3$ more easily since they can "add on" 3 to the 5 instead of counting out five and then counting three more.

The emphasis on cardinal forms of comparison goes back at least to the work of the Swiss developmental psychologist Jean Piaget, who focused on children's ability to discriminate among arrays of objects on the basis of the quantity of them present. The focus on quantity can also be seen in Gelman and Gallistel's [12] five principles of counting, which have been highly influential in mathematics education research. These are: (1) the importance of assigning only one counting tag to each counted object in the array (the principle of one-to-one correspondence); (2) number words should be provided in a constant order when counting; (3) repeated counting of the same set must always produce the same final number word (often called the cardinality principle; it presupposes (1) and (2)); (4) it does not matter whether the items in the set are identical; and (5) it does not matter in which order objects are counted. The first three deal with *how to count*, where the objective of counting is clearly to answer the "how many?" question. Indeed, the third "principle of counting" is itself the cardinality principle, namely that repeated counting of the same set must always provide the same final number. In other words, counting is seen as being aimed towards the saying of the final number word. More recently, neuroscientists have also focused on cardinality as being core to human (and, indeed, animal) number sense (e.g., Ncider & Dehaene [16]).

In the rush towards cardinality, then, what it means to be a number in primary school mathematics is very clearly about quantity. Depending on your interests and expertise, this may seem totally natural. (But like most aesthetic and perhaps even political choices, what is natural is largely a matter of enculturation.) But I would like to argue that there is much more to mean by a number—especially in terms of ordinality. Indeed, the philosopher Alain Badiou [1] has insisted, drawing on the work of mathematicians Richard Dedekind and Georg Cantor, that "ordinals constitute the absolute ontological horizon of *all* numbers" (p. 68). For Dedekind, number was to be thought of as a link in a chain; it was about sequence, or seriation. For Cantor, number was also about seriation, but his focus is on order (and the well-ordering principle for which the natural numbers provide an excellent model as well as a source of intuition). The latter describes the way in which: (1) given two numbers, one will always come after the other; (2) given a set of numbers, there will always be one that came first. Neither of these have anything, a priori, to do with answering "how many?". They may be thought of as tools that enable one to answer the "how many?" question, but might they also be tools for accomplishing something else? As with most tools, they take time to apply: ordinals are produced over time and questions about order are decided in time.

In discussing Cantor's work, Badiou [1] describes order in terms of which number will be greater and in terms of which number will be smaller. This may be evidence of some slippage into cardinality, which seems hard to avoid. My own re-wording into a temporal register is inspired by the work of Seidenberg, which can be helpful in reclaiming ordinality. The title of this section is taken from Seidenberg's paper [19]. It is relevant both to pedagogy and to mathematics. In terms of the latter, as Seidenberg notes, modern theories put forth by the logicians Gottlob Frege and Bertrand Russell analyse the concept of number without much attention to ordinality. They speak of "how many?" with little attention to counting. But theirs is a paradigmatic de-temporalisation of number, and it has found its way into the pedagogical realm.

In contrast to Dedekind and Cantor, who were developing mathematical theories of number, Seidenberg was interested in the origins of number and, in particular, the origins of counting. As he writes, most historical accounts of number trace its spread across cultures. But Seidenberg aimed to identify the special circumstance of its origin, which he argued cannot be indicated by its use or by its practical applications: "an application of a device (or idea) is an effect of the device, not a cause" (p. 2). He proposes a ritual origin for counting, one that speaks significantly of *myths* and *rites*. He draws on the work of Lord Raglan [17], who connects them thus: a *myth* "is a narrative linked with a rite" (p. 117), where a *rite* is the corresponding enacted practice. Taking these ideas of myth and rite into the specific context of numbering, Seidenberg observes, "the 'serializing' is the rite, and the 'counting' is the myth" (p. 10).

Based on extensive anthropological evidence, Seidenberg's contention is that the origin of counting lies in a creation ritual, in the bringing onto stage the correct deities in the correct order. These techniques, according to Seidenberg's account, were later adapted to counting and geometrical measurement. (There is also evidence that this creation ritual involved counting by 2s, much as in Noah's ark, which would mean that counting on one's fingers—either by 5s or 10s—is a later development, with hand gestures sometimes used to count silently under speaking taboos.) Thus, counting emerges first by associating number names with deities, which means that numbers did not exist prior to their ritual application. One might more appropriately speak of numbering as a verb, rather than number. The names of participants in the ritual, or the words announcing them, were the original number words. This ritual suffuses itself into all sorts of activities. Seidenberg again:

> The original intention is to mimic a portion of the Creation ritual. It is in this way that we envisage "counting" to have become detached from the ritual and to have acquired its abstract or general character. Higher counting may have started as a method of taking care of longer and longer processions (not with the idea of counting them, however). The base (which is not logically inherent in counting) corresponds to the number of persons in the basic ritual, and the higher counting derives from the continued repetition, with slight modifications, of this basic ritual. (p. 10)

The child who puts her stuffed animals one by one on a table, naming each one as it is placed there (with its proper name, of course, and not as 'elephant' or 'dog'), is thus involved in the rite of counting. These stuffed animals are ordered by time, and not by name. Her myth (the names of the stuffed animals) does not correspond with current English number myths, but that is a mere detail. In this counting, the question of how many stuffed animals are on the table is of little concern. However, the first one there will likely have meaning, and the difference between the third and the fifth ones too, *in terms of the order in which they appeared,* not in terms of the quantity of stuffed animals thus far named. I note that while it is customary to think of the counting myth in terms of named numbers ('three' or 'eight'), we can also think of them in their symbolic form ('3' or '8'). The child who writes a series of marks on the paper, 1 2 3 4, etc., can thus also be seen as participating in a creation ritual. No sense of quantity is required—no more, at least, than what might be required to be able to write *a b c.*

I am not arguing here that ontogeny recapitulates phylogeny; I am not making a developmental argument. I am simply pointing to some of the distinctive features of number that this counting myth entails. I am interested in it for two reasons: first, it may help elucidate some pedagogical concerns such as the relative importance of ordinality and cardinality in the teaching of counting and arithmetic; second, it is suggestive of an ontology of number that is deeply temporal, as well as rhythmic. In fact, I will argue in the next section that the power of ordinality—its capacity to function as a tool or a device—lies precisely in its temporality.

There are at least two strands of research that have attempted to reinstate ordinality into the teaching and learning of mathematics, which I mention briefly here because they support my argument that ontological questions are inseparable from pedagogical ones. The first concerns the work of Caleb Gattegno, who took ordinality as the primary dimension of the developing a concept of number [6, 11]. In Gattegno's curriculum, there is first a focus on associating number names with number symbols, along with finger gestures. The focus is less on determining the number of fingers that are unfolded than on coordinating different counting myths as well as coordinating counting myths and counting rites. Children then play with Cuisenaire rods (wooden blocks with 1 cm square faces and different lengths—each one associated with a specific colour) and work on relations (such as longer than, shorter than). The first number to be introduced is "2", to represent the action of placing two rods of the same size to match the length of a single rod. "2" is thus associated not with a single rod, but with a ratio between rods—no matter how long those rods are. But children are not asked "how many?": numbers are introduced as relations, rather than as denoting objects.[2] Further, there is little emphasis on the procedure of *how to count*. This gives rise to a sense of numbers as relational, rather than being ontologically distinct one from the other; three makes no sense in and of itself, as it might in a cardinal view of number—three *is a relation*, much as the meaning of one deity coming onto the stage is intertwined with who came before and who is to comes after.

Depending on how one approaches the use of the rods, the emphasis could be more cardinal than ordinal. (Interestingly, this is what often happens in primary school classrooms, and can be seen in the *Native Number* iPad App, which assigns a specific number to each rod.) If, for example, a teacher chooses to focus on one rod only, and the length is taken to be a discrete multiple of units, then the attention is to cardinality. If, on the other hand, the length is taken to be more continuous, then the quantity '2' is determined by how long it is in relation to another rod. In this latter case, it seems that length is providing the scene for the rite, rather than time.

The primacy of cardinality is also being challenged in the neuroscience literature, which has hypothesised that humans share an early (in evolutionary terms) Approximate Number System (ANS), which is used to judge the relative size of groups of objects in non-symbolic ways [15]. Some of this research suggests that cardinality is the key to learning early number in that it shows a link between ANS and a symbolic awareness of number. However, other researchers have argued that the experiments used in ANS research assume that cardinality is the primary aspect of number [14]. Based on the results of their own experiments, they argue that the "extraction of ordinal information from the ANS" (p. 257) and codifying relations between symbols for numbers may be key. Performing ordinal tasks with number symbols activates different areas of the brain from tasks involving collections of objects (with or without the link to number symbols) [15]. This means that an awareness of relations among number symbols (in an ordinal sense) and an awareness of how to link objects to numbers involve different processes. There is an argument to be made that the former is much more related to the ritual origins of counting. There is also an argument to be made that in developing a concept of number, ordinality is more significant yet much less emphasised in the curriculum.

In the next section, I consider a new App that has been developed to help support children's early experiences with counting, adding and subtracting. I will use it as

---

[2]This is in line with the Ancient Greek approach to 'ratio number', but markedly different from that of Descartes, for whom a line segment represents "an arbitrarily known number (a "unity") rather than a number to be measured in numbers" [10, p. 65]. Leibniz will take issue with Descartes' approach, in its stead creating a "baroque mathematics" that rejects the elemental notion of a unit of measure in favour of continuity and continuous repetition (see [8]).

an example of what it might look like to re-ontologise number (to move from asking what *is* number, to what can number be, or become), particularly a concept of ordinal number, in terms of its mobility. This application places children in the phylogenetic world of ordinal numbers.

2 Digital encounters with ordinality

The multi-touch device is a novel technological affordance in mathematics education. Through its direct mediation, it offers opportunities for mathematical expressivity by enabling children to produce and transform screen objects with fingers and gestures, instead of engaging and operating through a keyboard or mouse. This makes it highly accessible, in that children can act and manipulate with their fingers, rather than through symbols and words. It also opens the way for new, tangible forms of mathematical communication. In this section, I describe a new application called *TouchCounts*, which can be downloaded free of charge from the App Store [20].

TouchCounts is open-ended and exploratory, rather than practice-and level-driven and supports the development of number by offering modes of interaction with number objects that involve fingers and hands. Specifically, it aims both (1) to engage one-to-one correspondence by allowing every finger touch to summon a new sequentially-numbered object into existence, whose appearance is both spoken aloud and symbolically labelled; and (2) to enable gesture-based summing and partitioning—by means of pushing objects together and pulling them apart in a way that exposes very young children to arithmetic operations. Currently, there are two sub-applications in *TouchCounts*, one for Enumerating and the other for Operating. I will only describe the Enumerating world in this chapter, but readers are invited to visit the website (www.touchcounts.ca) or to download the App to see how the Operating world works.

In the Enumerating world, a user taps her fingers on the screen to summon numbered objects (yellow discs). In the terminology used by Seidenberg, this is the rite. The first tap produces a disc containing the numeral "1". Subsequent taps produce successively numbered discs. As each tap summons a new numbered disc, *TouchCounts* audibly speaks the English number word for its number ("one", "two", . . .). This is the myth that accompanies, that narrates, the rite. Fingers can be placed on the screen one at a time or simultaneously. With five successive taps, for instance, five discs (numbered '1' to '5') appear sequentially on the screen, which are counted aloud one by one (see Figure 2a). However, if the user places two fingers on the screen simultaneously, two consecutively numbered discs appear at the same time (Figure 2b), but only the higher-numbered one is named aloud ("two", if no previous taps have been made). The entire 'world' can be reset, to clear all numbered discs and return the 'count' of the next summoned disc to one. Note that the discs always arrive in order, pre-baptised, with their symbolic names imprinted upon them.

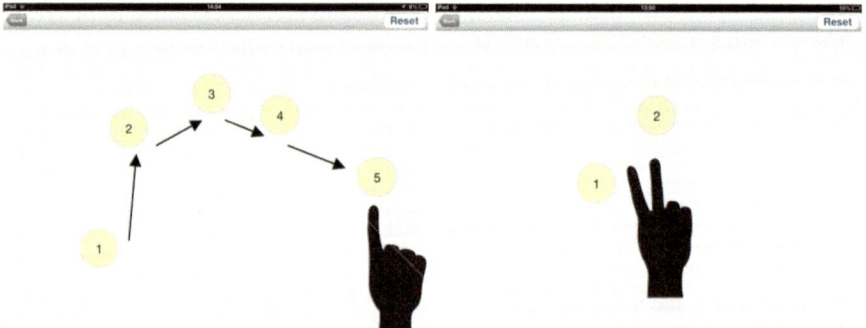

Fig. 2 (a) Five sequential taps—"one, two, three, four, five" is said (the arrows are only to indicate the sequence; they are not shown on the screen); (b) A simultaneous two-finger tap—only "two" is said (both discs appear simultaneously)

Earlier, I described the common practice of early school mathematics, which focuses attention on the "how many?" question, where children are asked to count a given set of objects. This question seems to provoke a routine of sequential counting in which the attention is focused on the last word said. In *TouchCounts'* Enumerating world, however, the child is engaged in a somewhat different practice—rather than counting a *given* set, she is actively *producing* a chain of objects with her finger(s) and these objects seem to count themselves (both aurally and symbolically) as they are summoned into existence. This echoes with Seidenberg's contention that the origin of counting lies in a creation ritual, in the bringing onto stage the correct deities in the correct order. At the end of the spoken count (the myth), no trace is left of what has been said. On the screen, however, each action leaves a visual trace, one (or more) discs bearing numerals of what has once been summoned into being.

If the 'gravity' option for this world is turned on in the App, then as long as the user's finger remains pressed to the screen, the numbered object holds its position beneath her fingertip. But as soon as she 'lets go' (by lifting that finger), the number object falls and then disappears "off" the bottom of the screen, as if captured by some virtual gravity. With 'gravity' comes the option of a 'shelf', a horizontal line across the screen. If a user releases her numbered object above the shelf, it falls only to the shelf, and comes to rest there, visibly and permanently on screen, rather than vanishing out of sight 'below'. (Thus, Figure 3 depicts a situation in which there have been four taps below the shelf—these numbered objects were falling—and then a disc labelled '5' was placed above the shelf by tapping above it.) Since each time a finger is placed on the screen, a new numbered disc is created beneath it and, once released by lifting the finger, begins to fall, one cannot "catch" or reposition an existing numbered object by re-tapping it.

Discs dropping away (under 'gravity') mirror the way spoken language fades rapidly over time, with no trace left—the impermanence of speech. Also, with discs disappearing, any sense of cardinality goes too: the disc labelled '2' is simply the second one to have been summoned by the rite, in the absence of the presence of '1'.

Fig. 3 After four sequential taps below the shelf and a fifth tap above the shelf (the yellow discs 1, 2, 3 and 4 will fall off the bottom of the screen)

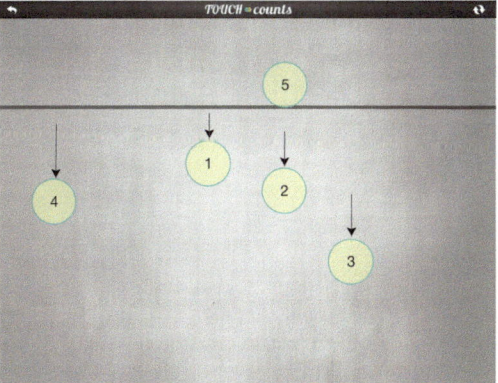

So the Enumerating world with 'gravity' enabled (it is an option) is almost entirely an ordinal one, with the shelf acting as a form of visible memory.

One of the characteristics, then, of *TouchCounts* is that the computer handles the counting (the iPad is the one who announces and manages the arrival of various figures onto the ritual scene). While the initial design of the App was to support children's one-to-one correspondence (between pointing and number naming), it quickly became apparent that children could also be corresponding finger touches with number naming, object-creating and symbol-making. This multiplicity of ordinal number production changes what number might mean for these children, as I will describe in these short episodes in which 5- and 6-year-old children play. I will present three episodes in all, each one aiming to highlight the way in which the logical and the ontological are in play.

2.1 "I made 18"

This first episode involves two children, Rodrigo and Grace. In the no-gravity setting, Rodrigo (aged four) was asked to "make five", which he did by tapping with his left index finger along an imaginary horizontal line from left to right, stopping exactly after *TouchCounts* said "five". Grace, aged three, who had not played with *TouchCounts* before, and who had been watching carefully, said, "I want to do it now". She started tapping but not saying anything (Figure 4a). She started on the left and then tapped towards the right of the screen along a horizontal, until she reached the edge of the screen at "six" (with *TouchCounts* counting along). Instead of stopping there, she tapped underneath her numbered disc 1 (thus mirroring writing conventions) and then continued tapping toward the right until she reached the screen edge again, at "twelve". She then placed the numbered disc 13 under the one showing '7', close to the bottom of the screen, and she tapped until she got to the right edge a third time at "eighteen" (Figure 4b) and then took her finger away from the screen.

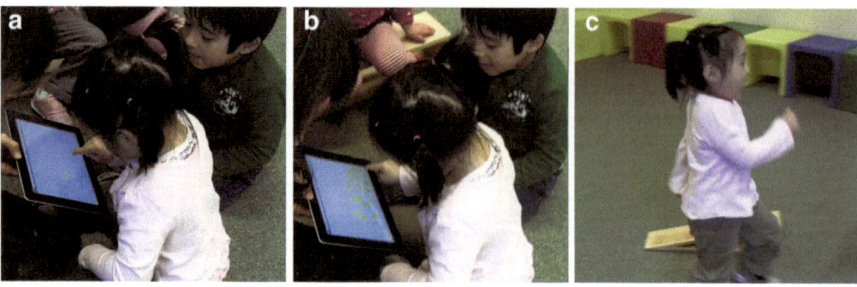

Fig. 4 (a) Grace tapping on the screen; (b) Grace tapping on the third row; (c) Grace strutting around the classroom shouting, "I made eighteen!"

As mentioned above, the classic exchange in this area of early number starts with the question: "How many?" In the initial prompts here, the teacher asks "How many did you make?", despite being well aware of the fact that to ask "How many?" presupposes a great deal of awareness and experience about counting. This point provides a strong example of how some of the apparent 'facts' about early number are apparently conditioned by the very questions used to ascertain them.

I: How many did you make?

Rodrigo: (*Answering for her*) Sixteen, seventeen, s'teen. (*The sounds before "'teen" are hard to hear, but the words are said in a rhythmic manner as he points in turn to discs on the screen*)

Grace: (*Looking up at the adult, smiling*) Eighteen! (*Pointing to the numbered disc 18 at the bottom right of the screen*)

I: You made eighteen. (*Presses Reset*) Now it's my turn.

Grace: (*Gets up and walks away from the group into the middle of the room*) I made eighteen! Eighteen, eighteen, eighteen, eighteen, eighteen, eighteen, eighteen. (*Marching around the room as she triumphantly calls out – see* Figure 4c)

This episode is interesting for several reasons, not least Grace's excitement at having made eighteen, which was not given as a task, but which emerged from her own exploration. Her pride in having made what she did (which is expressed in her excited "I made eighteen") underscores the fact that it was indeed *she* who *made* it, even though it was not she who initially proclaimed it (the iPad actually said "eighteen"). Her excitement shows how even an arbitrary number (not the more commonly-used ones of primary school, like 5 and 10) can be so interesting.

What is surprising, given her very young age, is the fact that she immediately says "Eighteen!", rather than counting all the numbered discs again. Perhaps Rodrigo's mumbling about "teens" oriented her attention to what she had just heard spoken aloud, which dissuaded her from counting them all up. Perhaps because she had made them with her finger, without having to keep count of them herself, she could attend more to the importance of the last spoken number and perhaps associate it with the quantity of discs on the screen.

It is possible to continue counting forever, to go beyond 5 (which Rodrigo did) or 10 (a number she has certainly heard before). This is the logic of seriation. It is important in this context because of the way Grace can count without having to know the number names. She can participate in the rite, while leaving the myth to *TouchCounts*. The more she does, the more the myth will become her own.

2.2 Putting 10 on the shelf

A group of three children are working together at an after school daycare—one is in kindergarten, another in grade one and the third is in grade two. The teacher has asked the children to place just 10 on the shelf (in the gravity mode). They have done this before, but she imposes an added constraint, which is that they should try to do it in a different way. Whyles goes first. He touches:

> with 5 fingers (*TouchCounts* says "five"),
> then 4 fingers (*TouchCounts* says "nine")
> then 1 finger (*TouchCounts* says "ten") above the shelf.

The only thing on the screen was a yellow circle labelled 10 sitting on the shelf. Benford, is next; he touches

> with 5 fingers (*TouchCounts* says "five"),
> then 1 finger (*TouchCounts* says "six"),
> then 2 fingers (*TouchCounts* says "eight"),
> then 1 finger (*TouchCounts* says "nine")
> then puts 1 finger above the shelf (*TouchCounts* says "ten").

The only thing on the screen is a yellow circle labelled 10 sitting on the shelf. Finally, Auden, goes next. He touches

> with 1 index finger (*TouchCounts* says "one")
> then 1 pinky finger (*TouchCounts* says "two")
> then 1 index finger (*TouchCounts* says "three")
> then 1 middle finger (*TouchCounts* says "four")
> then 1 index finger (*TouchCounts* says "five")
> then 1 middle finger (*TouchCounts* says "six")
> then 1 index finger (*TouchCounts* says "seven")
> then 1 middle finger (*TouchCounts* says "eight")
> then 1 index finger (*TouchCounts* says "nine")
> then 1 middle finger (*TouchCounts* says "ten")

There is nothing on the screen except the shelf. Auden presses reset and then touches:

> with 1 index finger (*TouchCounts* says "one")
> then 1 middle finger (*TouchCounts* says "two")
> then 1 index finger (*TouchCounts* says "three")
> then 1 pinky finger (*TouchCounts* says "four")
> then 1 index finger (*TouchCounts* says "five")
> then 1 middle finger (*TouchCounts* says "six")
> then 1 index finger (*TouchCounts* says "seven")
> then 1 middle finger (*TouchCounts* says "eight")
> then 1 index finger (*TouchCounts* says "nine")
> then puts 1 index finger above the shelf (*TouchCounts* says "ten")

It would be tempting to see these different ways in terms of decompositions of 10, but this would be a cardinal perspective on the number experience. Instead, the actions of the children resembles the setting of a digital clock: like Whyles, you hold the button for awhile, then hold it again, almost as long, until you are close, and then hold briefly for the final touch; Benford was more cautious, with a long hold followed by short holds during the second half; finally, Auden proceeds at a constant pace, alternating between different fingers as he proceeds. The time it takes to get to 10 changes, so that 10 now becomes a more plastic number, rather than a fixed extension.

A concept of number that is more temporal runs the risk of defying the Aristotelian conceit of a strict separation between the mathematical and the physical. In the next section, I draw on some contemporary work in philosophy that offers a different set of assumptions around the relation between the mathematical and physical, a set that not only eschews the banning of time and motion from mathematical concepts, but also rather, on the contrary, celebrates time and motion as significant elements in the creation of new mathematics, and not just as pedagogical devices that can be used to help scaffold novice mathematical understanding.

3 Virtual number

One of the assumptions in education is that children should engage in sensori-motor actions (like pointing at objects) in developing concepts. This view acknowledges the role of the human body in coming to know mathematics, but it does not help explain how these sensori-motor activities turn into mathematical concepts.[3] This is sometimes thought to occur through a process of mental abstraction in which, for example, number becomes a concept that is decontextualised, depersonalised and detemporalised. This dualist perspective, which erects a sharp divide between body and mind, can imagine a concept like number as being separate from any particular person, time and place. A contrasting monist perspective, put forth by philosophers such as Gilles Châtelet [5], sees such a concept as always intertwined with person, time and place, so that a concept is not fixed either historically or geographically; mathematics cannot be dissolved from 'sensible matter'—it is in a perpetual state of becoming. The advantage of this perspective is that it can account for the way a concept can begin, and continue to evolve, as a certain social practice in the physical world. Further, it also explains how concepts can remain alive, open to new forms.

---

[3]When embodied cognition theorists such as Lakoff and Núñez [13] describe the container metaphor from which the mathematical idea of 'set' emerges, they remain committed to an immaterial mathematical concept of set. Moreover, as I will argue, treating all concepts as metaphorical in relation to the 'real' reinforces the divide between the mathematically abstract and the physically concrete.

Châtelet posits the *virtual* as the necessary link that binds the mathematical and the physical together in mutual entailment. His use of the concept of virtual draws on the work of Gilles Deleuze [8], who writes, "[t]he virtual must be defined as strictly a part of the real object – as though the object had one part of itself in the virtual into which it is plunged as though into an objective dimension" (p. 209). It is the virtual dimension of this matter that animates mathematical concepts like number. Mathematical entities are thus material objects with both virtual and actual dimensions. In fact, doing mathematics involves both realising the possible (logical) and actualising the virtual (ontological).

Both realisation and actualisation bring forth something new into the situation (the possible and the virtual), but realisation plays by the rules of logic, while actualisation involves a different kind of determination, one that generates something ontologically new. Châtelet offers the example of mathematician William Rowan Hamilton's quaternions (which did not come about through logical deduction), as well as the point at infinity (which constituted a new ontological object). The virtual pertains to the indeterminacy at the source of all actions, whereas the possible pertains to the compliance of our actions with logical constraints. Thus novelty, genesis and creativity are fundamental concepts in a theory of actualisation. Actualising the virtual involves "an intrinsic genesis, not an extrinsic conditioning" [8, p. 154]. The virtual in sensible matter becomes intelligible, not by a reductionist abstraction or a "subtraction of determinations" (Aristotle's approach to abstraction), but by the actions that awaken the virtual.

Châtelet uses many examples from the history of mathematics to exemplify the way in which new kinds of entities are invented by actualising the virtual, such as Hamilton's quaternions and Abel's theorem (about indefinite integrals of a rational function). He underscores the important roles that these mathematicians' gestures and diagrams play in awakening the virtual—by creating new objects, new dimensions, new behaviour. Châtelet draws inspiration from the work of Deleuze, especially in terms of that latter's interpretation of Leibniz's "baroque mathematics". In contrast to modern mathematics, this is a mathematics that offers an alternative starting point for rethinking the relationship between immovable mathematics and movable matter. The term *baroque*, used in music and architecture, is in part about what is strange or bizarre, but is also about unpredictability and variation, where there might be a medley of contrasting themes instead of a timeless and uniform one. Deleuze considers Leibniz's approach to the differential calculus to be a baroque one, in that it is concerned with local variability, as is evident in Leibniz's notion of the family of curves: "Instead of seeking the unique straight tangent in a unique point for a given curve, we can go about seeing the tangent curve in an infinity of points with an infinity of curves: the curve is not touched, it is touching, the tangent no longer either straight, unique or touching, but now being curvilinear, an infinite, touched family" (quoted in [9, p. 19]).

Closer to the focus of this chapter, Deleuze also examines the baroqueness of the irrational number. Unlike the "fractional number", which has its particular value, "variation becomes presently infinite" with the irrational number as it is the "common limit of the relation between two quantities that are vanishing" (p. 17).

Fig. 5 Deleuze's triangle
diagram

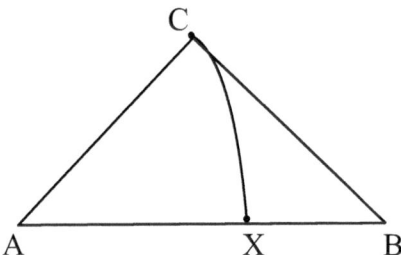

Deleuze provides a diagram of an isosceles triangle in which a curved line starting at the vertex C descends towards the straight segment of rational points AB (see Figure 5) and "exposes [AB] as a false infinity" (p. 18). Where the curved line intermingles with the straight line (at X) a point-fold is produced that forms the irrational number as the common limit of two convergent series. The irrational number is not a particular value, nor even a particular ratio (of AX to AB), but an infinite variation.

Philosophically, Leibniz sees the world as comprised of an infinite number of monads, each with its own distinct point of view and each presupposed by all the others. The ontology of monads feeds into Leibniz' theories of a relative space–time continuum (*spatium*) conceived as a fluid of relations and differentials. The monadology is a metaphysical counter to Descartes' view of nature's passivity. Within this fluid world of differential relations, actions of any kind are conceived as folds in the spatium. A mathematician's gesture or diagram creates a new fold on the surface, like the curve cutting the rational segment AB; it pleats and creases matter [9].

As Elizabeth de Freitas and I have shown in [7], novices can also be involved in such inventive acts, provided that there is opportunity for actualising the virtual. We provide the example of a group of grade 1 children doing just this in the context of deciding when two lines will intersect. They first had to extend the surface of the plane beyond the visible one on which lines had been drawn, and then make manifest a point of convergence on this extended plane. This enabled the children to come up with two theories for when two lines will intersect: when the "thickness" between the line changes and when one line is "slanted" more than the other. We underscore the kind of baroque mathematics at work in this example, which we see as significantly related to the dynamic geometry objects that the children were invited to interact with; moving points turned into lines, which also began to move on the screen.

For Grace, the action of repeatedly placing her finger on the screen in order to produce a new object/number/sound can be seen as realising the possible. The actualising of the virtual is in the particular spatial and temporal folds of the making of 18. Because of the way she taps on the screen, 18 is the natural conclusion of the screen size, in that her rows are limited to 6 and her columns to 3. This particular, and likely accidental, arrangement ends up giving 18 a kind of personality of its own that raises it above a mere stepping stone between 17 and 19. Like the Leibnizan monad, 18 gains local perspective from which different directions are possible—it could be the end, as it was in this case, or the number under 12, the launching point for 19, or any other new line of flight. The spacing of the tapping also renders the counting as 1, 2, 3, 4, 5, 6, pause, 7, 8, 9, 10, 11, 12, pause, 13, 14, 15, 16, 17, 18. This is baroque mathematics at work because it breaks free from the equi-ordering of counting, the assumption that each number is the same, modulo position. Grace disrupts this monotony by playing with time, until she pauses at 18. She invents a new way for time to control the count: at the end of the tapping, the count stops. This is what enables Grace to say that she *made* eighteen: it was not the biggest number, nor the total number, but the *last* number. Within time, the question of what comes before and after is sorted out. Time provides the order.

In the second example, the realising of the possible relates to a certain kind of order: each new tap will get closer and closer to ten (and might even surpass it). This may seem simple, but it is the very idea that the more you count, the bigger the number gets (where bigger is circumscribed in time), that provides a first way of making sense of the number song. When these children keep tapping, they often go past 100, and even try to get to 1000. The more time it takes for them to get to 100, the bigger their sense of that number. Indeed, in one Kindergarten class, where children played at getting to hundred by repeatedly tapping four times below the shelf and once above, until they got to 200, a child, said "I thought that two hundred was right after one hundred, but it's not". The use of "right after" echoes the fact that there were a long sequence of taps between putting 100 on the shelf and placing 200 there—this length of time will eventually become a question of size for the child, but for now it is a fact about ordinality.

In terms of actualising the virtual, it is evident in the children's new ways of counting. The one, which has been the motor of counting thus far, now flowers into a multiplicity: fives, fours and twos. If one deity at a time was coming on stage before, now we have several (and perhaps, in the Seidenberg sense in which counting by 2s is the origin myth, the real magic is in going from many to one). And not only can we have several, but also we can have different amounts of them (for Whyles and Benford at least). They can also come on stage in a different way (like for Auden, who baptises them with different fingers on different parts of the screen). The ontological function at work here can be seen as a folding of the continuum, where the previously continuous equivalent repetition varies into a continuous non-equivalent repetition.

If a concept is to function both logically and ontologically, it must sustain a certain vibrancy and vitality. In other words, a concept of this kind must be a multi-purpose device that resists reification while carving out new mathematical entities. Châtelet refers to these as 'allusive devices' that give rise to thought experiments that "penetrate closer to the heart of relation and operativity" by being deliberately productive of ambiguity [5, (p. 12)]. This seems relevant to the mathematician, who clearly wants concepts that that can be generative, as well as to the mathematics educator or teacher who sees learning as a creative enterprise. Seeing concepts in this way may also help explain some aspects of contemporary school mathematics, in which concepts are overly constrained towards the logical. Based on Châtetet's historical study of inventive mathematics, de Freitas and I argue that one way to redress the balance is by mobilising mathematics. Putting mathematics into motion reinstates the messier aspects of mathematics: its temporality, in particular, but also its indeterminacy. *TouchCounts* can be seen as an attempt to mobilise number, which is done in part by focusing on its ordinal nature.

4 Becoming baroque

Seidenberg's hypothesis about the ritual origins of counting need not be corroborated for it to help answer the question of what number can become. That is because it offers a practice (imaginary or not) in which numbers could come to matter, which may either anticipate or explain unknown practices. This was true in the context of *TouchCounts*, where the children's own rituals around tapping and chanting—and, especially, bringing objects on the stage of the shelf—can be seen as a new kind of ordinality that is often overlooked, both in the philosophy of mathematics and in mathematics education. Such an ordinality is intrinsically becoming in the sense that it is thoroughly temporal. And, at least in the context of *TouchCounts*, it is also tangible and aural, as well as symbolic. It is also becoming in the sense that it has built-in lines of flights; that is, it has both the logical and the ontological so that it is pregnant with the possibility of new meanings. The excerpts that I discussed in the previous two sections provided examples of how these new meanings have emerged in a very particular setting.

As mentioned above, I am not interested, in this chapter, in making a pedagogical argument about the effectiveness of *TouchCounts*. Instead, I use it as a particular case that enables us to see a kind of ordinality that may be insufficiently appreciated under the pressure of cardinality. There occurred a similar phenomenon with respect to the concept of ratio: the Euclidean approach to ratio effectively replaced the early Greek *anthyphairesis* (which means reciprocal subtracting in turn—it's a verb!). These pre-Euclidean ratios were actually a sequence of "repetition numbers", which effectively cut off a major phylogenetic root [10, p. 31]. Like cardinality, the Euclidean ratio obeys a more static, unitary form. In both cases, re-ontologising these concepts involves the kind of baroque focus on the local, the multiple and the temporal. Rather than argue that history should somehow be respected, and that our sense of what number can be or should be rooted in that history, I am interested in how those particular origins help us understand what number might become, particularly in a radically new digital era in which the hegemony of the alphanumeric (static, linear inscriptions) will give way to visual and dynamic mathematical expressions [18]. In this new world, might a baroque mathematics of the ordinal even give rise to a baroque mathematics of the cardinal?

Acknowledgments I would like to thank David Pimm for his many helpful comments and discussions about ordinality, the baroque and time. Thanks also to my research team and the children involved in the *TouchCounts* project.

References

1. A. Badiou, *Number and numbers,* Polity Press, 2008.
2. P. Benaceraff, "What number could not be", *Philosophical Review, vol 74*, no. 1 (1964) 47–73.
3. L. Lunney Borden, "The 'verbification' of mathematics: Using the grammatical structures of Mi'kmaq to support student learning", *For the learning of mathematics, vol 31, no.* 3 (2011) 8–13.
4. W. Burkert, *Lore and Science in Ancient Pythagoreanism* (trans: Minar, E.), Harvard University Press, 1972.
5. G. Châtelet, *Figuring space: philosophy, mathematics and physics*, (trans: Shore, R., & Zagha, M.), Kluwer, 2000.
6. A. Coles, Ordinality, neuro-science and the early learning of number, In C. Nichol, S. Oesterle, P. Liljedahl & D. Allen (Eds), *Proceedings of the joint PME 38 and PME-NA 36 conference* (Vol. 2, pp. 329–336), Vancouver, BC: PME, 2014.
7. E. de Freitas, & N. Sinclair, *Mathematics and the body: Material entanglements in the classroom*, Cambridge University Press, 2014.
8. G. Deleuze, *Difference and repetition* (trans: Patton, P.), Columbia University Press, 1994.
9. G. Deleuze, *The fold: Leibniz and the baroque* (trans: T. Conley), Athlon Press, 1993.
10. D. Fowler, *The mathematics of Plato's academy: A new reconstruction*, Clarendon Press, 1999.
11. C. Gattegno, *The common sense of teaching mathematics*, Educational Solution Worldwide Inc, 1974.
12. R. Gelman & C.R. Gallistel, *The child's understanding of number*, Harvard University Press, 1978.
13. G. Lakoff & R. Núñez, *Where mathematics comes from: How the embodied mind brings mathematics into being*, Basic Books, 2000.
14. I. Lyons & S. Beilock, Numerical ordering ability mediates the relation between number-sense and arithmetic competence, *Cognition, vol 121, no.* 2 (2011) 256–61.
15. I. Lyons & S. Beilock, Ordinality and the nature of symbolic numbers, *Journal of Neuroscience, vol 33, no.* 43 (2013) 17052–17061.
16. A. Nieder & S. Dehaene, Representation of number in the brain, *Annual Review of Neuro-science, vol 32* (2009) 185–208.
17. B. Raglan, *The hero: A study in tradition, myth and drama*, Greenwood Press, 1936/1975.
18. B. Rotman, *Becoming beside ourselves: The alphabet, ghosts, and distributed human beings*, Duke University Press, 2008.
19. A. Seidenberg, The ritual origin of counting, *Archive for History of Exact Sciences, vol 2, no 1* (1962) 1–40.
20. N. Sinclair & N. Jackiw. *TouchCounts* [software application for the iPad], 2011, https://itunes.apple.com/ca/app/touchcounts/id897302197?mt=8.

Enumerated entities in public policy and governance

Helen Verran

Abstract A poster promoting policy change in Australia's fisheries, which features a list of seven numbers in its effort to persuade, serves as occasion to do ontology. A performativist analytic is mobilized to show the mutual entanglements of politics and epistemics within which these numbers, agential in policy, come to life. Three distinct types of entanglement are revealed. Numbers have different political valences, and mathematicians need to attend to these in order to act responsibly as mathematicians.

1 Introduction

The poster reproduced below appeared in a document published in 2011. The paper in which it appeared, "Stocking Up: Securing our marine economy," proposed a radical new policy for Australia's fisheries [7]. For me the poster serves as occasion to focus-up numbers as they feature in contemporary environmental policy and governance. The analogy here is focusing-up particular elements in a field in microscopy; settling on a particular focal length for one's examination. The focal length I adopt, bringing numbers into clear view, is unusual in social and cultural analysis. But that focus is not the only way my commentary on the poster is unusual.

I work through a performativist analytic. And the first sign that mine is an unorthodox approach is that I will refer to enumerated entities rather than numbers from now on. From the point of view of my socio-cultural analysis, enumerated entities are very different from numbers, although a practitioner, a user, would certainly think that the difference, which I elaborate in my next section, rather exaggerated. For a user the terms number and enumerated entity can be interchangeable, but for a philosopher who wants to understand the mutual entanglements of epistemics and politics in contemporary life, a number is different than an enumerated entity.

A quick way to imagine my performativist approach to analysis is to remember John Dewey's description of his examination of "The State" in his analogous

H. Verran (✉)
School of Historical and Philosophical Studies, University of Melbourne, 18 Newgrove Road, Healesville, Melbourne, VIC 3777, Australia
e-mail: hrv@unimelb.edu.au

© Springer International Publishing Switzerland 2015
E. Davis, P.J. Davis (eds.), *Mathematics, Substance and Surmise*,
DOI 10.1007/978-3-319-21473-3_18

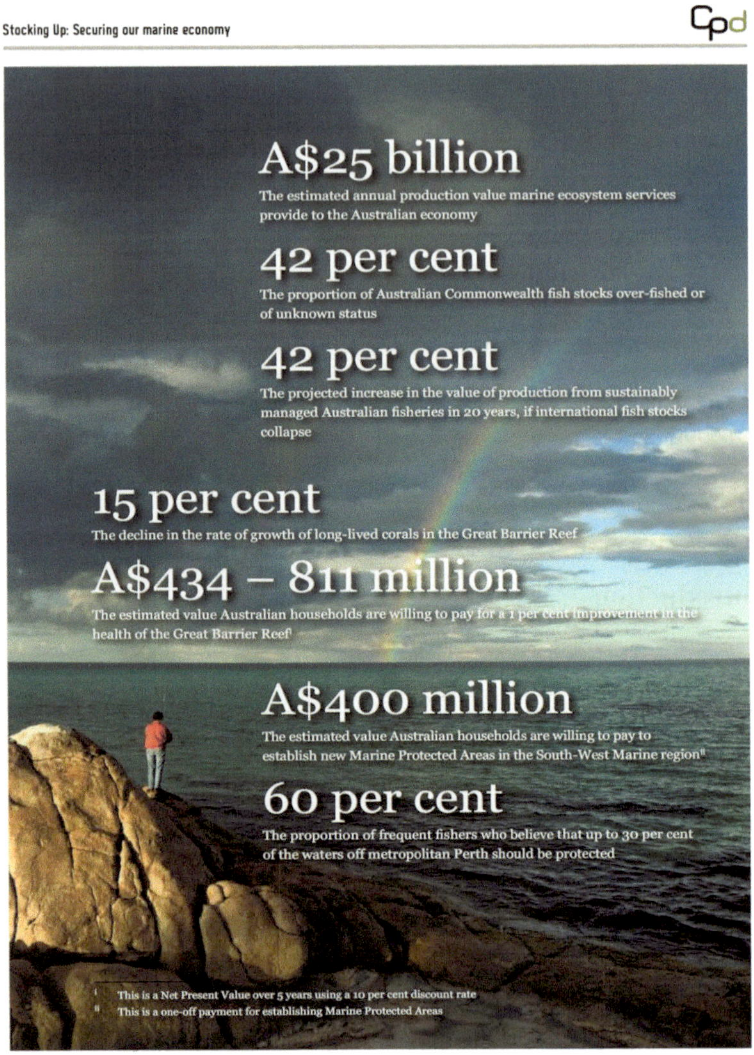

A$25 billion
The estimated annual production value marine ecosystem services provide to the Australian economy

42 per cent
The proportion of Australian Commonwealth fish stocks over-fished or of unknown status

42 per cent
The projected increase in the value of production from sustainably managed Australian fisheries in 20 years, if international fish stocks collapse

15 per cent
The decline in the rate of growth of long-lived corals in the Great Barrier Reef

A$434 – 811 million
The estimated value Australian households are willing to pay for a 1 per cent improvement in the health of the Great Barrier Reef

A$400 million
The estimated value Australian households are willing to pay to establish new Marine Protected Areas in the South-West Marine region

60 per cent
The proportion of frequent fishers who believe that up to 30 per cent of the waters off metropolitan Perth should be protected

ⁱ This is a Net Present Value over 5 years using a 10 per cent discount rate
ⁱⁱ This is a one-off payment for establishing Marine Protected Areas

attempt to understand how the concept of 'the public' is entangled in epistemics and politics [5]. "The moment we utter the words 'The State,' a score of intellectual ghosts rise up to obscure our vision," he wrote. The same happens with 'number.' Dewey recommended avoiding a frontal approach, and instead opting for "a flank movement." A frontal approach inevitably draws us "away from [the actualities] of human activity. It is better to start from the latter and see if we are not led thereby into an idea of something which will turn out to implicate the marks and signs which characterize political behavior. There is nothing novel in this method of approach," he asserts. "But very much depends on what we select from which to start . . . We must start from acts which are performed . . . " Accordingly, I consider the ordinary, everyday routines, the collective acts, by which enumerated entities come to life, routines which nevertheless entail metaphysical commitments. In doing this I am undertaking ontology. I argue that enacted in this way and not that, enumerated entities have particular constitutions; they are 'done' in collective actions with machines and bodies as well as in collective actions with words and graphemes (like written numerals). Commitments, enacted as part of those collective actions, to there being 'particular somethings' which justify the doing of those routines, add up to metaphysical commitments. Unearthing those commitments is ontology.

I am bringing the particular origins and character of the *political* aspects of enumerated entities into focus. Such characteristics are embedded in enumerated entities as they come into being with particular epistemic capacities. As Dewey notes of his enquiry, there is nothing new revealed here. I adumbrate the everyday routines in which the partialities lie (both in the sense of being implicated in certain interests and hence not impartial, and in being of limited salience and referring to specific aspects of the world), and thus show politicization of enumerated entities. In doing so, I suspect many readers will think 'So what? That's just ordinary.' Yes and that is the point. I am focusing up the political responsibilities that go along with being 'a knowledge worker'—a scientist or a mathematician. The rather unusual 'focal length' and framing of my study reveal that the work of scientists and mathematicians, contrary to popular understandings which proposes such work as 'objective,' actually bears "the marks and signs which characterize political behaviour," to use Dewey's words. Scientists and mathematicians of course are expected to behave responsibly when it comes to epistemic practices. But recognizing enumerated entities as *political* and their knowledge making as inevitably partial (both not disinterested and limited), implies that they have further responsibilities too. In my opinion, *these* responsibilities too, should be remembered by scientists and mathematicians as they participate in policy work.

2 How is an enumerated entity different than a number?

In doing ontology in a performativist analytic, Dewey's image of what he set out to do, is a good start in imagining how my analytic is constituted as a very different sort of framing. As Dewey has it, a performative analytic looks side-on at the conventional account of numbers. An enumerated entity as known in a performative social science analytic, is different from a number as understood by most ordinary modern number users. While the latter belongs in a stable state of being, the former exists in the dynamic of emergence realness.

Here's a definition of enumerated entities as I understand them. Enumerated entities as objects known in a performative analytic, are particular relational beings; they are events or happenings in some actual present here and now. Importantly there are quite strict conditions under which enumerated entities come into existence as relational entities in our worlds. Part of the wonder of enumeration is that sets of very specific practices can indeed happen just as exactly as they need to, again and again, as repetitions. Sets of routine practices must be collectively, and more or less precisely, enacted in order for an enumerated entity to come to life. So, in looking 'inside,' what most people would simply call numbers (and probably not imagine as having an inside), I elaborate the every day techniques by which they come to life. Like all entities in a performative analytic such as I mobilize here, the several kinds of enumerated entities I consider in asking about the family of enumerated entities displayed on the poster, are relational particulars named by singular terms.

On one side, the roots of the performative analytic approach lie with a conceptual turn in the ordinary language philosophy of Wittgenstein [19] and Ryle [13] as mobilized in the subsequent 'speech act theory' of Austin [1] and Searle [14]. These philosophies of language evaded the dead-hand of a posited external logical framework with which to fix the proper meaning of linguistic terms. Later in the century, working from the material world, the 'things end' of meaning making so to speak, the actor network theory of Latour, Callon and Law [10] mobilizing a version of French post structuralism offered what might be called 'materiality act theory.' As in the earlier linguistic shift, this approach too circumvented the dead-end of positing some external 'real physical structure' with which to fix meanings things have. Linking these two distinct 'turns' to performativity by making an argument for word using, including in predication and designation, as a form of embodiment [17] I constituted a means for recognizing the performativity of the conceptual across the full range of meaning making with words and things. This analytic frame of performative material-semiotics (to give it a formal name) is particularly generative when it comes to 'looking inside' numbers, which are first and foremost entanglements of linguistic terms (either as spoken words or inscribed graphemes) and things; entanglements of words and stuff.

Here's an example of particular metaphysical commitments being 'wound-in' as enumerated entities come to life in a state institution, and the sort of thing we can see when we 'look inside numbers.' Mathematicians in the state weather bureau are required to offer predictions of future temperatures and future rainfall. They use models and data sets they know to be reliable, and which have actually been constituted as 'clots' of differing materializing routines, and differing linguistic gestures (written or spoken). In the mathematicians' careful work, knowledge, or justified belief about the future is generated. The state's epistemic grasp is extended to the future although of course the future is unknowable. At the core of this explanation of this vast collective endeavor is a metaphysical commitment: the future will resemble the past. This commitment is embedded as those enumerated entities that are temperature and rainfall predictions are constituted. But now pay attention to the enumerated entities that make up the data—lists of actual past temperature and rainfall readings produced in the extensive and enduring institutional work of the state weather bureau, data which are fed into models. We all know that these enumerated entities are different to the predictions. There is a quite different set of metaphysical commitments buried in those assiduously assembled values, just as they are produced in quite differing routines than the predictions. The commitment here is to the past as real, an actual then and there that happened, with rainfall and temperature having these specific values. So the enumerated entities which are predictions of temperature and rainfall, have different constitutions than the readings of past temperature and rainfall, and of course the predictions fold in the readings and are more complicated; each step introduces particular metaphysical commitments. As a socio-cultural analytic, performativity takes quite seriously the minutiae of the actual generation of enumerated entities, details which applied mathematicians are quick to cast aside.

3   The Poster Displays Three Types of Enumerated Entities

Most of the enumerated entities listed on the poster exemplify the type generated in a style of reasoning that more or less simultaneously emerged in seventeenth century European experimental enumeration [16], in mercantile life [11], and in governance practices of the early modern state [15]. This empiricist type of enumerated entity 'clots' in an elaborate set of categorical shifts [16]. In making up a stabilized and self-authenticating style of reasoning [9] emergent in the epistemic shifts of the innovative laboratory practices of the likes of Galileo, these empiricist enumerated entities evaded the domain of the dominant idealist enumerated entities which came to life in the scholastic institutions of the established Christian church [6]. These empiricist enumerated entities, coming to life in the political and epistemic maelstrom of seventeenth Europe, are politico-epistemological units. And they became the mainstay of governance and policy practices in organizations doing administration in early modern states and in early modern capitalism [11].

What empiricist enumerated entities offer to experimentation, commerce, and state governance is a way of managing the sorites, or heap paradox with enough certainty. The sorites paradox arises in vague predicates like heap, pile, or territory. If one starts with a pile of rice and begins taking grains away, when does it stop being a pile? Or if one starts as a king with a territory and keeps giving bits of it away to supporters to defray costs, when does one stop being king ruling the vague territory of a kingdom, and become just another landowner with a specific area of land? Managing this paradox in an actual circumstance is a contingent matter radically dependent on context.

Empiricist enumerated entities, as they came to be in the so-called scientific revolution are a way of rendering the vague whole of the empirical world, of what is capitalized, or of what is governed (which in the case of the early modern state was territories and populations), into specific units in and through which experimental proof, capitalization, and governance, respectively, can proceed (natural and social facts). Rendering a vague whole into specific units to effect the last of these, governance, is necessarily a form of state politics. In the early modern state as much as the contemporary state, the empiricist enumerated entity establishes an interested order as the order of government—the so-called natural order in this case, which inevitably distributes goods and bads differentially amongst the peoples governed. Empiricist enumerated entities offer a means for doing distributive politics through entities which offer a flawless nonpolitical appearance as epistemic units. These enumerated entities embed propositions about an actual (although past) world, and proclaim the epistemic soundness of the policy they are adduced to support.

This poster offers evidence that enumerated entities of the type which first emerged in governance practices of early modern collective life still do that work managing and stabilizing the vague whole of what is governed, infusing the process with an epistemic seemliness. It shows also that several further forms of enumerated entity have emerged over the past several hundred years expanding possibilities of managing and stabilizing in governance through enumerated entities. Two of the numerical values displayed on the poster are types that emerged in the second half of the twentieth century. One emerged out of statistical reasoning as turned to modeling, and a felt need to seem to extend the epistemic reach of the state into the future [9]. In the process of their constitution these (statistical) enumerated entities enfold particular empiricist enumerated entities. In part this elaborate enfolding has been enabled by the exponential increase in processing power brought about by digital technology. While recognizable as rhetoric in that they embed propositions about a future that is in principle unknowable, these entities are persuasive as back up for the enumerated entities and like the entities from which they are derived, epistemically salient for policy and governance.

The other type of enumerated entity we see on the poster arose as a modification in late twentieth century governance practices in state organizations, as they shifted the focus of their administrative practices away from biophysical and social matters (territories and populations) to having economic infrastructure as what it is that they govern. In a profound ontological shift, a new vague whole—a financialized economic infrastructure, became as the object of government, as economizing, a style of reasoning previously confined to economics, came to life as the primary means of operationalizing governance practices of the state, at the end of the twentieth century. In part this occurred through the activism of the 'new public administration' social movement with its commitment to "accountingization" [12], but also in association with the enhanced operationalizing capacity that digital technologies offer. The workings of these two relatively recent forms of enumerated entities (statistical and economistic) in policy and governance, builds on the self-authenticating capacities of empiricist enumerated entities. In different ways they each offer further means for managing the sorites paradox in rendering the vague whole of what is governed into the specific units in and through which governance proceeds. The further enfolding, embedding as it does further metaphysical commitments, embeds further partialities, further politicization.

4   The Poster's Empiricist, Statistical, and Economistic Enumerated Entities

In this section I attend to three of the enumerated entities listed on the poster in detail, speculating on the actual activities in which they came to life, discerning metaphysical commitments embedded in them, and hence, what I call their political valence. This section elaborates the nitty-gritty work of doing ontology in a performative analytic. The first enumerated entity I consider in detail is of the empiricist type—a natural fact; then a statistical enumerated entity, and next an economistic enumerated entity. Before I begin this detailing of the everyday I remind you that I am drawing attention to things 'everybody' knows and which most of us have learned to ignore. As you find yourself getting impatient I ask you to also recall that in doing their epistemic work these enumerated entities are agential—they are 'empiricising,' 'staticising,' and 'economising.' Once they begin to circulate in collective life enumerated entities have effects—they begin to change our worlds, at a minimum this occurs by directing our attentions to some things and not others, although some would claim that they have other sorts of capacities too. These enumerated entities are politically agential.

If one had lived through the savagery of Europe's seventeenth century, I can imagine that finding oneself sharing one's world with an empiricizing enumerated entity, might come as something of a relief, as indeed I discovered during the eight years I lived in Nigeria. Being politically agential is not necessarily a bad characteristic in an enumerated entity. But it is bad politically if we fail to recognize that and how such entities are politically active. And while the staticizing style of the language of the weather forecast might irritate many, the style of reasoning is made clear, the nature of the claim is explicit. Those who heed (or fail to heed) the enumerated entity embedded in a weather forecast must accept that it is they who make the decision in accord with it, and they bear the responsibility. But what of economistic/economizing enumerated entities, which unlike staticizing enumerated entities carry no instructions on how they are to be engaged with? What are the responsibilities come along with advocating governance practices through these entities? This is not the place to pursue that question. But my hope is, that such questioning emerges from what follows.

i. An Empiricist Enumerated Entity The type of enumerated entity we are concerned with here presents truth claims about a specific aspect of the immediate past of Australian nature or Australian society:

- 42 per cent. The proportion of Australian Commonwealth fish stocks over-fished or of unknown status;
- 15 per cent. The decline in the rate of growth of long-lived corals in the Great Barrier reef;
- A\$434–811 million. The estimated value Australian households are willing to pay for a 1 % improvement in the health of the Great Barrier Reef;
- A\$400 million. The estimated value Australian households are willing to pay to establish new marine protected areas in the South-West Marine Region;
- 60 per cent. The proportion of frequent fishers who believe that up to 30 per cent of the waters off metropolitan Perth should be protected.

Making claims about 'a then and there' that once was 'a present' in which actual humans undertook specific sets of enumerating practices, in some cases these entities claim a precise value for a property of nature, and in others they claim a precise value for an attribute of society. In the history of the emergence of enumerated entities, it was nature, subject to the interrogations of the early experimentalists that first yielded empiricist enumerated entities. In part because of the success of this style of reasoning, society emerged as nature's other, and it became possible to articulate social empiricist enumerated entities.

Here, in excavating what is inside such enumerated entities, I focus on '15 percent. The decline in the rate of growth of long-lived corals in the Great Barrier Reef' as an enumerated fact of the natural world. In attending to the banal routines in which this natural empiricist enumerated entity came to life I will revisit a paper "Declining Coral Calcification on the Great Barrier Reef" published in the prestigious natural science journal *Science* in 2009 which the policy paper cites as the origin of the enumerated entity. We are looking for the metaphysical commitments, the 'particular somethings' which justify the collective doing of various routines and arrangements embedded in this entity.

Fifteen per cent is a way of expressing the fraction 15/100, which in turn is a simplification of the ratio of two values arithmetically calculated from three hundred and twenty eight actual measurements on coral colonies. Examination of the skeletal records embedded in some actual corals which grew at particular places in Australia's Great Barrier Reef, and were collected and stored in an archive in an Australian scientific institution, showed that they grew at an average rate of 1.43 cm per year between 1900 and 1970, but only at 1.24 cm per year between 1990 and 2005 [3]. Growth slowed by 0.19 cm per year, which might be expressed as the fraction (0.19/1.24) or 15 per cent which, represents the value of the decline in the growth rate of long-lived coral.

This '15 per cent' is a truth claim, it is an index of a situation understood as having existed 'out-there.' It precisely values a characteristic of the natural world. The decline in the rate of growth of long-lived corals in the Great Barrier Reef' has been put together through the doing of many precise actions on the part of many people.

> We investigated annual calcification rates derived from samples from 328 colonies of massive Porites corals [held in the Coral Core Archive of the Australian Institute of Marine Science] from 69 reefs ranging from coastal to oceanic locations and covering most of the >2000-km length of the Great Barrier Reef (GBR, latitude 11.5° to 23° south...) in Australia. [3]

Imagine the justifications that lie behind the institutional resources that have been assembled and which are drawn on in contriving this entity: the instruments (from extremely complicated and complex—X-ray machines and GPS devices, to simple—hammers and boats), the work of sun burned bodies of scientists swimming down to chop pieces off coral colonies, and of the disciplined fingers of technicians who prepare specimens, and so on. Speculatively retrieving the human actions we recognize what a vast set of relations this enumerated entity is, albeit a modest little enumerated entity. And, this statement of 'what we did' expresses a huge and well-subscribed presupposition: there is a found natural order 'out there' (in the Great Barrier Reef and other places). Depending on one's particular science, this general commitment to 'the natural order' cashes out in a specific set of commitments which in this case would go something like 'in the found natural order long lived corals grow at a particular rate depending on their conditions of growth.' In turning the vague general nature, a metaphysical commitment to one of reality's far reaching categories, into a specific valuation scientists undertake complicated sets of measuring practices and come up with formal values.

> The 1572–2001 data showed that calcification increased in the 10 colonies from $\sim$1.62 g cm$-$2 year$-$1 before 1700, to $\sim$1.76 g cm$-$2 year$-$1 in $\sim$1850, after which it remained relatively constant, before a decline from $\sim$1960. [3]

Why have I adumbrated all this? I want to reveal the partial nature of this enumerated entity. Here's an example of its partiality. Many of those who use, love, care for, and participate in owing the Great Barrier Reef as an Australian National Reserve are (usually heedlessly) committed to there being a found natural order. But one group of Australians is not so committed. As well as being owned by the Australian State, the territory that is Great Barrier Reef is, according to the Australian High Court, also owned by two rather distinct groups of Indigenous Australians. It is recognized in Australian law that those Indigenous territories are *not* constituted as a natural order, as territory of the modern Australian state is. An enumerated entity that mobilizes that (modern) order is, in Australian political life, inevitably partial. Of course most members of the Indigenous groups, Australian citizens who co-own the Great Barrier Reef with the Australian state, recognize the usefulness of such an enumerated entity and in a spirit of epistemic neighborliness are content enough to have it inhabit their world, and in some cases protect it. Those Australians are making a political decision there. And so is the Australian state, and Australian citizens in general, in choosing to work through empiricist enumerated entities, although few would be able to recognize that. Enumerated entities, in being partially constituted participate in the distribution of goods and bads; they enact a particular sort of environmental justice.

ii. A Statistical Enumerated Entity The second entity from the poster that I consider is the sort of enumerated entity that is currently causing much puzzlement amongst philosophers of science [8]. 'Forty two per cent. The projected increase in the value of production from sustainably managed Australian fisheries in 20 years, if international fish stocks collapse,' or 42/100 is, like that entity I have just considered, arithmetically speaking, a ratio of two values. The denominator A\$2.20 billion per year has been assiduously assembled in one of the vast counting and measuring exercises that go towards constituting Australia's national accounts. It is the total of value of four classes of Australian fish products.

The numerator has quite different origins. This value emerged not through counting and measuring by specific people in specific places at specific times, in generating empiricist enumerations that embed commitment to there being a vague nature, and a vague general entity 'society' in which 'the economy' can be found— all metaphysical commitments, but through calculation by a computer. To generate the prediction that the total value of Australia's fish products in 2030 would be

A\$3.11 billion per year, a carefully devised computer programme, plied with rich streams of data, only some of which is derived from actual measurements, enfolded the A\$2.20 billion per year value in a complicated set of calculative processes.

The difference between the actual value of the recent past (derived in actual counting and measuring actions) and the foreshadowed value of a carefully imagined future, would amount to an increase of A\$0.91 billion per year. We can thus render the computed increase as a ratio: 0.91/2.20, which is more felicitously rendered as 42/100, or forty two per cent. Whereas A\$2.20 billion per year (the denominator), indexing a collection of actual fishery products, an accumulation of the value of all fishery products landed each day in all Australian fishing ports for the year 2010, makes a claim about an actual situation, A\$3.11 billion in 2030, and hence A\$0.91 billion per year, cannot be such a claim, and so neither is 42 per cent, since 2030 has not yet happened. A\$2.20 billion and A\$3.11 billion purport to index the value of the fish products landed by Australian fishers in 2010 and which will be landed in 2030, respectively. They present as the same sorts of enumerated entity, and allow a seemingly trustworthy claim: we can expect, in certain circumstances, an increase of forty two per cent in the value of four national fisheries products. But even the cursory examination I have just made reveals there is a profound distinction between these two values. The distinction concerns the metaphysical commitments embedded in these two enumerated entities.

The particular program which generated the denominator of this ratio is called IMPACT (International Model for Policy Analysis of Agricultural Commodities and Trade) developed and maintained in the International Food Policy Research Institute in Washington DC. Among the categories included in the algorithms used to generate this value were the previously mentioned four classes of fish products, differentiated as captured or farmed, and including two categories of animal feedstuffs made from fishery products. Embedded in this numerator are values very like the denominator A\$2.20 billion per year [7].

The denominator A\$2.20 billion reports the past. It is a representation contrived in an elaborate set of institutional routines which start with actual people, perhaps wearing gumboots, weighing boxes of fish of one sort or another, and writing figures down in columns on a sheet attached to a clipboard, or perhaps in a few places, entering values into a computer tablet, that must be protected from splashes and fish scales. By contrast, A\$3.11 billion is connected to such material routines only very weakly, and only to the extent that the values generated in such routines are incorporated into data sets to be manipulated by computer models. The computers in which A\$3.11 comes to life need no protection from the wetness associated with actual fish and fishers. Similarly the institutional and literary routines in which this enumerated entity is generated are very different to those that give life to those like A\$2.20 billion per year. No longer indexical (assiduously assembled in the past through actual people quantifying actual stuff), it becomes an icon where category and value are elided, in which there is no distinction to be made between category and value. It now performs as an element in a particular order. In being enfolded into what will become the numerator in the working of the computer program, the denominator begins to enact a 'believed in' order rather than the quantitative valuing of its previous indexical life.

The processes of the computing undertaken with this newly constituted enumerated entity then contributes to processes that are constructing new entities: icons that articulate the future. By setting these new entities, calculated as possibilities appearing in twenty years, in proportion with the indexical 2010 total, a purported partial expansion in value of Australian fish stocks can be calculated. After twenty years of oceanic ecological collapse ("an exogenous declining trend of 1 % annually," as Delgado et al, describe it [4, p. 10]), this purported partial expansion will, it is said, increase substantially—by A$0.91 billion per year. This too, is an iconic entity where category and value are one and the same, and it has various parts and sub-parts, which could easily be read off from the computer output by an expert. This value collapses the present into a narrowly imagined future and are now pervasive in contemporary public life. They increasingly feature in politics and are crucial in generating policy. These enumerated entities now carry relations between economy and state. Indeed in the policy paper I exhibit here, A$0.91 billion per year, a (42 %) projected increase in the production value of fish, is crucial to the argument.

If anything will persuade politicians and the citizens they represent, it is this enumerated entity. It justifies taxpayers' money being spent on establishing and running marine parks and protected areas. By justifying permission for one group to fish 'over-there,' but not the other, and disallowing them both from 'here,' the entity excuses differential control of various classes of economic agents, distinguished as they are by the socio-technical apparatus they mobilize.

This entity appears similar to the one I considered above, it seems to index, however, when we recognize it indexes a future, we see that it carries with it the commitment to a future that resembles the present, so it is a prediction, a prophecy, albeit with the performative features that go along with being a valid prediction within statistics. Nevertheless the metaphysical commitments embedded in its constitution have become obvious.

iii. An Economistic Enumerated Entity The final enumerated entity I interrogate is 'A$25 billion. The estimated annual production value marine ecosystem services provide to the Australian economy.' It purports to value of the debt that, in the year 2011, Australian society accumulated in its utilizing of services provided by Australian nature—specifically its oceans. Fisheries and other marine economic activity of humans utilized these environmental services to generate wealth. Or to say this another way, environmental services which humans commandeered, and to which they added value through their activities, saw them indebted to nature to the extent of A$25 billion. What justifies the collective routines in which this value is calculated, is the metaphysical commitment to humans as debtors in some absolute sense, and to nature not as a biophysical entity but as economic infrastructure. A$25 billion is the debt owed to nature accumulated by Australians in 2011. We can classify this enumerated entity as economistic, a variety of statistical enumerated entities generated through statistical procedures using the so-called laws of economics. This is a particularly controversial example of the family of economistic enumerated entities in explicitly proclaiming itself as politically interested, while mimicking the form of empiricist enumerated entities which are popularly supposed to be non-political [2, 18]. This enumerated entity introduces politics as affect.

A\$25 billion is a large value, so large that it seems to elude the comprehension of most numerate people. In addition, the claim that this names the very roughly estimated debt owed by the economy to nature, challenges our imaginations in a different way. Perhaps it is not surprising that the authors of the paper in which this enumerated entity is displayed make a fuss about it. Of the fifty-two pages of argument and evidence in this paper, nine are devoted solely to this value. Six are allocated to describing its conceptual design and three to justifying the actual value claimed. Of these nine pages more than half are devoted to showing that the value is really much, much larger than A\$25 billion, but the authors have "been conservative" because in the time available it was not possible to quantify all the environmental services values provided by Australia's oceans: for example, with respect to things like rivers and wetlands, it is not possible to determine where oceans begin and end; neither is it possible to determine how to separate Australian ocean ecosystem services from global ocean ecosystem services [7].

The first concerns that might occur to a careful reader of the paper in which this poster appeared are conceptual. In measuring environmental services, the authors claim to have identified a debt that Australia's marine economy owes to nature—a certain percentage of the total use-value of Australia's oceans. They claim that this debt the economy owes to nature, can be added to the credit generated in economic production—the exchange value of Australia's ocean (A\$44 billion). Adding A\$25 billion (a vague estimate derived from modeling) to A\$44 billion (a rather precise value derived from the national accounts), they come up with the statement that "Our oceans provide an estimated A\$69 billion per year in value." The enumerated entity mimics empiricist enumerated entities, in presenting this value as an index of an 'out-there' reality. Here a second concern arises. The physical object that is "our oceans," which we are told gifts A\$69 billion per year in value to Australia, is ineluctably vague, spatially indeterminable and existing for eternity. The authors' claim that they had insufficient time to more precisely quantify the ecosystem services of Australia's oceans is, in fact, made in bad faith. In actuality, the object is not quantifiable because it is not a physical object. It is no more quantifiable than the world's unicorns. This enumerated entity has no epistemic salience to environmental policy.

This value A\$25 billion, is presented as an index, proposed as 'an indicator,' when in actuality it is an icon; as an outcome of human collective activity the enumerated entity has a status that is analogous to a religious painting in a church which believers might kiss. In this enumerated entity, category and value are inseparable, they are one and the same. As a particular, it stands-in for an order, a set of beliefs, just as the icon of Christ on the Cross stands-in for an order of belief. Just as a pencil stroke in a geometry exercise book *is* the sole materiality of a geometric line, the sole materiality of this economistic enumerated entity *is* the nine pages in the report.

I suggest that it is useful to think of A$25 billion as words uttered with political intention. We should take the appearance of this value at the head of a list on the poster as performing like a shouted slogan [1]. In being uttered as a collective political act, a slogan that those at a demonstration might chant, "Australians owe their oceanic nature twenty-five billion dollars!" as they occupy Bondi Beach, aims to convince and/or enlighten as such chants do. This enumerated entity is emotive, and as rhetoric aims to convince. Here it is not truth that matters, rather it is persuasion. We can only ask about the felicity of this entity as rhetoric. Does it constitute good rhetoric to proclaim the approximate value of a debt to nature in hopes of enlightening, or in order to elicit recognition so that obligations might be imposed upon those who acknowledge its utterance? Is this entity persuasive? How to define a felicitous utterance here? The criteria for good and bad here lie in the esthetics of the art of politics and are not concerned with truth claims about the oceans or their fish. This enumerated entity explicitly relates to the conduct of environmental politics as affect.

5   Conclusion: Empiricizing/Statistitzing/Economizing Nature, and the Responsibilities of Scientists and Mathematicians

Excavating the human activities embedded in the enumerated entities listed on this poster (including metaphysical commitments embedded in those activities), reveals that the vague whole the contemporary state governs is biophysical territory and its population, and the economic infrastructure that supports present and future economic activity. Both these vague wholes are rendered governable as policy by being particularized as empirical, or alternatively, through statistical and modeling techniques. My argument has been that this rendering governable in the process of articulating policy amounts to a politics of distribution of goods and bads. The enumerated entities that enable this, are in so far as they are epistemically salient to policy, also political. In addition we saw that one of the enumerated entities listed on the displayed poster has no epistemic salience to environmental policy, announcing itself as explicitly politically interested through affect.

The question I pose to scientists and mathematicians in the light of this demonstration of the inevitableness of political nature of enumerated entities involved in government policy—in this case environmental policy, concerns their responsibilities. I am not referring here to their responsibilities as citizens who just happen to be scientists and mathematicians. On the contrary I am referring to their professional responsibilities as members of a knowledge community; their responsibilities as experts. Here I return to Dewey who in 1927, in the aftermath of the lethal demonstration provided by the First World War that science and technology were a challenge to democracy, was uncompromising in voicing his expectations of publically employed scientists and mathematicians [5]. He saw the professional responsibilities of scientists and mathematicians as of three types. First, he wrote, recognizing that human acts have consequences upon others implies that those who undertake acts—such as inventing techniques to calculate new sorts of enumerated entities, some examples of which we see displayed on this poster, have an obligation to consider how their acts as enumerators might involve untoward effects on those not directly involved in those acts. Second, as 'officials,' scientists and mathematicians bear the responsibility to actively to prosecute efforts 'to control action [arising from their acts] so as to secure some consequences and avoid others.' Third, recognizing the possibility of unforeseen, even perverse, effects of their acts as enumerators, as 'officials' scientists and mathematicians must recognize their responsibility to 'look out for and take care of the interests affected.' Unfortunately in the ninety years since Dewey wrote, the dangerous fantasy that enumeration somehow renders analytic entities as untainted with human interests, has showed no sign of diminishing its hold on the collective imaginations of mathematicians and scientists, who in consequence are no closer to considering *how* they might assume collective responsibility for the sorts of enumerated entities that find their way into public policy.

References

1. J. L. Austin (1962). *How to do things with words: The William James Lectures delivered at Harvard University in 1955,* J. O. Urmson & M. Sbisà eds. Oxford: Clarendon Press.
2. N. Bockstael, A. M. Freeman, R. J. Kopp, P. R. Portney, V. K. Smith (2000) "On measuring Economic Values for Nature" *Environmental Science and Technology* 34 (2000), pp. 1384-1389.
3. G. De'ath, J. Lough, and K. Fabricius (2009). "Declining Coral Calcification on the Great Barrier Reef," *Science* 323: 116–119.
4. C. Delgado, (2003). "Fish 2020 Supply and Demand in Changing Global Markets", International Food Policy Research Institute, Washington D. C.
5. J. Dewey, *The Public and its Problems* (1927). Swallow Press, Athens: Ohio University Press, 1954
6. Pierre Duhem (1996) *Essays in the History and Philosophy of Science*, Roger Ariew and Peter Barker, trans & ed. Indianapolis: Hackett Publishing Company.
7. L. Eadie and C. Hoisington, (2011). "Stocking Up: Securing our marine economy" http://cpd.org.au/2011/09/stocking-up/.
8. R. Frigg and M. Hunter (eds) 2010. "Beyond Mimesis and Convention Representation in Art and Science", *Boston Studies in the Philosophy of Science*, Vol. 262. Springer: Springer Science+Business Media.
9. I. Hacking (1992). "Statistical Language, Statistical Truth, and Statistical Reason: The Self-Authentification of a Style of Scientific Reason" in *The Social Dimensions of Science*, Notre Dame: Notre Dame University Press.
10. J. Law (1999). "After ANT. Complexity, naming, and topology" in j. Law and J. Hassard, *Actor Network Theory and After*, Oxford: Blackwell Publishers.
11. M. Poovey (1998). *A History of the Modern Fact. Problems of Knowledge in the Sciences of Wealth and Society.* Chicago: University of Chicago Press.
12. M. Power and R. Laughlin, (1992). "Critical Theory and Accounting" in Alveson, N. & Willmott, H. (eds), *Critical Management Studies*, London: Sage.

13. G. Ryle (1953). "Ordinary language." The Philosophical Review, 62(2), 167–186. doi:10.2307/2182792
14. J. Searle (1969). *Speech acts*. Cambridge: Cambridge University Press. Law in Law and Hassard
15. S. Shapin and S. Schaffer (1985) *Leviathan and the Air Pump. Hobbes Boyle and the Experimental Life*, Princeton: Princeton University Press
16. I. Stengers, (2000). *The Invention of Modern Science*, trans. Daniel W. Smith, Minneapolis: University of Minnesota Press.
17. H. R. Verran (2001). *Science and an African Logic*. Chicago: University of Chicago Press.
18. H. R. Verran (2012). "Number" in *Inventive Methods. The happening of the social*, Lisa Adkins and Celia Lury (eds.) London: Routledge Books, pp. 110-124.
19. L. Wittgenstein, (1953). *Philosophical investigations*. London: Blackwell.